PHOTOREACTIVE POLYMERS

PHOTOREACTIVE POLYMERS
THE SCIENCE AND TECHNOLOGY OF RESISTS

Arnost Reiser
Institute of Imaging Sciences
Polytechnic University
Brooklyn, New York

WILEY

A WILEY - INTERSCIENCE PUBLICATION
JOHN WILEY & SONS
New York • Chichester • Brisbane • Toronto • Singapore

Copyright © 1989 by John Wiley & Sons, Inc.

All rights reserved. Published simultaneously in Canada.

Reproduction or translation of any part of this work
beyond that permitted by Section 107 or 108 of the
1976 United States Copyright Act without the permission
of the copyright owner is unlawful. Requests for
permission or further information should be addressed to
the Permissions Department, John Wiley & Sons, Inc.

Library of Congress Cataloging in Publication Data:

Reiser, Arnost.
    Photoreactive polymers: the science and technology of resists/
Arnost Reiser.
      p.   cm.
    "A Wiley-Interscience publication."
    Bibliography: p.
    Includes index.
    ISBN 0-471-85550-2
    1. Polymers and polymerization.   2. Imaging systems.
3. Photoresists.   I. Title.
QD382.P45R45 1988
621.36'7—dc19   88-26092
                                                     CIP

Printed in the United States of America

10 9 8 7 6 5 4 3 2 1

for Ruth

CONTENTS

Preface   xi

1   A Brief History of Resists   1

    Early History of Resists, 1
    Synthetic Photopolymers, 11
    Photofabrication, 16
    References, 20

2   Negative Photoresists   22

    Introduction, 22
    The Photochemistry of Crosslinking, 24
    The Physical Chemistry of Crosslinking, 46
    References, 61

3   Aspects of Photophysics and Photochemistry in Solid Polymers   65

    Excited States, 65
    Energy Transfer, 70
    Spectral Sensitization, 82
    Photochemistry in Amorphous Solids, 90
    References, 98

4 Photoinitiated Polymerization — 102

Photoinitiated Radical Polymerization, 103
Photogeneration of Radicals, 103
The Initiation Step, 127
Propagation versus Termination and the Kinetic Chain Length, 130
Photoinitiated Condensation Polymerization, 144
Photoinitiated Cationic Polymerization, 147
Practical Systems, 154
References, 172

5 Positive Resists Based on Diazonaphthoquinones — 178

Introduction, 178
The Chemistry of Diazonaphthoquinones, 181
Measurement of Dissolution Rates, 197
Resist Characteristics, 202
Phenomenological Description of Image Formation, 206
The Molecular Mechanism of Novolac-Diazoquinone Resists, 211
References, 223

6 The Rudiments of Imaging Science — 226

Sensitometry of Imaging Materials, 226
Image Degradation and the Spread Functions, 234
Resolution in Optical Pattern Transfer, 242
References, 250

7 Deep-UV Lithography — 252

The Physics of Deep-UV Lithography, 252
Negative Deep-UV Resists, 258
Positive Deep-UV Resists, 262
References, 291

8 Electron Beam Lithography — 294

The Physics of the Electron Beam, 294
A Few Notes on Radiation Chemistry, 302
Negative Electron Resists, 311
Positive Electron Resists, 319
References, 331

9 X-Ray and Ion Beam Lithographies 335

X-Ray Lithography, 335
X-Ray Resists, 343
Ion Beam Lithography, 349
References, 356

10 Multilayer Techniques and Plasma Processing 359

Multilayer Techniques, 359
Plasma Etching, 366
Resists for Use in Dry Etching, 371
Plasma Development, 383
References, 390

Index 395

PREFACE

This book is the outcome of a graduate course in resist science given at the Polytechnic University, Brooklyn, since 1984 and at Fudan University, Shanghai, in 1985. At the time, the audience were chemists, material scientists, chemical engineers, and device engineers from the universities as well as from industry. Their background was rather varied, but they shared a professional interest in the application of polymers to some branch of the imaging industry, the most important being the printing industry and the fabrication of solid state devices. The emphasis of the course was accordingly on the science and practice of these two technologies.

The presentation of the material follows the history of the subject, and the generic resist systems are described approximately in the order in which they made their appearance. This approach maintains the logic of technical development where a new material emerges in response to some deficiency in its historic precursor. The chapters dealing with the resists proper are interspersed with sections on the related background in photophysics, radiation chemistry, and image science. This proved to be successful in class, and it is hoped that it will be acceptable in the text.

Resist science is a fast-expanding field. New materials and procedures are reported almost every month, and any book on the subject is of necessity out of date on the day of publication. An honest effort was made to give a balanced view of the subject as it now stands and at the same time promote an understanding of the perennial challenge of material science: to meet the demands of new technology by the successful design of new materials.

This book would not have been written without the encouragement and help of friends and colleagues. I am particularly grateful to the Geheimrat,

Herman Mark, to Jan Rocek and Herbert Morawetz who read and commented on the whole manuscript, and to Peter Walker, Catherine Chang and Grant Willson who read parts of it.

I would also like to thank my copy editor, Jeannette Stiefel, for her friendly care and the staff of Wiley for their patience and help.

ARNOST REISER

Brooklyn, New York
January, 1989

PHOTOREACTIVE POLYMERS

1 A BRIEF HISTORY OF RESISTS

EARLY HISTORY OF RESISTS

The history of resists goes back to the very beginnings of photography. In fact the first photograph was taken on a resist material. The graphic artist Nicéphore Niépce coated a pewter plate with a natural photosensitive resin, bitumen of Judea [1] and exposed it on a sunny afternoon in a primitive camera from the attic of his house in Chalon, France. After exposure he washed the plate carefully in a mixture of oil of lavender and mineral spirits; the areas of the coating that had received light withstood this treatment, the unirradiated areas dissolved and a faint resin image was left on the plate. This first photograph, taken in 1826, is shown in Fig. 1-1.

The ambiguous shadows betray the long exposure time (several hours), and the quality of the image is poor. It is nevertheless a document of momentous significance: Here, for the first time, a scene had been recorded "spontaneously," without the intervention of an artist.[1]

Niépce, whose portrait is in Fig. 1-2, realized the importance of his finding and wrote a short memorandum, in lieu of a patent, claiming priority for his

---

[1] In the nineteenth century, drawing with pen and pencil and painting with watercolors was part of general education in middle class families. At the time small portable cameras were popular for the secure construction of outline sketches. The draftsman would point the camera at a scene, which would appear in reverse, but with magic clarity, on the translucent paper tacked to the back of the instrument. All that remained to do was to trace the outline of the objects with a pencil. Many a user of the camera wished for a kind of light-sensitive paper that would capture the camera image. Many tried to put this idea into practice; Niépce (1827) and, in a more practical way, Daguerre (1839) and Fox Talbot (1839) succeeded [2, 3].

Figure 1-1 The first photograph. It was taken in 1826 by Nicéphore Niépce from the attic of his house near Chalon, France, on a pewter plate covered with a light-sensitive resin, bitumen of Judea. (Reproduced with permission of the Gernsheim Collection, Harry Ransom Humanities Research Center, The University of Texas at Austin.)

Figure 1-2 Portrait of Joseph Nicéphore Niépce, by L. F. Berger, 1854. Courtesy of the curator, Musee Nicéphore Niépce, Ville de Chalon-sur-Saone, France.

process of "Heliography" [1]. Although he did not make a public announcement, the news spread fast in the community of French graphic artists. One of them, Daguerre, a well-known painter of theater sets and dioramas, joined Niépce, and together they established a company for the commercial development of the new discovery [3].

The new invention did not develop into a practical photography, but it was to be most useful as a plate making process for the reproduction of line drawings. In this second application, which Niépce called "Heliogravure," he coated the light-sensitive resin on an engraver's copper plate, exposed it in contact with a translucent line drawing, and removed the unexposed parts of the coating with solvent, as before. The plate, bearing the resin image, was then etched with acid and in this way an intaglio printing master was obtained.

It is worth noting some of the detail of this operation. Drawings on thin paper were made transparent by soaking them in oil. The bitumen was exposed through this transparency; it hardened where it was exposed to light and remained soluble where it was protected by the black lines of the drawing. After washing, the image was revealed on the plate as a pattern of fine open channels in a background of solid resin. The plate was then immersed in acid and the pattern transferred (etched) into the metal. The resin protected the metal in the unexposed areas by "resisting" the action of the etchant; hence the name resist.[2]

The intaglio plate obtained in this way is the photomechanical equivalent of the classical engraved or etched plate. It was used to print on paper as follows: The plate was liberally coated with printing ink, some of which penetrated into the etched lines (depressions) of the plate. Excess ink was wiped off the plate with a metal blade leaving ink only in the etched channels. Damp paper was then pressed against the plate and the ink was transferred by capillary forces from the plate depressions to the paper. Figure 1-3 shows a print pulled from one of Niépce's early plates. The result is surprisingly good and points to a useful future for the new invention.

Bitumen of Judea is a light Syrian asphalt, which had been used by engravers for a long time as a "ground" or masking material. As a photopolymer it had serious drawbacks: very long exposure times were required to produce an image, the solubility differences between exposed and unexposed areas were slight, the images were fragile, and the material was not available in reproducible quality. The quest for a more sensitive photographic medium led eventually to the silver halides. In photomechanical reproduction, bitumen of Judea was soon replaced by dichromated colloids. It appears that Mungo Ponton (Edinburgh, 1839) [4] was the first to use potassium dichromate imbibed in paper to produce images by exposure to sunlight. A year later, Becquerel [5] claimed dichromated starch as a light-sensitive material. The first really practical system was proposed, however, by Fox Talbot in 1843 [1]. He

---

[2] The term "resist" was coined very early on, apparently by Becquerel (ca. 1840) [1].

Figure 1-3 The Cardinal d'Amboise. An early contact photogravure by Nicéphore Niépce, 1826. (Reproduced with permission of the Gernsheim Collection, Harry Ransom Humanities Research Center, University of Texas at Austin.)

used dichromated gelatin as the resist material, developed the images in warm water and used ferric chloride as etchant.

Fox Talbot (Fig. 1-4 shows him in his later years) was an English country gentleman of independent means and a gifted artist whose life's work was the invention and the early development of silver halide photography. Dichromated gelatin for the making of printing plates was only one of his many sidelines. Nevertheless, he realized the commercial possibilities of the material and he applied for and was granted in 1852 an early British Patent (No. 565). Since then, dichromated gelatin brought about a spectacular expansion of the printing industry and dominated photomechanical reproduction for over a hundred years [6].

Gelatin did not remain the only substrate for dichromates; albumen, agar, casein, fishglue, shellac, and starch have been tried and used. Synthetic polymers such as poly(vinyl alcohol), poly(vinyl pyrrolidone), and poly(vinyl butyral) have partially displaced the natural materials, but gelatin and albumen have remained in use until quite recently. A typical modern recipe for dichromated albumen is given in Table 1-1 [6].

Originally, Fox Talbot had used dichromated gelatin for the preparation of intaglio plates that were suitable only for the reproduction of line drawings.

Figure 1-4 Portrait of William Henry Fox Talbot. Photograph by Moffat of Edinburgh, 1864. Courtesy of the curator, Fox Talbot Museum, Lacock Abbey, Wiltshire, England.

Before long, however, dichromated colloids had entered almost all traditional printing procedures. In the long term, the most important application of dichromate was to be in lithography. This historically youngest printing method was introduced by Senefelder (1771–1834) in the last years of the eighteenth century. He found that a serviceable printing plate could be prepared by simply writing or drawing on a perfectly flat "lithographic" stone

TABLE 1-1 Dichromated Albumen Coating

A. Dichromate Solution	
$(NH_4)_2Cr_2O_7$	758 g
Make up with water to	3758 g
B. Colloid Solution	
Egg albumen	454 g
Ammonia (20% in water)	60 mL
Make up with water to	2360 mL
C. Coating Solution	
A	946 mL
B	192 mL
Water	237 mL

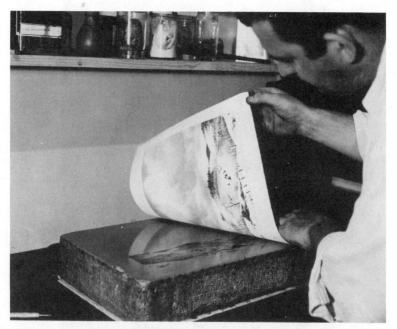

Figure 1-5 An ink image being transferred from the lithographic stone to a receiving sheet. (Reproduced with permission of the Metropolitan Museum of Art, New York.)

with a grease pencil [7]. When a greasy ink was applied to the drawing, it would stick only to the pencil lines and not to the somewhat hydrophilic background. By treating the stone with dilute sodium silicate, termed a "fountain solution," the difference between the ink accepting image areas and the ink rejecting background could be made more pronounced. The ink image was then transferred to a sheet of paper in a modified engraver's press. Figure 1-5 shows the ink image being transferred from the lithographic stone to a receiving sheet.

It was soon discovered that dichromated gelatin could substitute for the grease pencil. In the exposed gelatin coating, the irradiated areas remained on the stone, the unexposed areas could be washed off with warm water. After drying, the gelatin image became hydrophobic and accepted printing ink. This new method of photolithography allowed the photomechanical preparation of lithographic plates. Fox Talbot himself obtained some of the most delicate line reproductions from stones coated with dichromated gelatin.

The early days of dichromate were also the early days of photography [2, 3]. In the 1840s photography was at the center of popular interest, portrait studios and dioramas sprang up all over Europe and the United States, and photographic exhibitions drew huge crowds. The attraction of the new method of image making was of course its ability to produce almost effortlessly a "true and faithful" rendering of its subject. An important aspect of this was the fact

EARLY HISTORY OF RESISTS   7

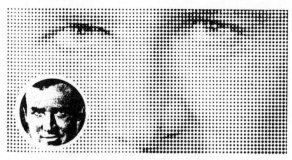

Figure 1-6 The simulation of continuous tone by halftone dots. (Reproduced with permission from M. Bruno [11]. © Gama Communications, Salem, NH, 1986.)

that the photographic plate was able to reproduce pictorial scenes not in the stark black and white of a line drawing, but in natural shades of gray. As a result, the printing industry was now faced with the practical problem of reproducing in print the "continuous" gray tones of original photographs.

Photolithography made tone reproduction possible. Although, like all other printing methods, lithography can print only one tone, say black, on a white background, it is possible to simulate grays in print by mixing small black areas with small white areas of different sizes. This technique, called halftone reproduction, is illustrated in Fig. 1-6. It has revolutionized the reproduction of images and it has made photolithography today's most important graphic reproduction method [8].

Almost all scientific publications, textbooks, and other materials with print runs up to 100,000 copies are now produced by photolithography. There have of course been changes since the early days. Lithographic stone has been replaced by grained aluminum sheets or other metals, the hydrophilic "fountain solution" has become more complex and more efficient, the plates are applied to printing cylinders rather than to flat-bed presses, and the pattern is not transferred from the inked plate directly to the paper, but is taken from the plate onto a flexible rubber blanket and then transferred from the blanket to the paper. This procedure, called lithographic offset printing, puts less wear on the printing plate and allows fairly long print runs [7]. Figure 1-7 is a schematic representation of a lithographic offset press [8].

Before long, attempts were made to reproduce not only continuous tone, but also color [9]. It had been known for a long time (Newton, 1666) that all natural colors were a mixture of the three primaries, red, green, and blue. In 1866, in a famous experiment, Maxwell demonstrated how this could be used to form colored lantern pictures [10]. He projected three monochrome transparencies (red, green, and blue) in exact overlay onto a white screen and found that the object appeared on the screen in vivid "natural" colors. With the advent of panchromatic photographic film, color separation by photographic means and color reproduction (not color photography!) became a practical possibility. In that process, the colored original is photographed in sequence

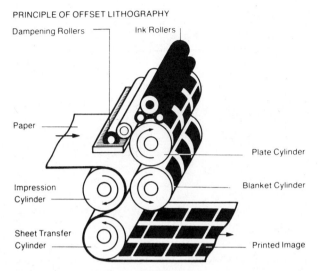

Figure 1-7 Schematic of a litho-offset press. (Reproduced with permission from *Pocket Pal*. © International Paper Company, Inc., 1983.)

through three filters for the three primary colors. From the three color separation negatives, which are the result of this operation, three separate printing plates are made. By overprinting these in accurate register (i.e., in exact overlay) with the appropriate inks, a color reproduction of the original is obtained. An example of modern color separations and their superposition is shown in Fig. 1-8.

Since color reproduction requires the imaging of areas (halftone dots) rather than lines, lithography is an appropriate medium for this purpose and the bulk of the color work in most print shops is carried out by halftone lithography [11]. The quality of color reproduction achievable by lithographic halftone is, however, limited because it approximates different color hues by using halftone dots of equal color intensity. An ideal color reproduction method would mix the three primaries by using image elements of equal area and different color intensity. This idea is realized in gravure printing (Klic, 1877), where printing cells of equal size but variable depth (see Fig. 1-9) dispense variable quantities of ink onto the paper.

Until recently, gravure cylinders were patterned by a rather improbable, but effective procedure, the "carbon tissue process." We describe it here to document the great versatility of dichromated gelatin in the hands of skilled craftsmen. Carbon tissue is a layer of dichromated gelatin tinted with a yellow dye and coated on a strong release paper. The yellow dye is added to the gel for the sake of controlling the depth of light penetration. The carbon tissue is

Figure 1-8 Color reproduction by color separation and subtractive color mixing. The yellow, magenta, and cyan color separation images are shown, as well as their superposition, the final color reproduction. Courtesy of Hell Graphic Systems, Inc.

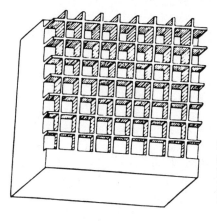

Figure 1-9 Printing surface of a gravure plate (schematic), showing cells of equal area but variable depth. (Reproduced with permission from M. Bruno [11]. © Gama Communications, Salem, NH, 1986.)

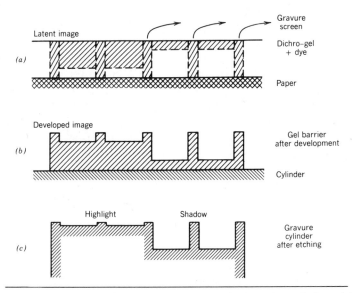

Figure 1-10 Schematic of the carbon tissue process. (*a*) The latent image is formed in dichromated gelatin on exposure through a cell raster and a positive transparency; large exposures occur in the highlights and weak ones in the shadows. (*b*) The developed gelatin image is applied upside down to the printing cylinder. (*c*) After etching, shallow highlight cells and deep shadow cells are formed on the surface of the gravure cylinder.

first exposed to a negative screen that divides the whole plate into cells of equal area. The prescreened plate is then exposed to a positive transparency. In the "shadow" areas of the positive original, little light is available for exposure and the gelatin will harden only on the surface, in the "highlights" of the original the coating is heavily exposed and the gelatin will harden deep into the layer. After exposure, the carbon tissue is applied, face down, onto the metallic printing plate or cylinder. The paper base is removed and the image is developed in warm water. After drying, the plate is immersed into the etchant; the gelatin image now acts as a diffusion barrier. Where the gelatin cover is thick, the etchant penetrates slowly and a cell of shallow depth will be formed in the metal, while in the shadow areas of the original deep cells will be etched into the plate. In the subsequent printing process large quantities of ink will be applied in the shadows, while in the highlights the white of the paper will shine through. Figure 1-10 illustrates the procedure.

Although the carbon tissue method is complex and its success very much depends on the skill of the operator, the quality of color reproduction obtainable in this way is outstanding. Gravure printing is used for the production of art books and of the more expensive color magazines. Today carbon tissue has been replaced by other, more controllable materials and indeed by electronic methods of producing the printing masters.

Finally, dichromated gelatin was and is used in serigraphy or "silk-screen printing." A fine screen, originally of silk, but today more often of nylon, is held by a frame and is impregnated with dichromated gelatin. The original image (a positive) is contact exposed to the screen. In the irradiated areas the gelatin hardens, the unexposed gelatin is washed away with warm water and the screen is dried. It can now serve as a permeable stencil through which printing ink can be applied to a receiving sheet (Fig. 1-11). Serigraphy is used extensively for the printing of posters. It is also used on a large scale in the manufacture of printed circuits for the electronics industry.

Figure 1-11 The principle of screen printing. The pattern is created in the form of a stencil on a (nylon) screen. It is transferred to the receiving sheet by squeezing printing ink through the holes of the screen. (Reproduced from *Pocket Pal*. © International Paper Company, 1983.)

SYNTHETIC PHOTOPOLYMERS

Although dichromated gelatin was highly successful as a photomechanical material, it did have serious shortcomings: First the coating tended to crosslink in the dark within a few hours, and as a result, dichromated materials could not be sold as precoated sheets or plates. Second, the gelatin images were not very resistant to strong etchants. This second problem was dealt with, somewhat brutally, by "burning in" the images at a high temperature (200–300°F). The carbonized coatings had a better etch resistance, but the process of carbonization was detrimental to image quality and, on the whole, did not provide a satisfactory solution. There was a genuine need for a hydrophobic photopolymer, which by its nature would resist an aqueous etchant. As the potential advantages of such a system became more widely recognized, several laboratories began to address the problem [12].

The route that eventually led to success started from the well-known solid state photodimerization of cinnamic acid.

$$\text{Ph-CH=CH-COOH} + \text{Ph-CH=CH-COOH} \longrightarrow \text{Truxinic acid}$$

Cinnamic acid   Truxinic acid

$$+ \quad \text{Truxillic acid} \qquad (1)$$

Truxillic acid

The dimers, truxillic acid and truxinic acid, are less soluble than cinnamic acid, and it was thought that the dimerization process would somehow affect the solubility of cinnamate containing compositions. A number of patents involving variously substituted cinnamates appeared in the literature around 1935, but it was left to Louis Minsk of Eastman Kodak (Fig. 1-12), to produce the first synthetic photopolymer. He functionalized a vinyl polymer with cinnamoyl groups, assuming that the dimerization of two groups belonging to different chains would produce a crosslink, and that a sufficient number of

Figure 1-12 Portrait of Louis Minsk the inventor of the first synthetic photopolymer, poly(vinyl cinnamate).

crosslinks would insolubilize the polymer. The embodiment of this idea was poly(vinyl cinnamate) [13].

$$\left[\begin{array}{c} CH_2-CH \\ | \\ O \\ | \\ CO \\ | \\ CH \\ \parallel \\ CH \\ | \\ C_6H_5 \end{array} \right]_n$$

Poly(vinyl cinnamate)

Poly(vinyl cinnamate) was an immediate success: There was no dark reaction and the coated material had a long shelf life; it could be sensitized to a very useful photographic speed; it was highly etch resistant and it had excellent resolution. The design of a whole new class of photopolymers, all based on photocycloaddition, was inspired by poly(vinyl cinnamate). These materials have their principal application as photosensitive coatings of presensitized (precoated) lithographic plates. Presensitized plates are a very large volume item; most scientific books and almost all educational literature are now being printed by offset lithography. In financial terms polymer plates are the most important photopolymer product on the market.

It is now time to define some of the terms commonly used in the context of photoresists [14]. Bitumen of Judea and dichromated gelatin are said to be negative working, or simply negative resists, because in a copying process using these materials the resist image is a "negative" of the transparent original. Where the original is dark, the polymer is not hardened and is washed away during development. In contrast, diazo papers produce an exact replica of the original; the black lines of the transparency appear as dark blue lines on the copy; diazotype is termed a "positive" reproduction system.

Producing a printing master with a negative working lithographic plate first requires the preparation of a photographic negative from the original artwork. Although the intermediate negative has some practical advantages since it allows various corrections to be made before the final plate is produced, it adds to the cost of the procedure. There is obvious interest in having a simpler, one-step reproduction process and a *positive polymer plate* was thought to be desirable. Such a plate was developed by the Kalle Company in Wiesbaden, Germany. Kalle had been in the graphic reproduction (reprographic) business for a long time. Their main line were diazo papers for the copying of engineering drawings [12]. In developing these materials they explored a large number of diazo compounds, among others also the diazoquinones. A chance observation led to the formulation of a positive resist: It was found that diazoquinones mixed with an alkali soluble novolac binder inhibited the dissolution of the binder in aqueous alkali [15]. Otto Suess (Fig. 1-13), a leading chemist at Kalle, realized what was happening. The diazoquinone, which is not water soluble, is transformed on irradiation into indene carboxylic acid, which is freely soluble in aqueous alkali [16].

$$\text{diazoquinone} \xrightarrow[H_2O]{h\nu} \text{Indene carboxylic acid} + N_2 \qquad (2)$$

If the *novolac–diazoquinone mixture* is coated, for example, on an anodized aluminum plate and exposed to a radiation pattern, the exposed areas of the coating dissolve much more rapidly in aqueous base than the unexposed areas. Since the novolac itself is oleophilic, the resinous pattern remaining on the aluminum support after development has lithographic properties and may be used as a positive lithographic plate. The Kalle positive plate came on the market around 1950 and was quite successful, its only drawback was a certain fragility and poor resistance to wear. In a different context, however, the diazoquinone–novolac mixture was later to become the dominant positive photoresist used in the production of today's semiconductor devices.

Figure 1-13 Portrait of Otto Suess the inventor of the novolac-diazoquinone resists. Courtesy of the Kalle Division of Hoechst AG, Wiesbaden, Germany.

So far we have not mentioned the photomechanical equivalent of the earliest printing method: Relief printing from the raised profile of an image, as is the case, for example, in printing from woodcuts or from metallic type. The reason for this is that relief printing from a polymer plate requires a high profile and therefore a thick polymer coating, and these cannot be patterned effectively by a crosslinking process. Thick polymer plates however, became a practical possibility with the advent of *photoinitiated chain polymerization*. Photoinitiation, that is, initiation by the action of light [or UV (ultraviolet) radiation] had been used early on as a means of inducing radical polymeriza-

Figure 1-14 Portrait of Louis Plambeck, Jr. the inventor of the first polymeric letterpress plate.

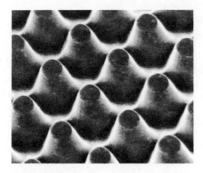 

Figure 1-15 Dot profiles and letters on a Merigraph flexographic printing plate. Courtesy of Hercules, Inc., Wilmington, Delaware.

tion for preparative purposes. In 1954, Oster at the Polytechnic Institute of Brooklyn demonstrated the high sensitivity of photoinitiated polymerization and suggested its use as a means of image reproduction [17]. The person who, independently, pioneered the development of this idea into a practical printing plate, the Dycryl plate, was Plambeck of Du Pont (Fig. 1-14).

The Dycryl plate was patented in 1956 [18] and was to be the first of a number of very successful letterpress plates. Time Incorporated (1957) and BASF with their Nyloprint plate followed soon after that (1958). The great advantage of the polymer plates was their cheapness and durability. Print runs of 500,000 and more could easily be achieved. Their principal application was and still is in the newspaper industry where they have completely displaced metal-type plates. Figure 1-15 shows detail of a modern polymer plate highlighting the depth of the printing profile.

Photoinitiated polymerization opened the way to a number of other uses and products. By a judicious choice of monomers and polymeric binders it is possible to arrange it so that irradiation affects not only the solubility of the film, but also its adhesion properties, its tackiness, its permeability to etchants or gases, or its refractive index. Each of these effects has been exploited for applications in image reproduction [19].

Photomodulation of adhesion has led to various drafting film transfers used in architects' offices, a combination of *solubility and adhesion modulation* is involved in the so-called dry resist films where the light-sensitive material is spread out between transparent covers. The dry resist film is sold in sheets that can be cut to size by the user and laminated to a support as required, eliminating the coating and drying operations of liquid resists. Dry resist films have now become the mainstay of printed circuit and wire board manufacture. *Changes in tackiness* are the basis of various color proofing systems, where a photogenerated tacky latent image is dusted with pigment and transformed into a visible color image. The *modulation of permeability* towards aqueous etchants is the basis of a modern substitute for the old "carbon tissue" method. These films, for example, "Cronavure" of Du Pont, are used as etch barriers in the preparation of gravure cylinders. Finally, the *modulation of the refractive index* of a polymer is used in the recording of holograms.

Photoinitiated polymerization has several important nonimage uses, such as the UV curing of varnishes and printing inks, the radiation curing of monomeric materials in precision molds, for example, in the replication of lenses and, more recently, in the copying of videodiscs [20].

PHOTOFABRICATION

The preparation of printing masters is part of the wider field of photofabrication. An overview of photofabrication is given in the following schematic.

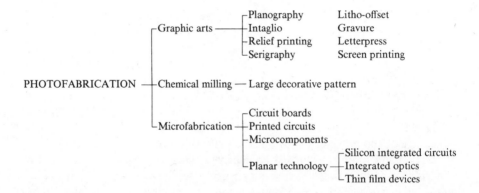

We have dealt with the applications of photosensitive polymers in graphic arts. The second item on this scheme, chemical milling, is concerned with the fabrication of large objects with the help of resist patterns and etchants. This macro method finds applications in the production of ornamental panels, embossed inscriptions, and so on.

Microfabrication

While chemical milling is often a convenient alternative to more traditional abrasive metalworking procedures, microfabrication is concerned with the making of small objects that could not be manufactured in any other way. We mention here the production of a variety of microcomponents such as small mechanical parts of photographic cameras, the grids of electric shavers, the contact frames ("spiders") in which computer chips are seated (Fig. 1-16) and many others. Microfabrication has assumed exceptional importance in modern technology as the miniaturization of technical devices progresses.

From the point of view of market size the most important part of microfabrication is the manufacture of printed circuits and wire boards, which carry

PHOTOFABRICATION   17

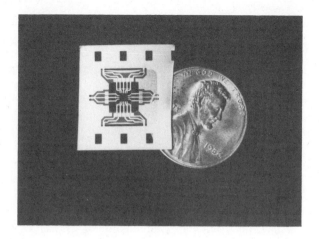

Figure 1-16 A contact frame ("spider") for the seating of an ultra-small integrated circuit, (1987). Courtesy of the Eastman Kodak Company.

and connect the various functional units of almost all electronic devices. Early printed circuits were made by screen printing appropriate conducting patterns onto an insulating substrate. Figure 1-17 shows such a low density printed circuit (ca. 1975). Later, dry resists were used as etching masks for the patterning of a thin copper (conductor) layer precoated on the insulator. Even more recently, the linewidth of the connectors and the size of other features

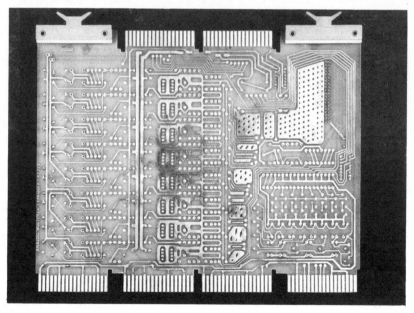

Figure 1-17 A low density printed circuit fabricated by screen printing the connector pattern onto an insulating board (ca. 1975).

has decreased to the point where liquid resists are required in the patterning process.

Planar Technology. The most significant contribution of photopolymers to modern technology is in the fabrication of integrated semiconductor devices. Although the quantities of resist used in the process are quite small, resists are an indispensable ingredient of semiconductor technology and play a crucial role in defining device patterns.

Early on, device fabrication was concerned with a two-dimensional pattern on a flat silicon wafer and the whole fabrication process was termed "planar technology." In the late 1940s the benefits of integrating a number of circuit elements on a single silicon chip were realized. On the urging of Bell Laboratories, where the semiconductor technology originated, polymer chemists at Eastman Kodak developed poly(vinyl cinnamate). This was a great advance over dichromated gelatin, but when the first experimental fabrication line was set up at Bell, device yields were low. The reason was the unsatisfactory adhesion of poly(vinyl cinnamate) to silicon and to silicon dioxide. In response, Kodak developed a resist based on bis-azides and rubber (Hepher and Wagner, 1954) [21] which had excellent adhesion and which, under the trade name Kodak Thin Film Resist (KTFR) became the preferred patterning medium of the industry until about 1972.

The use of resists in defining the patterns of integrated circuits by an essentially photographic method is termed "*optical microlithography.*" In the early days of the industry, the pattern required for the production of integrated circuits was transferred by contact printing from a negative template mask to the resist layer. The mask, in the form of a chromium or gold pattern on glass or quartz, was pressed firmly against the coated wafer in a mechanical device called a *mask aligner*. The procedure was optically very effective, but it led to a high incidence of mechanical defects in the mask as well as in the device. To overcome this difficulty, industry started to experiment with contactless printing by optical projection. *Projection alignment* machines made their appearance, first at Zeiss (1972) and soon afterwards at Perkin-Elmer and at other optical firms. The Perkin-Elmer "Micraligner" has since become the workhorse of the semiconductor industry. It is based on reflection optics and it is capable of imaging a whole 2 to 3 in. diameter wafer in a single exposure. (For an excellent discussion of these developments, see Reference 22.)

Projection exposure reduced defect densities drastically, but at the same time it created a new problem: At the resolution demanded from the instrument, the aerial image of the projector was poor and it needed an imaging material of high contrast to produce developed resist images with sharp edges. The negative azide resists were inadequate in this respect and they were soon displaced by the more contrasty and also otherwise more suitable positive diazoquinone–novolac resists. These were supplied in the United States under the name AZ resists by an affiliate of Kalle, the Azoplate Corporation [23]. The AZ resists are capable of producing integrated circuits with feature sizes

Figure 1-18 Detail of an advanced (1986) integrated circuit. It shows the center portion of the AT & T 256K dynamic random access memory (DRAM) chip. Photomicrograph at 1000 × magnification by Phillip Harrington, 1987. Courtesy of AT & T.

down to 1 μm and beyond. In fact, the resolving power of the AZ resists has not yet been a problem and the limiting factor in optical lithography are the optics of the printer. To improve printer performance one could either increase the numerical aperture of the lenses or use radiation of shorter wavelength as the information carrier.

Both strategies have been pursued: Lenses with higher numerical aperture have been produced, but these can image only a smaller field of view and so the whole wafer has to be exposed in several takes, square by square. This requires an accurate stepping mechanism in the projection table on which the wafer is located. The combination of high resolution optics with sophisticated table movements led to the so-called *"wafer steppers,"* which are currently used for most optical fine-line work. Figure 1-18 shows a detail from an integrated circuit produced by optical microlithography in 1986.

The other strategy, namely, the use of shorter wavelength radiation has led to deep-UV lithography and to the *electron beam, x-ray, and ion beam lithographies*. These new techniques require not only new instrumentation, but also new resist materials based on radiation chemistry rather than photochemistry. Electron beam lithography is now routinely used for the patterning of masks. X-ray lithography is being developed for the large scale manufacture of high density circuits. Ion beam techniques are used experimentally for mask repair and for specialized front line applications.

The quest for ever decreasing feature sizes has brought forth not only new optics and new radiation-sensitive materials, but also new imaging techniques. We mention the use of *multilevel resists*, where only a very thin top layer receives the original pattern and where this pattern is transferred to the lower layers by etching. Other advances in the direction of higher resolution are anisotropic pattern transfer techniques such as *reactive ion etching*, which uses the ions of reactive gases (O_2, F_2, etc.) accelerated in an electrostatic field. These techniques in turn necessitate materials that tolerate the corrosive plasma environment. In this way the physics and the chemistry of the imaging process are intimately interwoven and can no longer be considered in isolation. In earlier days the photosensitive material was central to the imaging process and the instrumentation (optics, mechanics, etc.) adapted itself to its requirements. Today the interconnections between hardware and material are so complex that a systems approach to the problem has become mandatory. This is the reason why in a text on resist materials the technology of their applications has to be considered.

Microlithography is still very much in flux and every year significant reductions in feature sizes are being reported. As the circuit designers demand for space economy becomes more stringent, novel production techniques are being developed. Surprisingly, optical lithography is still the most important production technology and has now been able to breach the 1 μm barrier and enter the submicron region. For some years x-ray lithography has been readied to take over as the mass production method of the next generation of devices. Whether this plan will materialize remains to be seen. With superconductors, high resolution ion implantation, laser exposure sources, and with integrated circuit feature sizes in the nanometer range, a new era of microfabrication is clearly about to begin.

REFERENCES

1. E. Courmont, *La Photogravure: Histoire et Technologies*, Gauthier-Villars, Paris, 1947.
2. H. Gernsheim and A. Gernsheim, *The History of Photography: From the Camera Obscura to the Beginnings of the Modern Era*, McGraw-Hill, New York, 1969.
3. N. Rosenblum, *A World History of Photography*, Abbeville Press, New York, 1984.

4. M. Ponton, *New Philos. J.*, p. 169 (1839).
5. E. A. Becquerel, *Compt. Rend.*, **10**, 469 (1840).
6. J. G. Jorgensen and M. H. Bruno, *The Sensitivity of Bichromated Coatings*, Lithographic Technical Foundation, New York, 1954.
7. J. E. Gogoli, *Photo-Offset Fundamentals*, 4th ed., Glencoe, Mission Hills, CA, 1980.
8. M. H. Bruno, Ed., *Pocket Pal: A Graphic Arts Production Handbook*, 13th ed., International Paper Co., New York, 1983.
9. R. W. G. Hunt, *The Reproduction of Color*, 3rd ed., Fountain Press, Windsor, England, 1975.
10. B. Coe, *Colour Photography*, Ash & Grant, London, 1978.
11. M. H. Bruno, *Principles of Color Proofing*, Gamma Communications, Salem, NH, 1986.
12. J. Kosar, *Photosensitive Systems*, Wiley, New York, 1965.
13. L. M. Minsk and W. P. VanDeusen, U.S. Patent 2,690,966 (1948).
14. C. G. Willson, in *Introduction to Microlithography* (L. F. Thompson, C. G. Willson, and M. J. Bowden, Eds.), *ACS Symp. Ser.*, **219**, 89 (1983).
15. Kalle AG, Ger. Patent 879,205 (1949).
16. O. Suess, *Annalen*, **556**, 65 (1944).
17. G. Oster, *Nature*, **173**, 300 (1954).
18. L. Plambeck, Jr., assigned to Du Pont, U.S. Patent 2,760,863 (1956).
19. E. Brinkman, G. Delzenne, A. Poot, and J. Willems, *Unconventional Imaging Processes*, Focal Press, London, 1978.
20. S. P. Pappas, Ed., *UV Curing: Science and Technology*, Technology Marketing Corp., Norwalk, CT, 1978.
21. M. Hepher and H. M. Wagner, Br. Patent 762,985 (1954).
22. "Dialogues in Optics: An Interview with Philip Blais on Ultrasmall Dimensions Lithographies," Optical Engineering Reports of SPIE, May 1984.
23. Azoplate Corp., U.S. Patent 2,766,118 (1956).

2 NEGATIVE PHOTORESISTS

INTRODUCTION

Chapter 1 briefly introduced the nomenclature of negative and positive resists. Here we describe the standard operations by which an image is formed in negative and positive working systems (see Fig. 2-1).

The material is coated on a substrate, dried, and exposed to a radiation pattern; it is then treated with a solvent termed the "developer." In negative resists the photochemistry that occurs on irradiation renders the material less soluble in the developer; the unirradiated parts of the coating are washed away and a negative polymer image of the original pattern remains on the substrate. In positive resists the photochemistry brings about an enhancement of solubility or of dissolution rate and it is the irradiated areas that are removed by the developer.

The early photomechanical materials such as bitumen of Judea, dichromated gelatin, and other dichromated colloids were all negative resists. At the time of their discovery the mechanism by which they functioned was not understood. Only in the 1920s was it recognized that gelatin and other natural colloids are macromolecules [1] and that rubber vulcanization is a crosslinking process. At that point the idea began to take hold that the hardening and the insolubilization that occur on exposure of dichromated gelatin are brought about by radiation induced crosslinking.

Exactly how crosslinks may be formed in dichromated coatings remained a mystery until Biltz and Eggert [2] established that the process involves the photoreduction of Cr(VI) to Cr(III).

$$Cr_2O_7^{2+} + 14H^+ + 6e^- = 2Cr^{3+} + 7H_2O \tag{1}$$

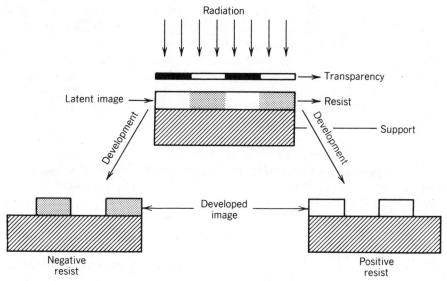

Figure 2-1 Schematic of negative and positive resist functions.

Datta and Sollern [3] have shown that the detailed fate of Cr(VI) depends somewhat on the colloidal medium, but that the final stage of the photoprocess always leads to Cr^{3+} ions. Since trivalent chromium is known to be a powerful coordination center, it was thought that crosslinks may be formed by the coordinative bonding of Cr^{3+} with the amide groups of the protein.

This view is supported by the fact that all polymers that have been used as gelatin substitutes, such as other proteins, starch, poly(vinyl alcohol), poly(vinyl pyrrolidone), and poly(vinyl butyral), carry ligands capable of forming complexes with Cr(III). More recently, coordinative bonding of gelatin to Cr(III) was investigated by Duncalf and Dunn [4] who found that crosslinking occurs only in dry coatings of dichromated gelatin, but not in concentrated aqueous solutions where Cr^{3+} preferentially coordinates water to form $CrCl_3 \cdot 6H_2O$

and does not interact with the protein. Crosslinking in gelatin films can also be suppressed by the addition of chelating agents such as EDTA (ethylenediaminetetracetic acid). Furthermore, the addition of chromic chloride hexahydrate does not have any effect on gelatin films, but when the films are heated high enough to drive off the water of crystallization, they become insoluble, as water in the coordination sphere of Cr(III) is being replaced by the amino and carbonyl ligands of the protein. This is convincing evidence that insolubilization is brought about by the reaction of substrate-bound ligands with a photogenerated coordination center.

Whatever the mechanistic detail, the most important conclusion from these studies is that polymer insolubilization can be achieved by crosslinking, and in fact, the great majority of negative resists are based on crosslink formation. However, this is not the only mechanism by which image and nonimage areas can be differentiated in negative resists. A number of materials have been designed on the principle of photoinduced polarity changes in polymer-bound functional groups. Such changes can have a profound effect on the solubility of the polymer. They are discussed in a later section of this chapter.

THE PHOTOCHEMISTRY OF CROSSLINKING

In general, the reactions that have been used to generate crosslinks in polymeric systems are of two types: Those where crosslinks are formed by the direct reaction of an excited molecule or group, and those where crosslinks arise through the action of a photogenerated reactive species in the ground state.

Reactions of Excited Chromophores

The classic example of crosslinking by an excited chromophore occurs in *poly(vinyl cinnamate)*.

Poly(vinyl cinnamate)

Minsk and co-workers conceived of the design of this polymer [5] while searching the literature for a potential crosslinking reaction. At the time the best known organic solid state photoreaction was the dimerization of cinnamic acid [6].

$$\text{HOOC—CH=CH—Ph} + \text{Ph—CH=CH—COOH} \xrightarrow{h\nu} \text{Truxillic acid} \quad (2)$$

Cinnamic acid → Truxillic acid

Poly(vinyl cinnamate) incorporates this reaction system. Crosslinks are formed by photoaddition between an excited (∗) cinnamoyl group of one chain with a ground state cinnamoyl group belonging to another chain.

$$(3)$$

The nature of the crosslinks in poly(vinyl cinnamate) has been confirmed by direct product analysis [7]. At low temperature up to 80% of the reaction products are cyclobutanes, the rest are the result of radical mediated side reactions.

[2 + 2] Cycloaddition in the ground state of two C=C double bonds is forbidden by orbital symmetry (Woodward–Hoffmann), but it is symmetry allowed if one of the reactants is in the excited state. The reaction is therefore eminently suitable as a photocrosslinking mechanism. It is particularly effective in the cinnamoyl group and its analogs where the carbonyl provides the desirable polarization of the reactive double bond, while the phenyl group increases the polarizability and enhances the light absorbing power of the chromophore. It is no coincidence that the cinnamoyl motif appears in the structures of a whole range of important photopolymers.

Minsk's procedure for introducing cinnamoyl groups into hydroxylic polymers is indicated in Eq. (4) and serves as a model for the synthesis of other resists [5].

$$\text{Photoreactive unit acid chloride} + \text{Hydroxylic polymer} \xrightarrow{\text{base}} \text{Photoreactive polymer} \qquad (4)$$

Although the crosslink-generating photoreaction is reasonably efficient (the quantum yield of crosslink formation is about 0.1), the practical photographic performance of poly(vinyl cinnamate), when exposed to a medium pressure mercury lamp is poor. This is caused by the mismatch between the absorption spectrum of the cinnamoyl group (its maximum absorbance is at 280 nm) and the spectral emission of the mercury arc (see Fig. 3-16). This difficulty can be overcome by spectral sensitization, or by the replacement of the cinnamoyl group with a chromophore that absorbs at longer wavelength.

Spectral sensitization of poly(vinyl cinnamate) is rather successful [8]. For example, the addition of 5% of Michler's ketone brings about a 300-fold enhancement of photographic speed. The reader is referred to Chapter 3 where the sensitization process is described in some detail.

An example of the second strategy is *poly(vinyl cinnamylidene acetate)*.

$$\left[\begin{array}{c} CH_2-CH- \\ | \\ O \\ | \\ C=O \\ | \\ CH \\ \| \\ CH \\ | \\ CH \\ \| \\ CH \\ | \\ C_6H_5 \end{array} \right]_n$$

Poly(vinyl Cinnamylidene Acetate)

This material, introduced by Leubner and Unruh [9] in 1966, absorbs at 360 nm and has a photographic speed 200 times higher than that of unsensitized poly(vinyl cinnamate). It is noted that only the double bond adjacent to the carbonyl group is reactive [10]. Cycloaddition is here reversible, the cyclobutanes formed on irradiation at 365 nm can be cleaved by irradiation with the 254 nm mercury line [11].

$$Ph-CH=CH-CH=CH-\overset{\vdots}{C}=O$$
$$+$$
$$Ph-CH=CH-CH=CH-\underset{\vdots}{C}=O$$

$$\underset{254 \text{ nm}}{\overset{365 \text{ nm}}{\rightleftarrows}} \quad \begin{array}{c} Ph-CH=CH-CH-CH-\overset{\vdots}{C}=O \\ | \quad | \\ Ph-CH=CH-CH-CH-\underset{\vdots}{C}=O \end{array} \quad (5)$$

Other photoreactive chromophores of interest are the *chalcones* of the general structure [12]

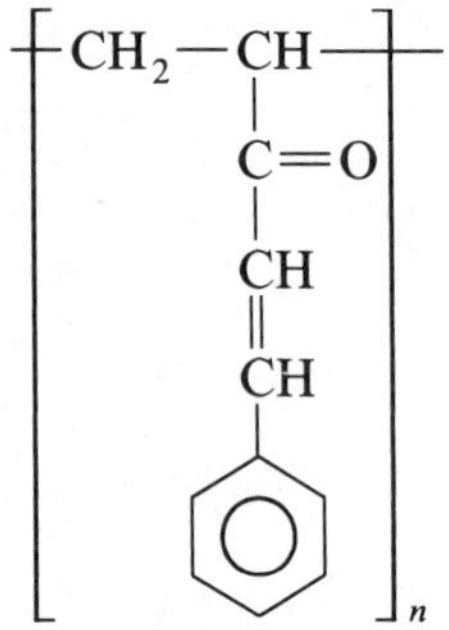

$$\left[\begin{array}{c} -CH_2-CH- \\ | \\ C=O \\ | \\ CH \\ \| \\ CH \\ | \\ C_6H_5 \end{array} \right]_n$$

and the *phenylene diacrylates* [13]. These can be used either as pendant groups, for example, attached by an ester linkage to poly(vinyl alcohol) [14],

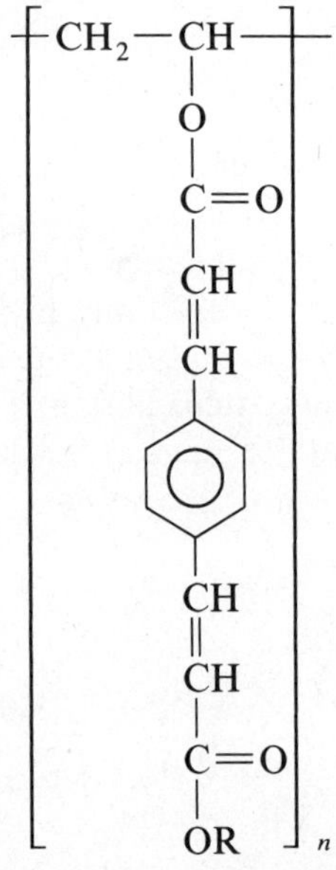

$$\left[\begin{array}{c} -CH_2-CH- \\ | \\ O \\ | \\ C=O \\ | \\ CH \\ \| \\ CH \\ | \\ C_6H_4 \\ | \\ CH \\ \| \\ CH \\ | \\ C=O \\ | \\ OR \end{array} \right]_n$$

or in the backbone of a polyester [15]. Particularly useful is a polyester of *p*-phenylenediacrylic acid (PPDA) with the not too flexible glycol shown in the following structure [13].

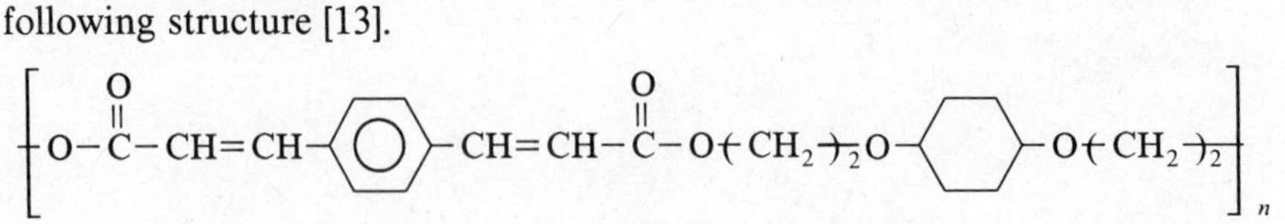

$$\left[O-\overset{O}{\overset{\|}{C}}-CH=CH-\langle\bigcirc\rangle-CH=CH-\overset{O}{\overset{\|}{C}}-O(CH_2)_2O-\langle\bigcirc\rangle-O(CH_2)_2 \right]_n$$

PPDA

This polyester is used on a very large scale as the photosensitive material of precoated lithographic plates (Eastman Kodak). The bifunctional chromophore absorbs strongly at 365 nm. The crosslinking reaction has a quantum yield of about 0.1.

The physical properties of condensation polymers can be altered by changes in the nonphotosensitive components [16, 17]. For example, the introduction of a rigid glycol component such as bisphenol A leads to heat-resistant materials [16]. Recently, Li et al. [18] have described a group of polyesters based on *styrylpyridine*, which are thermally stable up to 450°C (see also Reference 19).

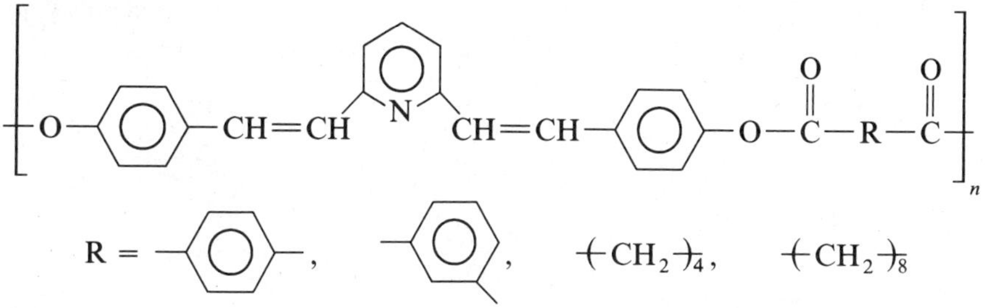

$$\left[O-\langle\bigcirc\rangle-CH=CH-\langle\bigcirc_N\rangle-CH=CH-\langle\bigcirc\rangle-O-\overset{O}{\underset{\|}{C}}-R-\overset{O}{\underset{\|}{C}} \right]_n$$

$$R = -\langle\bigcirc\rangle-, \quad -\langle\bigcirc\rangle, \quad +CH_2 \mathrel{\mathop:}_4, \quad +CH_2 \mathrel{\mathop:}_8$$

Such resists are of particular interest as fine-line solder masks or when the polymer is intended as a permanent component of the final device, be it as insulator, interlayer dielectric, or α-particle barrier.

Polyimides are particularly suitable for these functions and photoimageable polyimides have been developed in several laboratories. Some of these are based on polyamic acids to which photopolymerizable units have been added, and these are mentioned in Chapter 4. A recently disclosed photoreactive fully imidized material is described in a later section of this chapter on page 41.

Water Processable Resists. Environmental considerations have increased demand for water processable materials. These can be prepared by incorporating ionic or ionizable groups (acids or bases) in the polymer structure. An example is shown in the following reaction, where a commercial copolymer of styrene and maleic anhydride is hydrolyzed in the presence of a photosensitive graft unit [20].

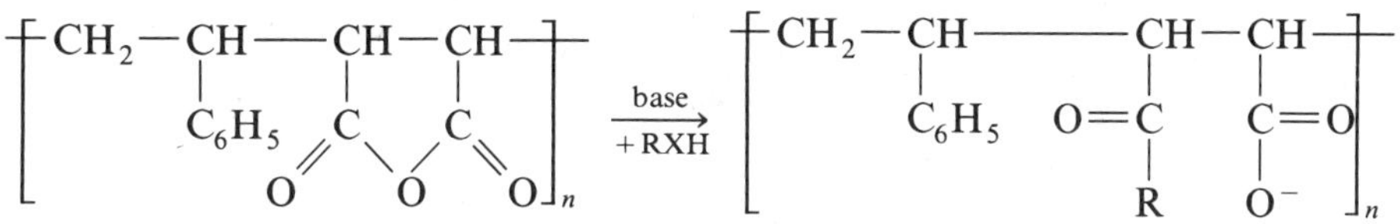

$$\left[CH_2-CH-CH-CH \atop \underset{O}{\overset{C_6H_5}{|}} \underset{O}{\overset{C}{\underset{\|}{\diagdown}}} \underset{O}{\overset{C}{\underset{\|}{\diagup}}} \right]_n \xrightarrow[+RXH]{base} \left[CH_2-CH-CH-CH \atop C_6H_5 \quad O=\underset{R}{\overset{C}{|}} \quad \underset{O^-}{\overset{C=O}{|}} \right]_n$$

Another approach to water processable photoresists is via polymeric quaternary pyridinium salts, which can be made light sensitive by reaction with an aldehyde [21].

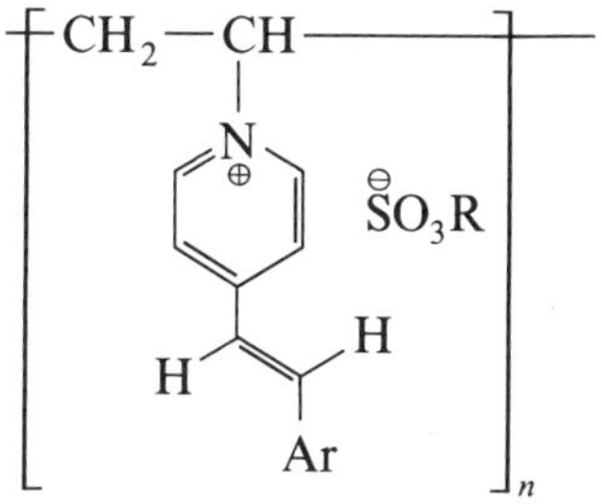

$$\left[CH_2-CH- \atop \underset{}{\overset{}{|}} \underset{}{\overset{N^\oplus}{\langle\bigcirc\rangle}} \overset{\ominus}{S}O_3R \atop \underset{H}{\overset{}{\diagdown}}\underset{Ar}{\overset{H}{\diagup}} \right]_n$$

The use of pyridinium or other heteroaromatic bases has been widely explored [22, 23], most recently by Ichimura and Watanabe [19, 24, 25] who used styrylpyridinium and styrylquinolinium side chains as photoreactive chromophores. Table 2-1 lists the sensitivities of some water processable vinyl polymers compared to that of poly(vinyl cinnamate).

Finally, two chromophores that lead to polycyclic dimers must be mentioned: dimethylmaleimide and diphenylcyclopropene. *Dimethylmaleimide* dimerizes in solution to form the tricyclic dimer shown in reaction (6) [26].

$$\text{Dimethylmaleimide} \xrightarrow{h\nu} \text{tricyclic dimer} \tag{6}$$

While most earlier photopolymers were prepared by the functionalization of a preexisting polymer, Baumann and Zweifel [27] found that bifunctional monomers containing dimethylmaleimide and another group, for example, an acrylate, can be polymerized via the acryloyl group without affecting the maleimide. (The steric hindrance of the methyl groups protects the double bond of maleimide from radical attack in the thermal polymerization process.) This allows much greater flexibility in the synthesis of polymers and copolymers. Accordingly, Berger and Zweifel, and co-workers [28] have incorporated dimethylmaleimide as a side chain in various acrylic copolymers and prepared resists of the general structure.

(a) R = CH$_3$ R' = OH
(b) R = CH$_3$ R' = CH$_3$
(c) R = H   R' = C$_2$H$_5$

Photographic sensitivities as high as 1 mJ/cm^2 have been achieved in some of these systems.

An interesting photopolymer was discovered by De Boer in 1973 when he introduced *diphenylcyclopropene* as a side chain into substituted poly(vinyl alcohol) [29].

TABLE 2-1 Water Processable Crosslinking Photoresists

Photosensitive Group R	Contents of R (mol%)	Relative Sensitivity	Reference
$[-OOC-CH=CH-C_6H_5]$	95	1	4
$-O-C_6H_4-CH=CH-(Py^+CH_3)$	1	5	24
$-O-C_6H_4-CH=CH-(Quinolinium^+CH_3)$	0.8	5	24
$-N^+(Py)-C_6H_4-CH=CH-C_6H_5$	31	225	21
$(N^+(CH_3)Py)-CH=CH-C_6H_5$	50	10,000	22
$-O-(CH_2)_2-N^+(Py)-C_6H_4-CH=CH-C_6H_5$	55	2,150	25
$-CH(CH_2-O-)_2 CH-C_6H_4-CH=CH-(Py^+CH_3)$	2	7,500	25

$$\left[\begin{array}{c} -CH_2-CH-CH_2- \\ | \\ O \\ | \\ C=O \\ | \\ CH \\ \diagup \diagdown \\ Ph-C=C-Ph \end{array} \right]_n$$

The conjugated π-system is that of *cis*-stilbene, except that here the conformation is fixed and cis–trans isomerization cannot occur. As a result, the lifetime of the excited triplet state of diphenylcyclopropene is longer than that of any other dimerizing chromophore. This has an important consequence: even in dilute solution triplet sensitized diphenylcyclopropene reacts with a quantum yield of 0.6 to form the tricyclic dimer shown in Eq. (7).

$$2 \underset{\text{Ph} \quad \text{Ph}}{\overset{\text{H} \quad \text{H}}{\triangle}} \longrightarrow \text{tricyclic dimer} \quad (7)$$

Diphenylcyclopropene

The same cyclization reaction occurs also in the solid matrix [29]. Although at best 40% of the side chains can be substituted with diphenylcyclopropene groups, the material has the highest photographic speed of any crosslinking resist. With triplet sensitizers such as thioxanthone the quantum efficiency of crosslink formation is almost unity. The key to this performance is the migration of the excitation quantum in the polymer matrix and its trapping at the few reactive sites. The wide migration range is a result of the long lifetime of the excited triplet state of the rigid chromophore (see Chapter 3).

Crosslink Formation by a Photogenerated Reactive Species

Nitrenes. In a number of successful resists systems crosslinks are formed by the thermal reaction of a photogenerated reactive species. Examples are resists based on the decomposition of organic azides, where the primary products of the photoprocess are nitrenes, electron deficient species analogous to carbenes

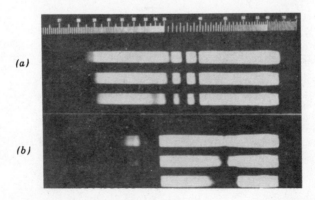

Figure 2-2 Absorption spectrum (a) of 1-azidoanthracene and (b) of the corresponding nitrene. The nitrene was produced by photolysis of a solid solution of the azide in an organic glass (diethylether: pentane: ethanol = 5:5:2) at 77 K. [Reproduced from *Nature*, **211**, 410, (1966)].

[30, 31]. The real existence of aromatic nitrenes and their role as reaction intermediates was established by Wasserman et al. (1964) [32] and by Reiser and Frazer (1965) [33] in matrix isolation experiments. Figure 2-2 shows the absorption spectra of 1-azidoanthracene and of the corresponding nitrene produced by photolysis in an organic glass at 77 K [34].

The photoprocess and the subsequent thermal reactions of azide photolysis in a polymeric substrate are described by the following scheme.

$$RN_3 \xrightarrow{h\nu} {}^1RN + N_2 \qquad (8)$$

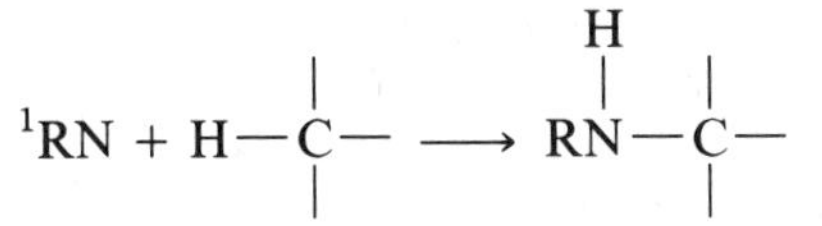

$$^1RN + H-\underset{|}{\overset{|}{C}}- \longrightarrow RN-\underset{|}{\overset{H}{\underset{|}{C}}}- \qquad (9)$$

$$^1RN \longrightarrow {}^3RN \qquad (10)$$

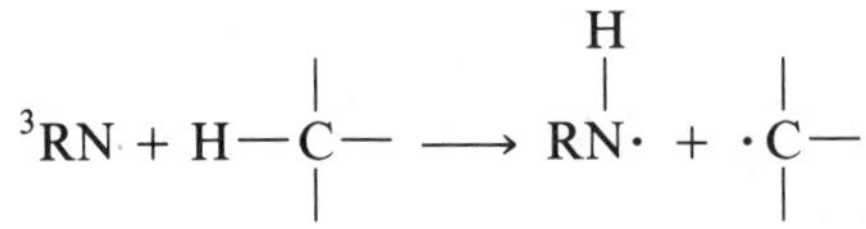

$$^3RN + H-\underset{|}{\overset{|}{C}}- \longrightarrow RN\cdot + \cdot\underset{|}{\overset{H}{\underset{|}{C}}}- \qquad (11)$$

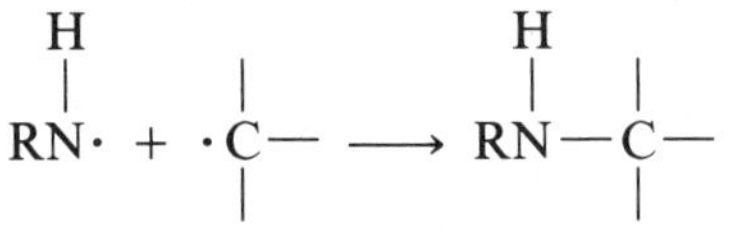

$$\overset{H}{\underset{|}{RN}}\cdot + \cdot\underset{|}{\overset{|}{C}}- \longrightarrow RN-\underset{|}{\overset{H}{\underset{|}{C}}}- \qquad (12)$$

$$\overset{H}{\underset{|}{RN}}\cdot + H-\underset{|}{\overset{|}{C}}- \longrightarrow RNH_2 + \cdot\underset{|}{\overset{|}{C}}- \qquad (13)$$

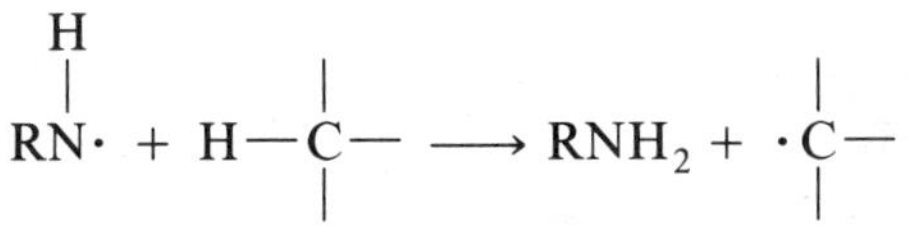

The key to this chemistry is the fact that nitrenes have a triplet ground state. In the first step, which is the only photoreaction in the scheme, the azido group loses nitrogen and is transformed into a (singlet) nitrene. The singlet nitrene inserts readily into H—C bonds of the polymeric substrate and forms, in a single step [Eq. (9)], a secondary amine that acts as a link between the azide fragment and the substrate. More often, however, the singlet nitrene decays to the ground state triplet nitrene [Eq. (10)] before the singlet reaction can occur. The triplet nitrene abstracts hydrogen and forms an amino radical [Eq. (11)] and a carbon radical at the same time. The two have correlated spins and can couple only after spin inversion. When coupling occurs, a secondary amine crosslink is again formed [Eq. (12)]. The amino radical, however, may also abstract a second hydrogen from a different site and produce a primary amine rather than a crosslink [Eq. (13)]. It is the competition between reactions (12) and (13) that determines the efficiency of the thermal crosslinking process.

Resists based on nitrene chemistry are prepared by attaching azido groups to a polymer chain [35].

$$\left[\text{CH}_2-\text{CH} \left(\text{C}_6\text{H}_4\text{-N}_3 \right) \right]_n \quad \text{and} \quad \left[\text{CH}_2-\text{CH}-\text{O}-\text{C}(=\text{O})-\text{C}_6\text{H}_4\text{-N}_3 \right]_n$$

TABLE 2-2 Quantum Efficiency of Nitrene Formation by Aromatic Azides [35a]

Aromatic Azide	ϕ
Ph–N$_3$	0.53
2,4,6-trimethylphenyl azide (Me, Me, Me on ring with N$_3$)	0.96
3-azidobiphenyl	0.37
1-azidonaphthalene	1.00
N$_3$–C$_6$H$_4$–CH=CH–C$_6$H$_4$–N$_3$	0.37
N$_3$–C$_6$H$_4$–CH=CH–C(=O)–(CH$_2$)$_2$–C(CH$_3$)H–CH=CH–C$_6$H$_4$–N$_3$ (ABC)[a]	0.33
N$_3$–C$_6$H$_4$–(CH=CH)$_3$–C$_6$H$_4$–N$_3$	0.20

[a] The abbreviation for 2,6-bis(4-azidobenzal)-4-methylcyclohexanone is ABC.

TABLE 2-3 Grafting Efficiency of Naphthylnitrene to Various Substrates [36]

	Yield of Amine	
Substrate	Secondary	Total
Polystyrene	0.93	0.95
PMMA[a]	0.82	0.95
Polyisoprene (cyclized)	0.67	0.90
Paraffin wax (hard)	0.33	0.90
Paraffin wax (soft)	0.03	0.35
Paraffin oil	0.04	0.07

[a] Poly(methyl methacrylate) is abbreviated as PMMA.

The quantum yield of crosslink formation for such a system is the product

$$\Phi = \phi\chi \tag{14}$$

where ϕ is the quantum yield of nitrene formation in reaction (8) and χ is the efficiency of the thermal reactions (11) and (12), which form the link between nitrene and substrate [35a]. The two parameters depend quite differently on structural features. The quantum yield of photolysis depends almost entirely on the structure of the azide and may vary within an order of magnitude, as can be seen in Table 2-2. The efficiency of the grafting reaction, measured by the relative yield of secondary amine in the overall process, depends equally on the properties of the substrate (see Table 2-3). In the solid environment the grafting efficiency is also a function of local rigidity: The more rigid the matrix, the higher the chances that hydrogen abstraction and radical coupling occur at the same carbon center and that a crosslink is formed.

Instead of attaching azido groups directly to the polymer, a low molecular weight bis-azide may simply be added to a nonphotoreactive substrate so that crosslinks are formed on irradiation when both azido groups react with sites of different polymer chains [30]. Since two azido groups must react before a crosslink can be formed, the overall efficiency of the crosslinking process depends now on the square of the efficiencies of the individual steps.

$$\Phi = \phi^2\chi^2 \tag{15}$$

This makes for lower photographic speed (both factors are smaller than unity), but that is compensated for by the simplicity and convenience of the system, which allows great flexibility in the choice of substrate. For example, the incentive for developing the bis-azide/polyisoprene resists was the desire to confer photosensitivity on a material with good adhesion to semiconductor surfaces; and for good adhesion, rubber seemed a natural choice.

The bis-azide 2,6-bis(4-azidobenzal)-4-methylcyclohexanone, (ABC), is used most often in near UV resists. Its quantum yield of photolysis in the solid matrix is about 0.4. Its popularity is partly due to the easy synthesis.

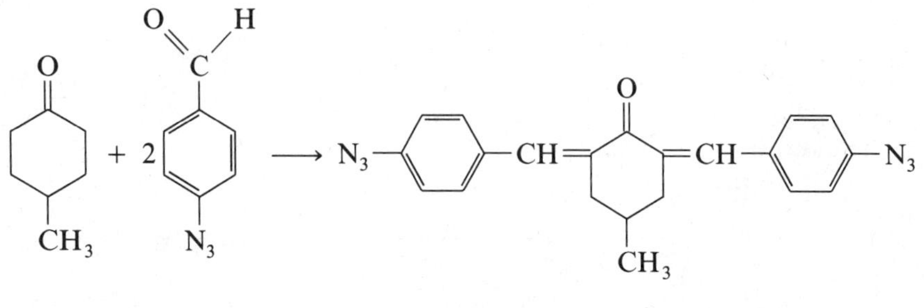

**2,6-Bis(4-azidobenzal)-
4-methylcyclohexanone ABC**

Some other bis-azides in commercial use are shown here [37]. They have been chosen for their compatibility with common substrates as well as for their spectral absorption.

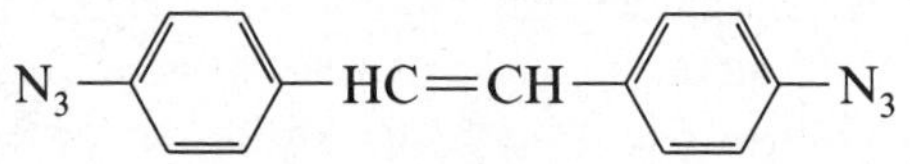

4,4′-Diazidostilbene

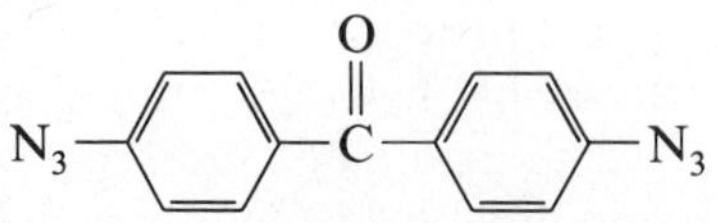

4,4′-Diazidobenzophenone

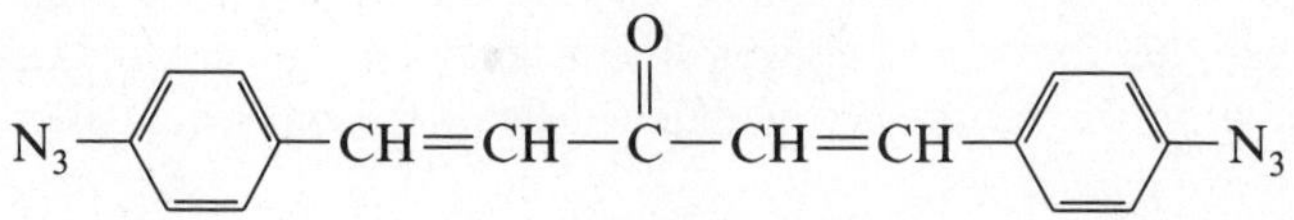

4,4′-Diazidobenzalacetone

The most frequently used substrate in bis-azide resists is poly(*cis*-isoprene), cyclized by treatment with a mineral acid. The cyclization reaction is reminiscent of cationic polymerization. In the first reaction step the acid cation adds to a double bond and produces a carbocation. The reaction of a nearby double bond with this ion leads to cyclization [38].

$$\text{[structural formulas]} \quad (16)$$

Cyclization first of all reduces the degree of unsaturation of the polymer and this alleviates the problem of spontaneous thermal crosslinking; it introduces alicyclic units, which make the polymer less flexible and increase its softening temperature. The acid treatment also severs some of the C—C bonds of the chain and lowers the molecular weight and with it the intrinsic viscosity. This allows the use of coating solutions with a higher solids content. Table 2-4 compares some of the properties of uncyclized and cyclized poly(*cis*-isoprene) [38].

Cyclization is a delicate process, the outcome of which determines the properties of the final resist. Cyclization procedures are therefore proprietary to the various manufacturers, and great care is taken to characterize the product. The principal criterion is the degree of unsaturation, a secondary characteristic is the "cyclicity," distinguishing between monocyclic, bicyclic,

TABLE 2-4   Properties of Uncyclized and Cyclized Poly(*cis*-isoprene) [38]

Property	Uncyclized	Cyclized
$M_w{}^a$	10^6	10^4
Density	0.92	0.99
Softening point (°C)	28	50–65
Intrinsic viscosity	3–4	0.49–0.36
Unsaturation (mmole/g)	14.7	8–4

$^a M_w$ is the weight average molecular weight.

and tricyclic motifs in the polymer

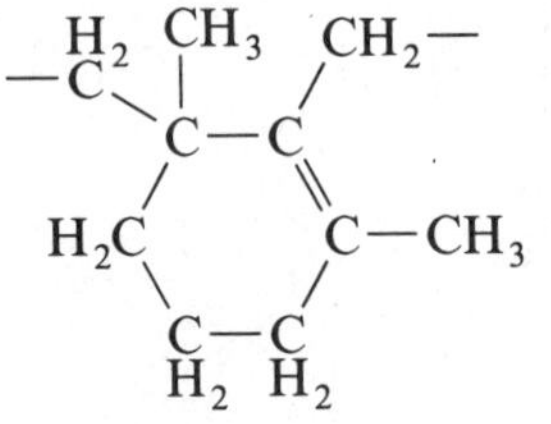

Monocyclic

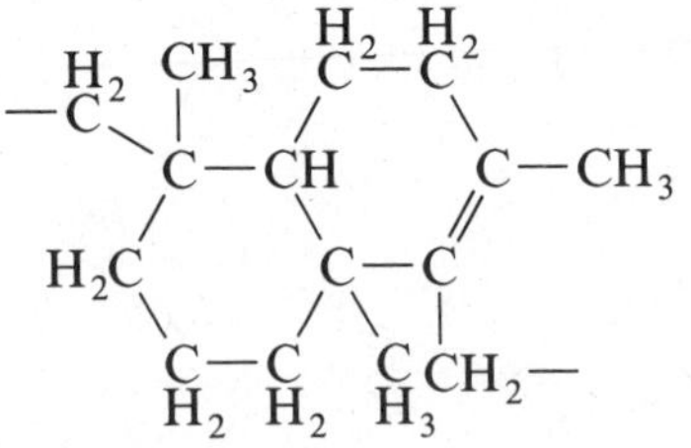

Bicyclic

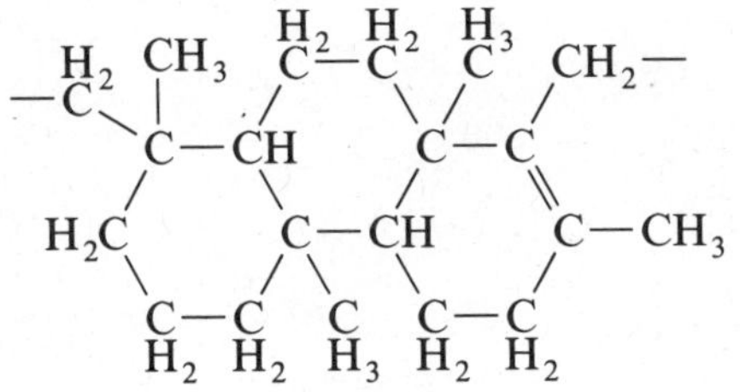

Tricyclic

Double bonds can be located inside (endo) or outside (exo) of the rings, and the ratio of endocyclic to exocyclic double bonds has some effect on physical properties. A representative value of the endo–exo unsaturation ratio is 0.8. These data are obtained by careful NMR (nuclear magnetic resonance) and IR (infrared) spectrometry [38].

The generic bis-azide resist from which most others have been derived is Kodak Thin Film Resist (KTFR), which is composed of a cyclized poly(*cis*-isoprene) with an approximate molecular weight of 150,000, using 2,6-bis(4-azidobenzal)-4-methylcyclohexanone (ABC) as crosslinking agent. The KTFR resist was the workhorse of the semiconductor industry from 1957 till about 1972 [39].

A revival of interest in azide resists has occurred as a result of the work of the Hitachi Laboratories in Japan [40–44]. Nonogaki and co-workers have recently described a water processable azide resist based on poly(acrylamide) or polyvinylpyrrolidone together with water soluble bis-azides such as

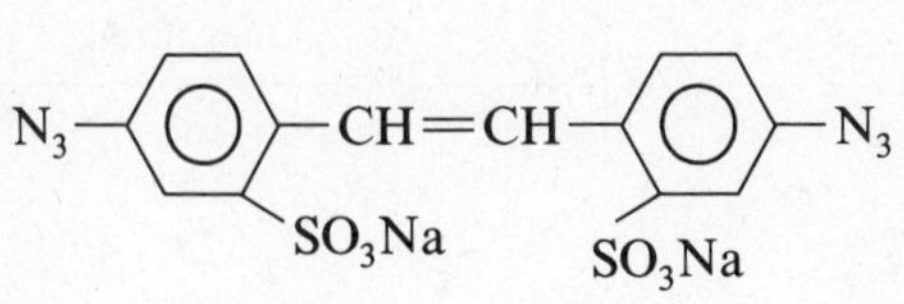

$$N_3-\underset{SO_3Na}{\bigcirc}-CH=CH-\underset{SO_3Na}{\bigcirc}-N_3$$

Even more interesting are some new azide resists based on poly(vinyl phenol) as the polymeric substrate and on various mono-azides as the photosensitive component such as 4-azidochalcone and its homologs [43, 44, 46].

$$N_3-\text{C}_6H_4-CH=CH-\overset{O}{\underset{\|}{C}}-C_6H_5$$

$$N_3-\text{C}_6H_4-CH=CH-\text{(cyclohexenone)}$$

$$N_3-\text{C}_6H_4-CH=CH-CH=CH-\text{(cyclohexenone)}$$

The azido compounds are added to the resin in large quantity (20% by weight). On irradiation the photogenerated nitrenes abstract hydrogen from the polymer backbone and produce carbon radicals. The recombination of these provides the crosslinking mechanism. The remarkable feature of these materials is that they do not swell in the developer, which is aqueous alkali. Submicron resolution has been achieved.

An important reaction that competes with crosslink formation in azide resists is the *reaction of nitrenes with oxygen* [34].

$$^3RN + {}^3O_2 = RNO_2 \qquad (17)$$

This reaction is very efficient because oxygen, like the nitrene, has a triplet ground state. In fact, the oxygen reaction is so efficient that azide resists have to be used in a nitrogen or carbon dioxide atmosphere or in a mild vacuum.

The reaction with oxygen may also be used to achieve complete *image reversal* by the following procedure. The resist film is first exposed in an oxygen atmosphere to a positive original. At this stage all the azido groups in the irradiated areas will be destroyed without producing crosslinks. Subsequently, the film is placed into a vacuum frame and its whole area is exposed. Those parts that had been shielded during the first image exposure and therefore still contain azido groups will now crosslink. After development, a positive image of the original is formed.

Carbenes. In analogy with the nitrenes, carbenes are potentially useful as crosslinking agents, but only recently have suitable carbene precursors been reported. Vicari [45] was able to use the photoisomerization of acylsilanes to siloxycarbenes as a crosslinking mechanism. Under irradiation, the acylsilanes are in photostationary equilibrium with siloxycarbenes [45a]. These in turn can insert into OH bonds or add to the pyridine nitrogen, and so on.

$$Me_3Si-\underset{O}{\overset{\|}{C}}-Ph \underset{}{\overset{h\nu}{\rightleftarrows}} Me_3Si-O-\underset{..}{C}-Ph$$

$$\downarrow ROH, pyridine$$

$$Me_3Si-O-\underset{OR}{\overset{H}{\underset{|}{C}}}-Ph \quad (18)$$

The reaction can be used either by grafting acylsilyl groups onto suitable polymers, or by adding bifunctional acylsilanes, such as, for example,

[structure of bifunctional acylsilane with Me₃Si-C(=O)-C₆H₄- groups on both sides of a methylcyclohexanone]

to poly(vinyl alcohol), poly(vinylpyridine), or similar polymers. Both methods lead to efficient resists with good sensitivity.

Radicals. Radicals are usually associated with polymerization reactions, but the coupling of radicals can be used to form crosslinks. Reactions of this type are common in radiation chemistry (see Chapter 8). In photochemistry, hydrogen abstraction by excited triplet ketones (e.g., benzophenone) and subsequent radical coupling is one of the classical processes [46]. It has been shown to occur also in solid polymers [47].

$$Ph_2C=O^* + H-\underset{CH_3}{\overset{CH_3}{\underset{|}{C}}}-OH \rightarrow Ph_2\dot{C}-OH + \cdot\underset{CH_3}{\overset{CH_3}{\underset{|}{C}}}-OH$$

Excited Triplet Benzophenone   **iso-Propanol**

[products: Ph₂C(OH)-C(OH)Ph₂, (CH₃)₂C(OH)-C(OH)(CH₃)₂ (Acetone pinacol), Ph₂C(OH)-C(OH)(CH₃)₂]   (19)

Smets et al. [48] have used this process in photocrosslinking copolymers of vinylbenzophenone and 4-dimethylaminostyrene. The reaction, in this case, passes through an exciplex (see Chapter 3), which the authors were able to detect by fluorescence spectrometry.

$$\text{Vinylbenzophenone} + \text{4-Dimethylaminostyrene} \xrightarrow[\text{transfer}]{h\nu \text{ electron}} [\cdots]^* \xrightarrow{\text{proton transfer}} [\cdots] \longrightarrow \text{Crosslink} \quad (20)$$

Recently, a photoreactive polyimide has been reported [49, 50], where crosslinking is brought about by photogenerated radicals. The repeat unit of this material has the structure where the R groups may be methyl or other alkyl substituents. The key step in the functional mechanism is hydrogen abstraction by the carbonyl group of the benzophenone moiety from nearby alkyl substituents and the formation of two radicals. The recombination of these radicals on neighboring chains leads then to the formation of crosslinks [51].

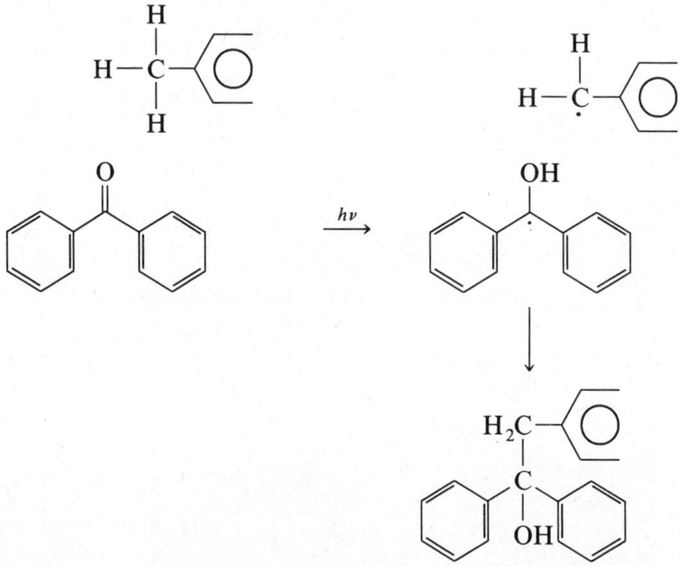

(21)

The photoreactive polyimide has fairly high photographic speed (a typical exposure requirement is 25 mJ/cm^2), good contrast, and high resolution. There are no volatile products on heating and the images are dimensionally stable up to 400°C. Finally, the resist can be coated and imaged in thick layers, which is important for its use as an α-particle barrier, as well as for other applications.

It is of some interest to note that the hydrogen abstraction reaction, which is so efficient in fluid solution, ($\phi = 0.6$), is reversible in the rigid polymer and leads to a low net quantum yield of crosslink formation ($\Phi = 0.03$).

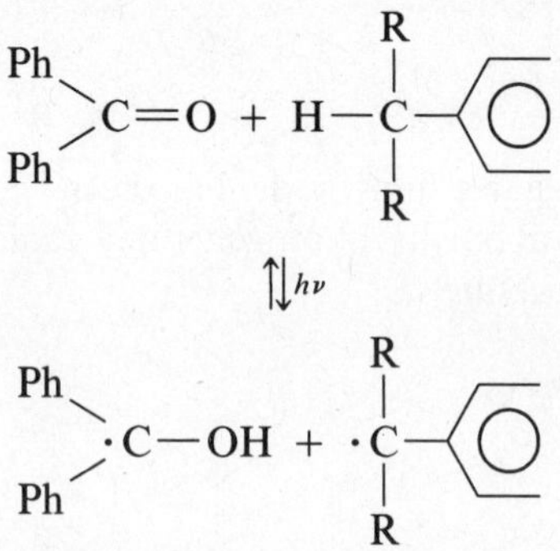

(22)

A Hybrid System. An interesting new crosslinking system based on the reaction of radicals was recently disclosed by Plambeck et al. [52]. It is a hybrid

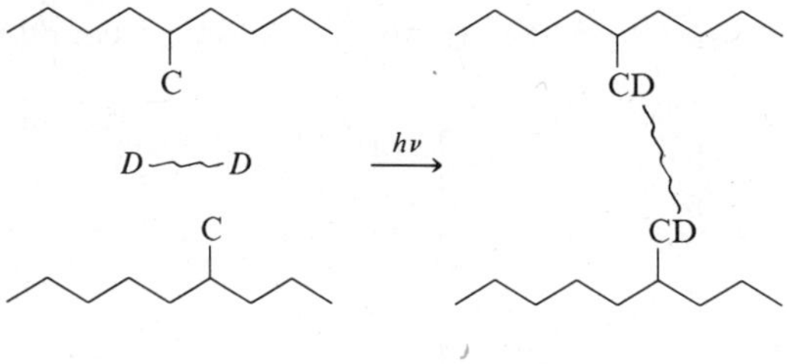

Scheme 1

system that combines the photographic speed of silver halide photography with the desirable physical properties of a polymer. It makes use of the catalytic development of exposed silver halide to generate reactive intermediates capable of forming crosslinks in a high molecular weight polymer.

The idea was realized by designing polymers with pendant color coupler groups and using bifunctional color developers to act as crosslinking agents in the presence of a silver latent image. The process is illustrated in Scheme 1.

The reaction between developer and silver ions is described by reaction (23) [53].

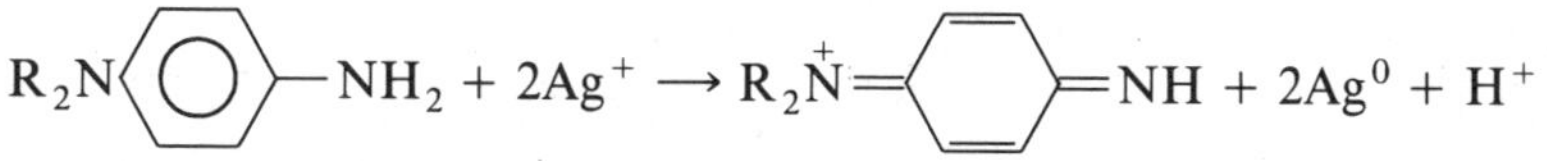

$$R_2N-\langle\bigcirc\rangle-NH_2 + 2Ag^+ \rightarrow R_2\overset{+}{N}=\langle\bigcirc\rangle=NH + 2Ag^0 + H^+ \quad (23)$$

The diimine cation reacts with the active methylene of a color coupler to form a dye.

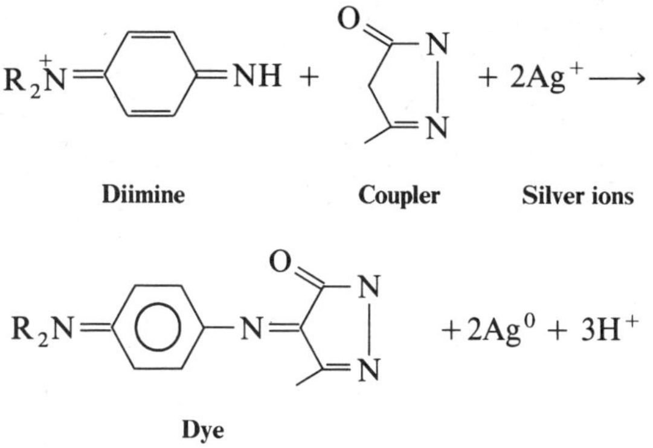

Diimine   Coupler   Silver ions

$$\quad (23a)$$

Dye

The first reaction is strongly catalyzed by the silver nuclei of an exposed silver halide emulsion. The practical problem is to design a polymer that will disperse silver halide grains.

Coupler polymers compatible with silver halide were prepared by copolymerization of acrylates with methacrylates and with polymerizable coupler units well known in color photography. A magenta forming pyrazolone coupler was found to be particularly useful. A typical copolymer composition is shown in the following formula.

$$\left[\text{CH}_2-\text{CH} \atop \substack{| \\ \text{CO} \\ | \\ \text{OEt}} \right]_{0.43} \left[\text{CH}_2-\overset{\text{CH}_3}{\underset{|}{\text{C}}} \atop \substack{| \\ \text{CO} \\ | \\ \text{OMe}} \right]_{0.35} \left[\text{CH}_2-\overset{\text{CH}_3}{\underset{|}{\text{C}}} \atop \substack{| \\ \text{CO} \\ | \\ \text{NH} \\ | \\ \underset{\text{H}}{\underset{|}{\text{N}}}\!\!-\!\!\overset{\text{N}}{\underset{}{\text{N}}} \\ \text{O}} \right]_{0.12} \left[\text{CH}_2-\overset{\text{CH}_3}{\underset{|}{\text{C}}} \atop \substack{| \\ \text{CO} \\ | \\ \text{OH}} \right]_{0.10}$$

The bifunctional color developer was modeled on the Kodak CD4 developer.

$$\text{H}_2\text{N}-\!\!\bigcirc\!\!-\text{N}-(\text{CH}_2)_4-\text{N}-\!\!\bigcirc\!\!-\text{NH}_2$$
$$\phantom{\text{H}_2\text{N}-\!\!\bigcirc\!\!-}\underset{\underset{\underset{\text{OH}}{|}}{\underset{\text{CH}_2}{|}}}{\underset{\text{CH}_2}{|}}\phantom{-(\text{CH}_2)_4-}\underset{\underset{\underset{\text{OH}}{|}}{\underset{\text{CH}_2}{|}}}{\underset{\text{CH}_2}{|}}$$

To prepare a negative working lithographic plate, equal amounts of coupler copolymer and sensitized silver halide are mixed together at a carefully controlled pH of 9.5 to form a combined emulsion. This is coated on anodized aluminum at a dry coating weight of between 10 to 40 mg/dm^2. The plates are exposed in a process camera and developed, subjected to a conventional spray washout, and rinsed in 2% acetic acid. Good quality images are obtained and the plates are not visibly degraded after 100,000 impressions [52].

Resists Based on Photoinduced Polarity Changes

Photoinduced changes in the polarity of polymer-bound functional groups can profoundly affect the solubility of the polymer. To be effective, polarity changes have to be quite dramatic to transform the resist from a material soluble in organic solvents to one that dissolves in an aqueous medium. This idea has been used extensively in the design of electron resists (see Chapter 8), but it can be applied equally to photopolymers sensitive in the visible and near UV.

The classical example of a photoinduced polarity change is the photodecomposition of diazoquinones, which transforms a moderately polar compound, the diazoquinone, into an ionizable compound, the indene carboxylic acid. This reaction is the basis of the important positive working photoresists described in Chapter 5. If diazoquinone units are attached to a polymer backbone, for example via acrylic side groups, the polymer will be soluble in organic solvents before exposure and become soluble in dilute aqueous alkali after exposure.

Schwalm et al. have used the photochemical transformation of *N*-iminopyridinium ylides (I) to 1,2-diazepins (II) as the basis of a negative working resist [54].

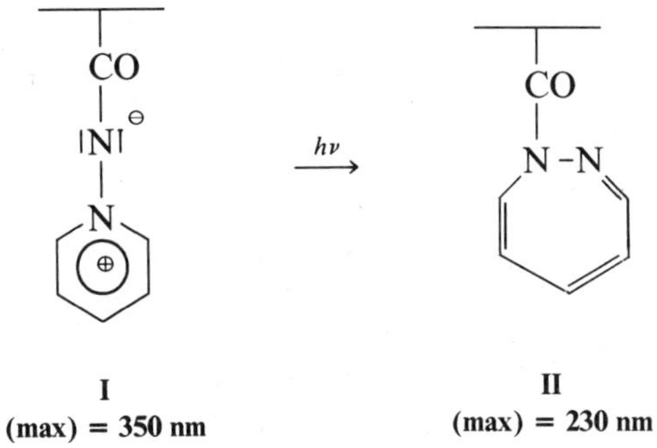

(24)

I
(max) = 350 nm

II
(max) = 230 nm

The polymer-bound ylides are water soluble while the diazepins are hydrophobic. The system works well as a negative resist with an aqueous developer. The positive tone function expected on development with organic solvents is less satisfactory because of some adventitious crosslinking [54].

In diazoquinone chemistry the driving force of the photoreaction is the great stability of N_2, which makes the $=N_2$ substituent such an effective leaving group. A rather popular negative photoresist is based on the same principle [55–57]. This so-called diazoresin is a low molecular weight polymer produced by the diazotization of an anilin–formaldehyde resin. Its structure follows:

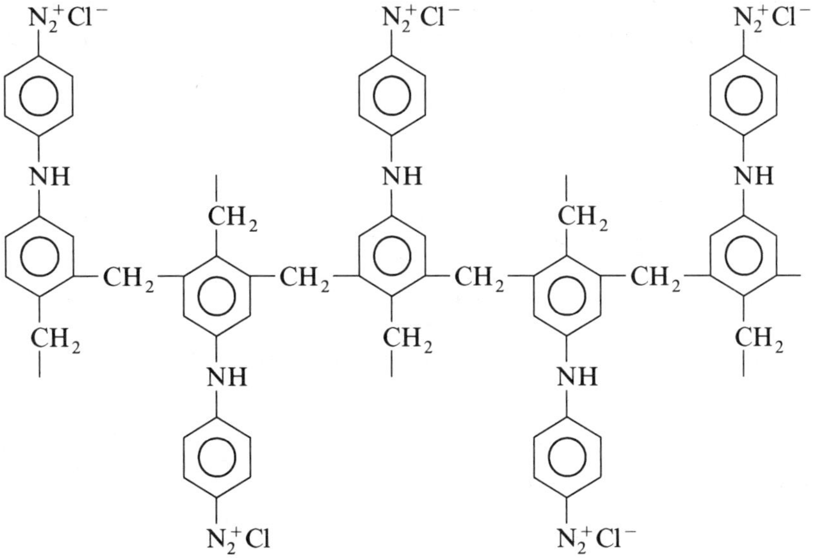

On irradiation the diazonium ion eliminates nitrogen, leaving behind a carbocation that then reacts by substitution with the chloride ion.

$$-\text{HN}-\langle\text{O}\rangle-\text{N}_2^+ \text{ Cl}^- \xrightarrow{h\nu} -\text{HN}-\langle\text{O}\rangle-\text{Cl} + \text{N}_2 \quad (25)$$

The chlorinated resin is no longer water soluble and will act as a negative resist with an aqueous developer. The material is widely used in the printing industry for negative lithographic plates and for the preparation of silverless negatives. In the diazoresin material some crosslinks are also formed on exposure, which make the image mechanically and thermally more resistant. This occurs at sites where a diazonium ion is located near a secondary amine and where the substitution reaction may produce a quarternary ammonium ion and lead to a crosslink.

$$\sim\!\langle\text{O}\rangle-\text{N}_2^+ \text{ Cl}^- + :\overset{|}{\underset{|}{\text{NH}}} \xrightarrow{h\nu} \sim\!\langle\text{O}\rangle-\overset{|}{\underset{|}{\overset{+}{\text{NH}}}} \text{ Cl}^- \quad (26)$$

Recently, the photolysis of diazonium salts was used in a different way to produce a negative working resist [58]. An aqueous acetic acid solution of 4-dimethylaminophenyldiazonium chloride, zinc chloride, and the binder poly-N-vinyl pyrrolidone is coated over a cresol novolac film on a silicon wafer. The diazonium compound diffuses into the phenolic layer and is eventually partitioned between the top layer and the phenolic resin; a uniform two-layer resist is formed. On exposure, the diazonium compound in the top layer bleaches and acts as a contrast enhancing dye (see Chapter 6). In the bottom layer it decomposes to the nonpolar 4-dimethylaminochlorobenzene, which acts as a dissolution inhibitor of the phenolic resin on development in aqueous alkali. A resolution of 0.5 µm, lines and spaces, was achieved with this material.

THE PHYSICAL CHEMISTRY OF CROSSLINKING

In crosslinking resists the irradiated areas are made resistant to solvent by the interconnection of polymer chains into a network. In the preceding sections we have presented the chemical mechanisms by which this is brought about; here we are concerned with the physical consequences of the chemical process.

Let us start with a qualitative description [59]. Crosslinks are formed at the sites of individual chromophores. In the early stages this leads only to an increase in the average molecular weight of the polymer, but it has a dramatic effect on the molecular weight distribution in the system. Since all chromophores (polymer repeat units) have *a priori* the same probability of excitation, in a polydisperse system large molecules are more likely to react than smaller ones. As a consequence, the largest molecule in the ensemble will rapidly outstrip all others. When this many-branched supermolecule permeates the whole irradiated area, it forms there a continuous three-dimensional network that the solvent may penetrate, but which it can no longer disperse: An insoluble residue or gel has been formed. This gel constitutes the image in polymer photography. Image formation in crosslinking resists is thus identified as a process of photoinduced gelation.

The point at which the gel makes its first appearance is called the gel point of the system. At the gel point only an infinitesimally small fraction of the material is part of the network. If crosslinking continues, more and more of the original molecules become linked to the network and the weight fraction of gel gradually increases. A plot of the gel fraction against the exposure (radiation dose) that caused it is termed the gel curve of the photopolymer. Fig. 2-3 shows the gel curve of a resist film of low optical density.

In Fig. 2-3 the weight fraction of gel (W) is plotted as a function of incident exposure in terms of photons per square centimeter. These so-called quantum exposures are often expressed in einsteins per square centimeter, an einstein being one mole of quanta (or photons). Since the gel constitutes the image in crosslinking resists, the gel curve is also the photographic characteristic curve of the resist. The progress of the crosslinking process can be expressed by the weight fraction of the gel, or more commonly by the relative (or normalized) thickness of the film remaining in the exposed areas after development. Following the custom in silver halide photography, the normalized thickness is often plotted as a function of the decadic logarithm of the exposure dose, measured in millijoules per square centimeter. Such a plot is

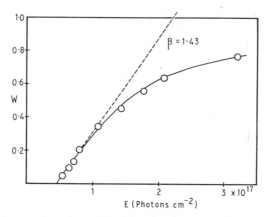

Figure 2-3 Gel curve of a polyester of *p*-phenylenediacrylic acid, sensitized with 3% Michler's ketone, irradiated at 395 nm. β is the dispersity of the molecular weight of the polymer, W the gel fraction, E the exposure dose in photons/cm^2. [Reproduced with permission from *J. Chim. Phys.*, **77**, 469 (1980).]

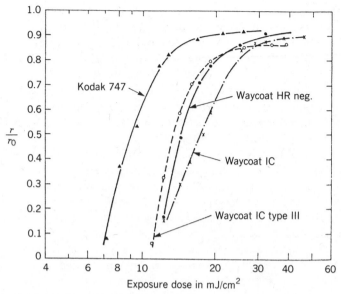

Figure 2-4 Sensitometric curves of a group of commercial negative photoresists. [Reproduced with permission from L. F. Thompson and R. E. Kerwin, *Annu. Rev. Mater. Sci.*, **6**, 272 (1976).]

shown in Fig. 6-2. Equation (27) may be used to convert einsteins per square centimeter into millijoules per square centimeter.

$$D(\text{mJ/cm}^2) = \frac{11.96 \times 10^{10} \, E(\text{einsteins/cm}^2)}{\lambda \, (\text{cm})} \qquad (27)$$

Sensitometric curves for a few typical negative resists [59a] are shown in Fig. 2-4.

Gel Point Exposure and Photographic Sensitivity

In Fig. 2-3 the curve originates abruptly at the gel point, which marks the minimum exposure required for incipient image formation [60]. The gel point exposure or gel dose E_G is a measure of the inherent photographic sensitivity of the system without reference to any particular application. The exposure that satisfies the requirements of a specific application is higher than E_G and defines a work point on the gel curve. For example, in Fig. 2-3 the work point might be reached when about half of the resist in the exposed areas has been insolubilized. The position of the work point varies with the application; it is higher for etching masks, lower for photolithography.

Because gel point and work point differ, the practical speed of a resist depends not only on the position of the gel curve, but also on its initial slope,

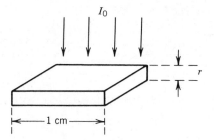

Figure 2-5 Unit area of a resist film of thickness r, exposed to a photon flux I_0.

or more generally, on its shape. The initial slope of the gel curve is related to the photographic contrast of the resist. The two parameters, gel dose and initial slope of the gel curve, provide in most respects an adequate characterization of the photographic performance of a resist.

A relation between gel point exposure E_G and the molecular properties of the photopolymer can be derived from gelation theory [61] (see Fig. 2-5).

Consider unit area of a resist film of thickness r and specific gravity d. If the molecular weight of the monomer unit is M_0, the number of moles of photoreactive groups in unit area of the film is rd/M_0. In the process of irradiation a fraction ρ of the available groups has become crosslinked and the number of crosslinked groups in unit area is therefore, $rd\rho/M_0$. If during irradiation a fraction A of the incident photons is absorbed, the photochemical balance equation can be written in the form

$$EA = \frac{rd\rho}{M_0 \Phi} \qquad (28)$$

where E is the incident radiation dose in einsteins per square centimeter and Φ is the quantum yield of crosslink formation. This quantum yield is defined as the number of groups or chromophores taking part in intermolecular crosslinks, for each photon absorbed in the film.[1]

Equation (28) relates the crosslink density ρ to exposure energy at any stage of the crosslinking process. At the gel point the crosslink density is such that on the average every macromolecule carries one crosslink. This is expressed by Stockmayer's rule

$$\rho_G = \frac{M_0}{M_w} \qquad (29)$$

where M_w is the weight average molecular weight of the polymer. The minimum exposure that brings the photopolymer to the gel point is obtained

---

[1] Only intermolecular crosslinks contribute to network formation and hence to the polymer gel. Intramolecular links between two groups on the same chain produce loops and do not connect with another macromolecule.

by combining Eqs. (28) and (29).

$$E_G = \frac{rd}{AM_w\Phi} \doteq \frac{d}{2.303\varepsilon m M_w \Phi} \qquad (30)$$

On the right of Eq. (30) the absorbed fraction of radiation, A, has been approximated, for $D \ll 1$ by the following expression.

$$A = 1 - 10^{-D} = 2.303D = 2.303\varepsilon mr \qquad (31)$$

In Eqs. (30) and (31) ε is the molar extinction coefficient of the chromophore (reactant group), which can be determined from the absorption spectrum, and m is here the molarity of the chromophore in the solid film. The weight average molecular weight M_w may be obtained either by gel permeation chromatography (GPC) or from light scattering experiments.

Equation (30) is the fundamental equation of crosslinking resists. It relates a photographic characteristic, namely, the gel dose E_G, to molecular properties of the polymer. It shows that the photographic sensitivity is proportional to the quantum yield of the crosslinking reaction, to the weight average molecular weight M_w of the polymer, and to the light-gathering power, εm, of the chromophores. In optically thin films the gel point exposure E_G is independent of film thickness and becomes an intrinsic property of the material.

Equation (30) allows an estimate of the maximum photographic speed that can be achieved with a crosslinking resist. In the absence of a chain mechanism the maximum value of Φ is 2. Assuming realistic values for ε and m and for the molecular weight a value of $M_w = 10^5$, the minimum gel dose required in an optically thin film is found as

$$E_G = 0.25 \times 10^{-9} \text{ einsteins/cm}^2$$

for UV radiation of 360 nm. This corresponds to an energy density of 1×10^{-4} J/cm^2 = 0.1 mJ/cm^2. This value constitutes an absolute benchmark in the evaluation of resist performance. In Table 2-5 it is compared with the practical exposure requirements of several Kodak resists. It can be seen that the sensitivity of commercial crosslinking resists has not yet reached the theoretical maximum.

Equation (30) applies to optically thin films. In many practical applications, however, comparatively thick films are used. Under these conditions an image is formed only when the gel point has been reached in the resist layer adjacent to the support. For that case, a minimum image exposure E_i is readily derived from Eq. (30). It is noted first, that the absorption of the critical sublayer is low and can be approximated by Eq. (31). Furthermore, the fraction of light reaching this layer is determined by the overall transmittance of the film, $T = 10^{-D}$. Thus, for frontal irradiation of a resist layer coated on a nonre-

TABLE 2-5 Exposure Requirements of Photoresists
and of Other Photographic Materials [59]

Photographic Material	E (mJ/cm^2)
Positive AZ resist	100
Kodak Photo Resist (KPR II)	70
Diazo paper	30
Kodak Thin Film Resist (KTFR)	20
Kodak Ortho Resist (KOR)	4
Equation (30)	0.1
Dry silver (3M Company)	0.02

flecting support the minimum exposure is given by [61]

$$E_i = \frac{rd}{AM_w\Phi} \frac{1}{T} \tag{32}$$

If the resist is coated on a transparent support and irradiated through that support, the minimum image exposure and the gel point exposure coincide and Eq. (30) applies directly. This method of irradiation is convenient if the quantum yield of crosslinking of the polymer is to be determined [62].

Resist Contrast. The photographic contrast of a resist film is related to the initial slope of the gel curve. Gelation theory makes it possible to link this to molecular properties of the polymer. The starting point of the argument is a general expression derived by Flory [63], which connects gel fraction W with crosslink density ρ.

$$W = 1 - \sum_{y=1}^{y=\infty} yf(y)[1 - \rho W]^y \tag{33}$$

Here y is the degree of polymerization ($y = M/M_0$), and $f(y)$ is its distribution function in the system.

It so happens that in the majority of crosslinking photopolymers the distribution of molecular weight, and hence of the degree of polymerization, y, is well approximated by a so-called lognormal distribution. That is a distribution, which on a logarithmic scale has the bell-shaped appearance of the classic Gaussian or normal distribution (see Fig. 2-8). It can be represented by a distribution function of the form

$$f(y) = \exp\left\{-\frac{(\ln y - \ln y_0)^2}{2\beta^2}\right\} \tag{33a}$$

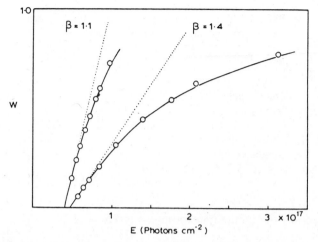

Figure 2-6 Gel curves of two photopolymers with similar gel point exposure doses, but different dispersities of molecular weight. [Reproduced with permission from *J. Photogr. Sci.*, **29**, 187 (1981).]

Here β is the half-width or dispersity of the distribution and y_0 is the most probable value of y located at the apex of the distribution curve.

If this distribution function is introduced into Eq. (33), the initial slope of the gel curve is obtained by differentiation [59] as shown below.

$$\left(\frac{dW}{dE}\right)_G = 2\frac{M_w}{M_0}\frac{M_0 A\Phi}{rd}e^{-\beta^2} = \frac{2}{E_G}e^{-\beta^2} \qquad (34)$$

Introducing the customary expression for resist contrast (γ) with respect to a log E scale one finds

$$\gamma = \left(\frac{dW}{d\log E}\right)_G = 2 \times 2.303\exp(-\beta^2) = 4.606\frac{M_n}{M_w} \qquad (35)$$

It can be seen from Eq. (35) that resist contrast depends only on the width of the molecular weight distribution, measured here by the ratio M_w/M_n [see Eq. (36)]. This point is illustrated in Fig. 2-6; it shows the gel curves of two photopolymers that have similar gel point exposures, but differ in the dispersity of molecular weight.

Shape of the Gel Curve. The behavior of the resist film at all stages of exposure is described implicitly by the Flory function, Eq. (33). If the distribution function of molecular weight is known, the relation between W and ρ or between W and E can be made explicit and the gel curve established. Figure 2-7 shows the Flory functions for lognormal distributions of molecular

THE PHYSICAL CHEMISTRY OF CROSSLINKING   53

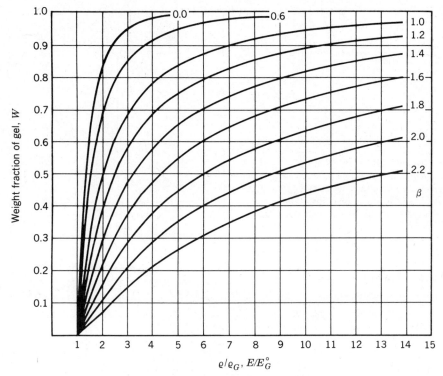

Figure 2-7 Generalized, dimensionless chart of the Flory function $W = \mathscr{F}(E/E_G, \beta)$. In this equation W is identical with the normalized film thickness r/r_0, E and E_G are the exposure dose and the gel point exposure dose, respectively, and β is the dispersity of molecular weight of the original polymer.

weight in a generalized dimensionless plot where the gel fraction W is plotted as a function of the ratio ρ/ρ_G or E/E_G for a range of values of the dispersity β.

The dispersity, which is again defined as the half-width of the bell-shaped lognormal distribution function of molecular weight, can be obtained directly from Gel Permeation Chromatography, as it is shown in Fig. 2-8. It is also related to the ratio of weight average to number average molecular weight by the expression

$$\frac{M_w}{M_n} = \exp(\beta^2) \tag{36}$$

Since the molecular weight distributions of most resists is adequately represented by a lognormal distribution, the Flory function in Fig. 2-7 may quite generally be used to predict resist performance if the dispersity of molecular weight of the material is known.

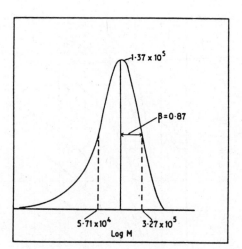

Figure 2-8 Molecular weight distribution in poly(vinyl cinnamate). The dispersity of molecular weight is measured by the half-width of the lognormal distribution curve. [Reproduced with permission from *J. Chim. Phys.*, **77**, 469 (1980).]

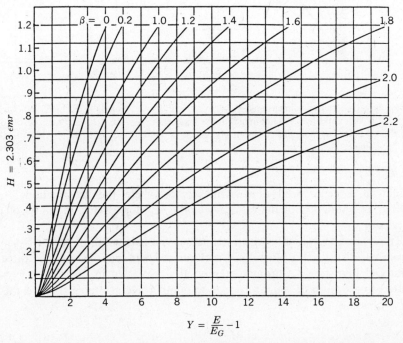

Figure 2-9 Flory function for an optically thick film irradiated through a transparent support. The dimensionless quantities are defined as $H = 2.303\ \varepsilon m r$, $Y = (E/E_G - 1)$ where ε is the molar extinction coefficient of the light absorbing chromophore (either the reactant group or a sensitizer), m its molarity in the solid film, r the film thickness. The chart is useful for the accurate determination of the gel point exposure and the quantum yield of crosslinking. [Reproduced with permission from *Photogr. Sci. Eng.*, **21**, 145 (1977).]

The practical procedure is as follows: The generalized curve corresponding to the dispersity β of molecular weight of the polymer is identified on Fig. 2-7. If the intended application of the material requires a final normalized thickness of, say, $r/r_0 = 0.7$, the corresponding dimensionless exposure ratio E/E_G is found on the axis of abscissas, and from that the actual exposure E is calculated. If E_G is not known from previous experiments, it may be calculated via Eq. (30) from the quantum yield of crosslinking, from the optical properties of the chromophore, and the weight average molecular weight of the polymer.

The Quantum Yield of Crosslink Formation. While gel point exposure is a property of the resist film and depends on film thickness, except for optically thin films, the quantum yield of crosslink formation is a property of the material, irrespective of its shape. It is defined as the ratio of the number of reactant groups involved in crosslinks to the number of photons absorbed in the system.

To determine the quantum yield, the material is coated as a thick film on a transparent support and irradiated through this support so that gel formation starts at the polymer-support interface. By exposing only a small area at a time, and by varying the exposure dose as well as the position of the exposing beam, a series of resist images is produced, corresponding to a photographic step wedge. After development and drying, the thickness of the resist images is measured by interferometry or by other means. A plot of image height against incident dose represents the gel curve of the thick film.[2] The value of the gel point exposure may be obtained from this curve by using a generalized Flory function derived for this purpose by Pitts and co-workers [62] and presented in dimensionless form in Fig. 2-9.

In this figure, the dimensionless quantity H

$$H = 2.303 \varepsilon m r \qquad (37)$$

is plotted as a function of the dimensionless exposure Y

$$Y = \frac{E}{E_G} - 1 \qquad (38)$$

This function depends again on the dispersity β of molecular weight. In using the chart of Fig. 2-9 for every value of the film thickness, r, the value of H is determined and a corresponding value of Y is read off the appropriate curve. These values are plotted against the incident exposure E that produced the film thickness r (see Fig. 2-10). A straight line is the result, the slope of which

---

[2] This is not identical with the gel curve of a thin film as represented in Fig. 2-7, where it was assumed that the intensity of the radiation flux is uniform throughout the depth of the coating.

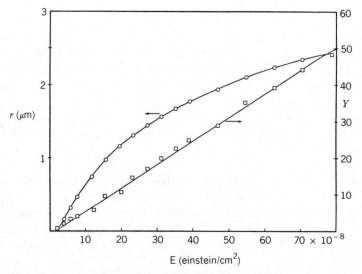

Figure 2-10 Gel curve, a plot of developed film thickness (r) versus exposure dose, of a thick film of poly(vinyl cinnamylidene acetate) irradiated through a quartz support with 365 nm radiation. The straight line is the function $Y = F(E)$ derived from the gel curve via the chart in Figure 2-9. The slope of the straight line is the reciprocal of the gel point exposure dose. [Reproduced with permission from *J. Chim. Phys.*, **77**, 469 (1980).]

is the reciprocal of gel point exposure.

$$\frac{dY}{dE} = \frac{d(E/E_G)}{dE} = \frac{1}{E_G} \tag{39}$$

Once the critical gel dose has been obtained, the quantum yield of crosslinking is calculated from Eq. (30).

Resist Processing

The term "resist processing" denotes all the operations in which the resist takes part and which, together, produce the final result, for example a finished printing plate or an integrated circuit pattern. While the specifics vary from system to system, the general sequence of operations is the following [64].

1. *Substrate preparation.* This involves the degreasing and drying of the surface and sometimes the precoating with an adhesion promoter. Degreasing is carried out with solvents or with aqueous (alkaline) detergents, and it is followed by a pH adjustment and rinse.

2. *Resist coating.* This is carried out by distributing a coating solution (5–15% solids) over the surface of the substrate with a coating knife, a roller, a coating hopper, or a (slow) rotary whirler. In microlithography the resist is

applied to the wafer by spin coating on a fast-rotating turntable or spinner. The thickness of the coating is controlled either by the setting of the gap between knife (hopper) and substrate, or by the viscosity and the speed of flow of the solution over the surface. In microlithography the relation between viscosity, spin speed measured in rpm (revolutions per minute), and final coating thickness is an important characteristic of the resist and is usually supplied in the form of a chart (see, e.g., Fig. 5-21).

3. *Prebake*. The air-dry films are subjected to a thermal treatment to remove the last traces of coating solvent and to release the mechanical strains in the coated films. In some systems the prebake conditions are rather critical and can make the difference between success and failure of the whole imaging operation. (KMR 747, e.g., is heated for 30 min to 85–90°C in an oven with intensive air flow.)

4. *Exposure*. This is of course the central imaging operation and the final result depends very much on the quality of the radiation pattern that is impressed onto the resist (see Chapter 6). At this point, the important quantity is the exposure dose. For example, the recommended exposure for KMR 747 films of 1 to 2 μm thickness is 20 to 30 mJ/cm^2.

5. *Development*. In a negative resist this is the treatment of the exposed resist film with solvent to remove the unexposed areas. It can be carried out by simply immersing the work piece in the developer (puddle development) or by washing or scrubbing it in a highly agitated developer bath. Alternatively, the resist may be exposed to a fine spray of developer under pressure (spray development). To be practical in a manufacturing process, development should not take more than 1 min. Exposed KMR 747, for example, is spray developed for 30 s in n-decane or in petroleum ether 80 to 100°C and then rinsed for 20 s in butyl acetate. It is finally dried in a rapid air jet (air knife).

6. *Postbake*. The developed resist pattern is heated to remove traces of developer and anneal the polymer pattern. (KMR 747 is heated in postbake to 120–130°C for 30 min.) The resist pattern is now ready to be used as an etch mask, for the patterning of metal, and so on.

7. *Stripping*. After the resist pattern has been used in the manufacturing operation it is removed from the surface of the substrate (stripped) by soaking or scrubbing in a *strong* solvent or by exposure to an oxygen plasma (ashing).

All these operations have a bearing on the chemical and physical properties, which the resist must have. The properties of the material, however, are most critically involved with the development step.

Development of Crosslinking Resists. In crosslinking resists development of the latent image after exposure entails not only the removal of the unexposed areas of the film, but also in the exposed areas the removal of polymer chains that have not become attached to the network of the gel. For this to happen,

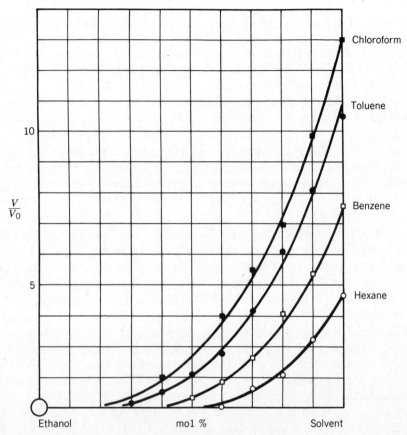

Figure 2-11 The swelling of a lightly crosslinked resist film in several mixed solvents as indicated. The resist is poly(*cis*-isoprene) crosslinked with 1% by weight of 4,4'-diazidostilbene. The swelling ratio V/V_0 is plotted against the composition of the mixed solvents. V is the swollen volume, V_0 is the unswollen volume of the sample.

the resist must swell in the developer. This, however, changes the dimensions of the image and leads to image degradation.

The extent of swelling that may be encountered in crosslinked films is shown in Fig. 2-11. The data in the figure were obtained by coating a lightly crosslinked film of poly(*cis*-isoprene) on a release plate, and floating it off into solvent to let it expand freely. When the liquid medium was gradually changed from ethanol (a nonsolvent) to toluene (a thermodynamically good solvent) the volume of the sample increased about ten-fold, corresponding to a linear expansion by a factor of 2.2. If this film is transferred directly from toluene to ethanol it contracts rapidly to one tenth of its swollen volume. Such a large change cannot be made reversibly in a single step, but if the composition of the solvent is changed gradually, the expanded film will return reversibly to its original dimensions [65].

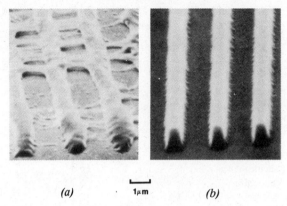

Figure 2-12 A negative electron resist, poly(styrene-co-chlorostyrene), exposed and developed (*a*) 10 s in toluene, (*b*) 30 s in methyl isobutyl ketone. [Reproduced with permission from E. D. Roberts, *Solid State Technol.*, June, p. 135 (1984).]

Different solvents cause swelling to a different degree: In the system of Fig. 2-11, chloroform expands the poly(*cis*-isoprene) network greatly, a poorer solvent such as hexane has a much smaller effect. What this can mean in practice is shown in Fig. 2-12 where the same resist image is developed in a thermodynamically good solvent (toluene) and, for comparison, in a poor solvent (methyl isobutyl ketone).

Several lessons are drawn from these results: First, to keep swelling reversible it is advantageous to use several solvents (developers and rinses) in succession, and to vary solvent power gradually. Second, to minimize swelling, the developer should only be a moderately good solvent in the thermodynamic sense. Third, for dissolution to proceed at a reasonable rate, the developer has to be a kinetically good solvent with a fast rate of penetration into the polymer.

Novembre et al. [66] have shown how the choice of developer and rinses may be optimized by using the solubility parameters of the polymer and the solvents. Solubility parameters were originally introduced by Hildebrand and Scott [67] as a measure of the cohesive energy density of liquids. They defined the solubility parameter δ by the expression

$$\delta = \left[E_{\text{vap}}/V \right]^{1/2} \tag{40}$$

where E_{vap} is the molar energy of vaporization and V is the molar volume of the liquid. It was thought that liquids with similar cohesive energy densities will mix freely and that in general two substances with similar values of δ will be compatible with each other, the degree of compatibility being measured by the difference in their solubility parameters. Hansen [68] has developed this idea further by assigning separate δ values to the three principal forces

TABLE 2-6 Solubility Parameters for Potential Developers and Rinses to be Used with a Particular Electron Resist [66]

	δ_v	δ_h
Polymer	18	7
Acetone	16.3	11.0
Methyl butyl ketone	16.2	7.1
Toluene	18.2	1.6
Ethylene glycol diacetate	18.2	10.6
n-Butyl acetate	16.5	6.8
Isobutyl acetate	16.5	5.1
sec-Butyl acetate	15.7	5.0
n-Amyl acetate	16.1	5.7
n-Hexyl acetate	16.5	6.4
n-Heptyl acetate	16.0	4.9
2-Propanol	8.7	8.0

responsible for cohesion: the dispersive or van der Waals forces, the permanent dipole forces, and hydrogen bonds.[3]

$$\delta^2 = \delta_d^2 + \delta_p^2 + \delta_h^2 \qquad (41)$$

Solubility parameters are tabulated for a great number of materials [68]. Table 2-6 gives values for a few solvents [69].

Novembre has combined the dispersive and polar parameters,

$$\delta_v = \left(\delta_d^2 + \delta_p^2\right)^{1/2} \qquad (42)$$

By plotting δ_v against δ_h a two-dimensional solubility map is established in which every solvent is represented by a point. Such a map is shown in Fig. 2-13. The solubility parameters of the polymer are defined by the center of gravity of the map points of those solvents in which the polymer is soluble. For the polymer of Fig. 2-13 [66], a polystyrene based electron resist, the coordinates of this point are $\delta_v = 18$ and $\delta_h = 7$.

In keeping with the idea of least possible polymer–solvent interaction, a developer will be chosen near the circumference of the solubility region, say with $\delta_v = 16$ and $\delta_h = 6$. For the resist considered in Fig. 2-13, methyl butyl ketone, n-butyl acetate, and n-hexyl acetate fit these requirements most closely (see Table 2-6); toluene and ethylene glycol diacetate are clearly too powerful solvents for this polymer and would induce excessive swelling.

---

[3]Although the generality of a hydrogen bonding parameter for a particular component is questionable, since the interaction depends critically on the hydrogen donor or acceptor character of the other components, the three-dimensional representation of solubility has been found useful in practice.

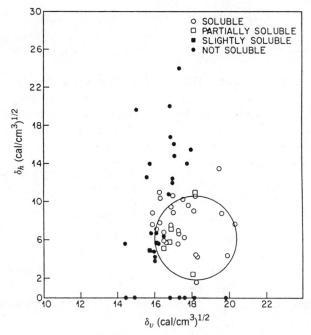

Figure 2-13 Map of solubility parameters for an experimental polystyrene-based electron resist. The image points represent solvents with their δ_h and δ_v parameters as coordinates. The solvents are coded according to their ability to dissolve the polymer. The solubility parameters of the polymer are given by the center of gravity of the image points of the liquids that dissolve it (indicated by the circle). [Reproduced with permission from A. Novembre et al., *Tech. Pap. Reg. Tech. Conf. Soc. Plast. Eng.*, Ellenville, NY, Oct. p. 223 (1985).]

As was mentioned earlier, the overall suitability of a solvent depends also on its kinetic properties, that is, on its rate of penetration into the polymer. Diffusion coefficients and molecular sizes are therefore important. Liquids with almost identical thermodynamic parameters may behave quite differently in this respect. For example, the isomers of butyl acetate are thermodynamically very similar (see Table 2-6), but only *n*-butyl acetate is a kinetically good solvent and may be used as developer, isobutyl acetate is only a partial solvent, and *sec*-butyl acetate is practically a nonsolvent. With this information the final choice for the polymer was as follows. Developer: *n*-butyl acetate (a moderately good solvent); first rinse: *n*-hexyl acetate (a somewhat poorer solvent); and second rinse: 2-butanol (a nonsolvent).

REFERENCES

1. H. Morawetz, *Angew. Chem., Int. Ed. Engl.*, **26**, 93 (1987).
2. M. Biltz and J. Eggert, *Wiss. Veroeff. AGFA*, **3**, 294 (1928).
3. P. Datta and B. R. Sollern, *18th SPSE Fall Symp.*, Nov. 1978.

4. B. Duncalf and A. S. Dunn, *Print. Technol.*, **14**(3), 125 (1970).
5. L. M. Minsk and W. P. Van Deusen, U.S. Patent 2,690,966 (1948); L. M. Minsk, J. G. Smith, W. P. Van Deusen, and J. F. Wright, *J. Appl. Polym. Sci.*, **11**, 302 (1959).
6. H. Stobbe, *Ber.*, **45**, 3396 (1912); H. Stobbe and A. Lehfeldt, *Ber. B*, **58**, 2415 (1925).
7. P. L. Egerton, E. Pitts, and A. Reiser, *Macromolecules*, **14**, 95 (1981).
8. E. M. Roberston, W. P. Van Deusen, and L. M. Minsk, *J. Appl. Polym. Sci.*, **11**, 308 (1959).
9. C. W. Leubner and C. C. Unruh, U.S. Patent 3,257,664 (1966).
10. H. Tanaka, M. Tsuda, and H. Nakanishi, *J. Polym. Sci., Part A-1*, **10**, 1729 (1972).
11. H. Tanaka and K. Honda, *J. Polym. Sci., Polym. Chem. Ed.*, **15**, 2685 (1977).
12. C. C. Unruh and A. C. Smith, *J. Appl. Polym. Sci.*, **3**, 310 (1960).
13. Farbenfabricken Bayer A.G., Br. Patent 838,547 (1968).
14. J. L. R. Williams, S. Y. Farid, J. C. Doty, R. C. Daly, D. P. Specht, R. Searle, D. G. Borden, J. J. Chang, and P. A. Martic, *Pure Appl. Chem.*, **49**, 523 (1977).
15. P. L. Egerton, J. Trigg, E. M. Hyde, and A. Reiser, *Macromolecules*, **14**, 100 (1981).
16. J. A. Arcesi and F. J. Rauner, U.S. Patent 3,640,722 (1972).
17. D. G. Borden, *Polym. Eng. Sci.*, **14**, 487 (1974).
18. M.-Y. Li, E. M. Pearce, A. Reiser, and S. Narang, *J. Polym. Sci., Polym. Chem. Ed.*, **26**, 2517 (1988); M.-Y. Li, E. M. Pearce, and S. Narang, *Proc. SPIE* **77**, 46 (1987).
19. K. Ichimura and N. Oohara, *J. Polym. Sci., Polym. Chem. Ed.*, **25**, 3063 (1987).
20. J. L. R. Williams, in *Polyelectrolytes* (E. Selegny, Ed.), Reidel, Dordrecht, Netherlands, 1974.
21. G. Leubner, J. L. R. Williams, and C. C. Unruh, U.S. Patent 2,811,510 (1957).
22. J. L. R. Williams and D. G. Borden, *Makromol. Chem.*, **73**, 203 (1964).
23. D. G. Borden and J. L. R. Williams, *Makromol. Chem.*, **178**, 3053 (1972).
24. K. Ichimura and S. Watanabe, *J. Polym. Sci., Polym. Chem. Ed.*, **18**, 891 (1980).
25. K. Ichimura, *J. Polym. Sci., Polym. Chem. Ed.*, **20**, 1411 (1982); K. Ichimura and S. Watanabe, *J. Polym. Sci., Polym. Chem. Ed.*, **20**, 1419 (1982).
26. G. O. Schenk, W. Hartmann, S. P. Mansfeld, W. Metzner, and C. H. Krauch, *Chem. Ber.*, **95**, 1642 (1962); see also F. C. DeSchryver, N. Boens, and G. Smets, *J. Polym. Sci., A-1*, **10**, 1687 (1972).
27. M. E. Baumann and H. Zweifel, assigned to Ciba-Geigy, U.S. Patent 4,079,041 (1978); assigned to Ciba-Geigy, Eur. Patent 21,019 (1981).
28. J. Berger and H. Zweifel, *Angew. Makromol. Chem.*, **115**, 163 (1983); J. Finter, E. Widmer, and H. Zweifel, *Angew. Makromol. Chem.*, **128**, 71 (1984); J. Finter, A. Haniotis, F. Lohse, K. Meier, and H. Zweifel, *Angew. Makromol. Chem.*, **133**, 147 (1985).
29. C. D. De Boer, *J. Polym. Sci., Polym. Lett. Ed.*, **11**, 25 (1973).

29a. M. V. Mijovic, P. J. Beynon, T. J. Shaw, K. Petrak, and A. Reiser, *Macromolecules*, **15**, 1464 (1982).
30. M. Hepher and H. M. Wagner, Br. Patent 762,985 (1954).
31. A. Reiser and H. M. Wagner, in *The Chemistry of the Azido Group* (S. Patai, ed.), pp. 461 ff., Interscience, New York, 1971.
32. E. Wasserman, G. Smolinksy, and W. A. Yager, *J. Am. Chem. Soc.*, **86**, 3166 (1964).
33. A. Reiser and V. Frazer, *Nature*, **208**, 682 (1965).
34. A. Reiser, G. C. Terry, and F. W. Willets, *Nature*, **211**, 410 (1966).
35. S. H. Merrill and C. C. Unruh, *J. Appl. Polym. Sci.*, **7**, 273 (1963).
35a. A. Reiser and R. Marley, *Trans. Faraday Soc.*, **64**, 1806 (1968).
36. A. Reiser, L. J. Leyshon, and L. Johnston, *Trans. Faraday Soc.*, **67**, 2389 (1971).
37. W. S. De Forest, *Photoresists; Materials and Processes*, McGraw Hill, New York, 1975.
38. A. Stein, *The Chemistry and Technology of Negative Photoresists*, A Waycoat Tutorial, Philip A. Hunt Chemical Co., Palisades Park, NJ 07650, 1984.
39. R. K. Agnihotri, D. L. Falcon, F. P. Hood, L. G. Lesoine, C. D. Needham, and J. A. Offenbach, *Photogr. Sci. Eng.*, **16**, 443 (1972).
40. M. Agaki, S. Nonogaki, T. Kohashi, Y. Oba, M. Oikawa, and Y. Tomita, *Photogr. Sci. Eng.*, **17**, 353 (1977).
41. T. Kohashi, M. Agaki, S. Nonogaki, N. Hayashi, and Y. Tomita, *Photogr. Sci. Eng.*, **23**, 168 (1979).
42. T. Iwayanagi, M. Hashimoto, and S. Nonogaki, *Polym. Eng. Sci.*, **23**, 935 (1983).
43. S. Nonogaki, M. Hashimoto, T. Iwayanagi, and H. Shiraishi, *Proc. SPIE* **539**, 189 (1985).
44. M. Hashimoto, T. Iwayanagi, H. Shiraishi, and S. Nonogaki, *Tech. Pap., Photopolym. Conf. SPE*, Ellenville, NY, Oct. 1985.
45. R. Vicari, Ph.D. Thesis, Polytechnic University, Brooklyn, NY, 1986.
45a. J. M. Duff and A. G. Brook, *Can. J. Chem.*, **51**, 2869 (1973).
46. N. J. Turro, *Modern Molecular Photochemistry*, pp. 262 ff., Benjamin-Cummings, Menlo Park, CA, 1978.
47. K. Horie, H. Audo, and J. Mita, *Macromolecules*, **20**, 54 (1987).
48. G. J. Smets, S. N. El Hamouly, and T. J. Oh, *Pure Appl. Chem.*, **56**, 439 (1984).
49. J. Pfeifer and O. Rohde, *Proc. 2nd SPE Int. Conf. Polymides*, Ellenville, NY, Oct.–Nov. 1985.
50. O. Rohde, M. Riediker, A. Schaffner, and J. Bateman, in *Advances in Resist Technology and Processing II, Proc. SPIE* **539**, 175 (1985).
51. A. A. Lin, V. R. Sastri, G. Tesoro, A. Reiser, and R. Eachus, *Macromolecules*, **21**, 1165 (1988).
52. L. Plambeck, A. Cairncross, W. J. Chambers, C. S. Cleaver, D. S. Donald, and D. F. Eaton, *J. Imaging Sci.*, **30**, 221, 224, 228 (1986).
53. T. H. James, in B. H. Carroll, G. A. Higgins, and T. H. James, "*Introduction to Photographic Theory*," Chap. 10, Wiley, New York, 1980.
54. R. Schwalm, A. Bötcher, and H. Koch, *Proc. SPIE*, **920**, 3 (1988).

55. M. P. Schmidt and R. Zahn, Ger. Patent 596,731 (1934).
56. Kalle A. G., Belg. Patents 630,565 and 630,566 (1963).
57. M. Hepher, *J. Photogr. Sci.*, **12**, 181 (1964).
58. S. Uchino, T. Iwayanagi, and M. Hashimoto, *Proc. SPIE*, **920**, 13 (1988).
59. A. Reiser and E. Pitts, *J. Photogr. Sci.*, **29**, 187 (1981).
59a. L. F. Thompson and R. E. Kerwin, *Annu. Rev. Mater. Sci.*, **6**, 267 (1976).
60. A. Reiser, *J. Chim. Phys.*, **77**, 469 (1980).
61. A. Reiser and E. Pitts, *Photogr. Sci. Eng.*, **20**, 225 (1976); J. Finter, Z. Haniotis, K. Meier, and H. Zweifel, *J. Imaging Sci.*, **30**, 259 (1986).
62. N. J. H. Parham, E. Pitts, and A. Reiser, *Photogr. Sci. Eng.*, **21**, 145 (1977).
63. P. J. Flory, *Principles of Polymer Chemistry*, Chaps. IV and IX, Cornell University Press, Ithaca, New York, 1953.
64. L. F. Thompson and M. J. Bowden, in *Introduction to Microlithography*, *ACS Symp. Ser.* **219**, Chap. 4 (1983).
65. L. F. Leyshon and A. Reiser, unpublished results, 1973.
66. A. E. Novembre, L. M. Masakowski, and M. A. Hartney, *Tech. Pap.*, *Photopolym. Conf. SPE*, Ellenville, NY, Oct. 1985.
67. J. Hildebrand and R. Scott, *Solubility of Non-Electrolytes*, 3rd ed., Reinhold, New York, 1949.
68. C. M. Hansen, *J. Paint Technol.*, **39**, 104, 505 (1967).
69. A. F. Barton, in *CRC Handbook of Solubility Parameters*, Chaps. 7 and 8, CRC Press, Boca Raton, FL, 1983.

3 ASPECTS OF PHOTOPHYSICS AND PHOTOCHEMISTRY IN SOLID POLYMERS

EXCITED STATES

Most industrial applications of organic photochemistry are concerned with reactions that occur in thin polymer films. A short section of this text is therefore devoted to the photochemistry of amorphous solid systems. We can supply here only the most rudimentary information required for an understanding of the subject. For a comprehensive exposition of the underlying photochemistry the reader is referred to the excellent textbook by Turro [1] and to other monographs, for example that in reference [2].

A brief introduction to the terminology is in order here. The excited states of organic molecules that play the central role in organic photochemistry are identified by their multiplicity (the overall spin of the state), by their orbital character (indicating the involvement of π, n, σ, or other orbitals in the excitation process), and by their energy in relation to the energy of the ground state. The relevant excited states of a molecule are often represented in a state diagram of the type shown in Fig. 3-1 for 1-chloronaphthalene.

The excited singlet state S_1 of a molecule is produced either by direct absorption of a photon or by energy transfer from another excited molecule. The fate of the excitation energy after the initial event can be read off the diagram in Fig. 3-1. The excited singlet state S_1 can either emit fluorescence (rate constant $k_F = 6 \times 10^6 \text{ s}^{-1}$) or it may be deactivated by a nonradiative transition to the ground state ($k < 10^4 \text{ s}^{-1}$).[1] The fastest process originating

---

[1] Nonradiative transitions between states of the same multiplicity (i.e., singlet to singlet or triplet to triplet) are called *internal conversions*, transitions between states of different multiplicity are termed *intersystem crossings*.

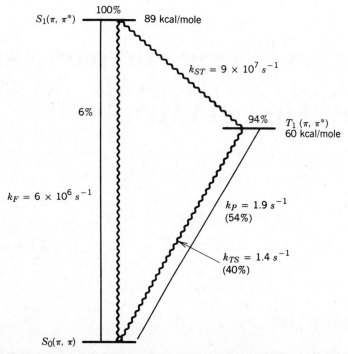

Figure 3-1 State diagram of 1-chloronaphthalene, showing the energy levels of the lowest excited singlet and triplet states and indicating the rate constants for the transitions between states. (After N. J. Turro [1]. © Benjamin-Cummings, 1978.)

in S_1 is the transition to the lowest triplet state T_1, an intersystem crossing that proceeds here with a rate constant of $k_{ST} = 9 \times 10^7 \text{ s}^{-1}$. Intersystem crossings are spin forbidden transitions that are made possible by the intervention (in aromatic systems) of out-of-plane vibrations or by the state-mixing effects of heavy atoms, such as, for example, Cl in chlorobenzene. The fraction of excited S_1 states that emit fluorescence (6% in this case), is determined by the competition between the radiative fluorescence transition, internal conversion to S_0 and intersystem crossing to T_1.

The excited triplet state in turn is deactivated by a radiative process (phosphorescence, rate constant $k_P = 1.9 \text{ s}^{-1}$) and by a nonradiative intersystem crossing to the ground state ($k_{TS} = 1.4 \text{ s}^{-1}$). The data in Fig. 3-1 refer to chlorobenzene at 77 K when both processes are about equally likely; 57% of triplet states emit phosphorescence, 43% return nonradiatively to the ground state. Figure 3-2 shows the emission spectra of 1-chloronaphthalene, fluorescence and phosphorescence, taken in solid solution at 77 K. The wavelengths of the 0,0-vibronic transitions measure the energies of the electronic levels S_1 and T_1 as shown in Fig. 3-1.

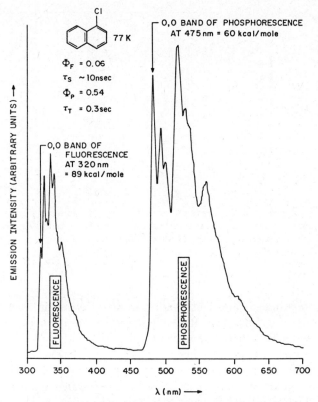

Figure 3-2 The emission spectrum of 1-chloronaphthalene dissolved in an organic glass of ether, pentane, ethanol 5:5:2 at 77 K. In these conditions both fluorescence and phosphorescence are observed. (Reproduced with permission from N. J. Turro [1]. © Benjamin-Cummings, 1978.)

At room temperature, the nonradiative intersystem crossing process T_1–S_0, has a rate constant of the order of 10^4 s^{-1} and far outruns the radiative process so that no phosphorescence is observed under these conditions. It will be noted, however, that even at room temperature the return to the ground state from the triplet state is several orders of magnitude slower than the deactivation of the singlet state. This means that triplet states have in general longer lifetimes than excited singlet states. As a result, most photochemical reactions in fluid solution originate in the triplet state, or more precisely, triplet state reactions have usually higher quantum yields in fluid systems than singlet state reactions.

Excimers. In photopolymer systems an important type of excited state are excimers [3]. Excimers are formed by pairs of aromatic or heteroaromatic molecules that do not significantly interact in the ground state, but are weakly

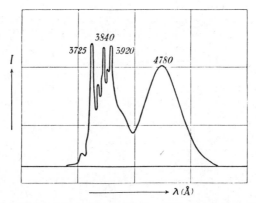

Figure 3-3 The fluorescence emission of pyrene in 2×10^{-3} M solution in benzene, showing the structured monomer fluorescence and the structureless excimer fluorescence, $\lambda_{max} = 4780$ Å. This figure is taken from a paper by Förster and Kaspar [4], where the discovery of excimers was first announced. (Reproduced with permission from Oldenbourg Verlag.)

bonded in the excited state. This bonding occurs between an excited molecule and a ground state molecule of the same species. It has its origin in the change of orbital symmetry that accompanies excitation and leads to cooperative (positive) orbital overlap and hence to bonding between the two systems. Excimers were discovered by Förster and Kasper [4] in 1954 when they observed two kinds of fluorescence in fairly concentrated solutions (10^{-3} M) of pyrene. Figure 3-3, taken from their original paper, shows the "normal" fluorescence of monomeric pyrene, which has the characteristic vibrational structure of the emission spectrum of an aromatic molecule, and the strong, red shifted, completely structureless emission of the excimer.

The red shift and the loss of vibrational structure can be understood from the shape of the potential energy surfaces of the two molecules as they approach each other broadside (Fig. 3-4). In the ground state the two molecules repel each other when they come within the distance of their van der Waals radii, but in the excited state the attractive force of positive orbital overlap creates a potential energy well that defines the excimer. The depth of the well is the excimer binding energy B; that is, that amount of energy by which the excimer is stabilized relative to the energy of the excited state of an isolated (monomeric) molecule. Fluorescence emission returns the pair molecule to the shoulder of the repulsive branch of the potential energy curve. The energy change corresponding to the fluorescence transition from the excimer, ΔE_{FD}, is smaller than the energy change in the fluorescence of the monomer, ΔE_{FM}, hence the red shift in the emission. Furthermore, it covers a continuous range of values and obliterates any vibrational fine structure of the molecular skeleton. In solid solutions of organic molecules pair sites do occur that have the correct geometry for excimer formation. On irradiation, these sites emit excimer fluorescence. Figure 3-5 shows the gradual change in the fluorescence

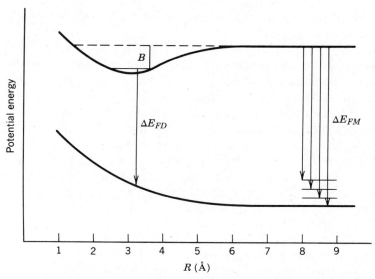

Figure 3-4 Potential energy diagram of an excimer forming pair of molecules. Lower curve: both molecules in ground state. Upper curve: excimer formation on the approach between an excited molecule and a molecule in the ground state. ΔE_{FM} is the excitation energy of monomer, ΔE_{FD} is the excitation energy of excimer, and B is the excimer binding energy.

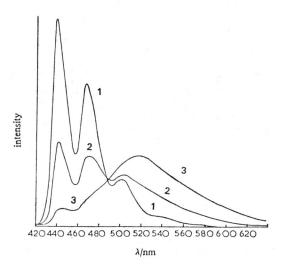

Figure 3-5 Fluorescence emission of perylene dissolved in a matrix of poly(methyl methacrylate) showing the gradual appearance of excimer emission as the concentration of perylene increases. Perylene concentrations (mole/kg^1): (1) 5×10^{-3}; (2) 5.33×10^{-2}; (3) 2.04×10^{-1}. (Reproduced with permission from J. A. Ferreira and G. Porter [5].)

spectrum of perylene as its concentration in a solid matrix of poly(methyl methacrylate) is increased.

Exciplexes. When two molecules of similar, but not identical structure, for example, anthracene and tetracene, interact on excitation, heteroexcimers are formed. If, however, the two molecules differ significantly in their electron affinities so that one may be regarded as an electron donor and the other as an electron acceptor, the bonding process is accompanied by a partial transfer of charge. In that case the interaction is stronger than in conventional excimers and the transient excited species is termed an exciplex (excited complex), [6, 7]. Exciplexes emit a structureless, red-shifted fluorescence, or phosphorescence, similar to that of an excimer. Figure 3-6 shows the formation of an exciplex as the electron acceptor biphenyl is gradually added to a solution of the electron donor, dimethylaniline [8].

Exciplexes are often stages in the complete transfer of an electron from one molecule to another leading to the formation of radical ion pairs and finally of separated radical ions. An important example is the photoreaction of benzophenone with tertiary amines [9].

$$\begin{array}{c} Ph \\ Ph \end{array}\!\!C\!=\!\overline{O}^* + \begin{array}{c} CH_3 \\ | \\ |N\!-\!CH_3 \\ | \\ CH_3 \end{array} \longrightarrow \left[\begin{array}{cc} \overset{\delta -}{|\overline{O}|} & \overset{\delta +}{} \\ | & | \\ -C\!\cdots\!N- \\ | & | \end{array} \right]^{*}$$

$$\downarrow$$

$$\begin{array}{c} Ph \\ Ph \end{array}\!\!\dot{C}\!-\!\overline{\underline{O}}|^{-} + \begin{array}{c} CH_3 \\ | \\ H\!-\!\overset{+}{N}\!-\!CH_3 \\ | \\ \cdot CH_2 \end{array} \longleftarrow \begin{array}{c} Ph \\ Ph \end{array}\!\!\dot{C}\!-\!\overline{\underline{O}}|^{-} + \begin{array}{c} CH_3 \\ | \\ \cdot \overset{+}{N}\!-\!CH_3 \\ | \\ CH_3 \end{array}$$

ENERGY TRANSFER

Energy flows not only between the different states of one molecule, it can also be transferred between molecules that are in proximity of each other. Indicating the molecule that carries excitation energy as the donor (D), the other molecule as the acceptor (A), the process can be represented by reaction (1)

$$D^* + A \longrightarrow D + A^* \qquad (1)$$

Whatever the precise mechanism, energy transfer is an electronic process and

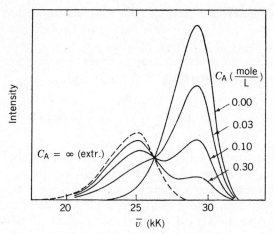

Figure 3-6 Fluorescence emission of the electron donor dimethylaniline in toluene solution, in the presence of increasing concentrations (c_A) of the electron acceptor biphenyl. (Reproduced with permission from H. Knibbe and A. Weller [8].)

therefore essentially adiabatic; it will occur with reasonable probability only if the excitation energy of D^* is equal to or greater than that of A^*.[2]

$$E_{D^*} \geq E_{A^*} \qquad (2)$$

At the moment of transfer, donor and acceptor are coupled and form a single quantum mechanical entity. Two distinct coupling mechanisms are important in this connection: coulombic or dipolar interaction (Förster) and electron exchange or orbital interaction (Dexter).

Dipole Resonance Transfer

Energy transfer by coulombic interactions is in essence a dipole resonance effect. Förster [11, 12] has shown that under favorable circumstances the electronic transitions of two molecules may couple in a way similar to the coupling of two oscillating dipoles where energy is transmitted from one to the other. In the molecular case the rate of energy transfer for this kind of coupling is given by the expression

$$k_{ET}(\text{coulombic}) = \frac{\mu_D^2 \mu_A^2}{R_{DA}^6} \qquad (3)$$

where k_{ET} is a rate constant ($M^{-1} s^{-1}$), μ_D and μ_A are the transition dipole

---

[2] See, however, Sandros and Bäckstrom [10] on reversible energy transfer when E_{A^*} is slightly larger than E_{D^*}.

moments of the fluorescence transition (D*–D) and the absorption transition (A–A*), respectively, and R_{DA} is the separation between the two molecular centers at the moment of transition.

By expressing the transition moments in terms of measurable quantities the Förster transfer rate can be written in the form [13]

$$k_{ET} = 8.8 \times 10^{-25} \frac{\kappa^2 \phi_F(D)}{n^4 R^6 \tau_D^0} \int f_D(\nu) \varepsilon_A(\nu) \frac{d\nu}{\nu^4} \qquad (4)$$

Here n is the index of refraction of the medium, $\phi_F(D)$ is the quantum yield of fluorescence of the donor, τ_D^0 is the fluorescence lifetime of the donor in the absence of the acceptor, κ^2 is a geometric factor, which in fluid systems (where the rotational relaxation time is shorter than the lifetime of the excited state) has a value of 0.67. In a rigid medium the value of κ^2 is 0.457 [13]. The fluorescence spectrum of the donor normalized to unity is $f_D(\nu)$, $\varepsilon_A(\nu)$ is the absorption spectrum of the acceptor (not normalized!), and ν are wave numbers.

It can be seen that in dipole resonance transfer the rate of transfer depends on the fluorescence intensity and on the fluorescence lifetime of the donor, as well as on the spectral overlap between the fluorescence of the donor and the absorptance of the acceptor. The most important feature of this transfer mechanism is the sixth-power dependence on the separation of the two molecules. On closer scrutiny it is found that this should make it possible to transmit energy by resonance transfer over distances of up to 50 Å.

The general propensity for energy transfer between a pair of molecules is characterized by a critical distance R_0 at which the rate of energy transfer and the rate k_d of spontaneous deactivation of the donor are equal [14]. This condition corresponds to a concentration of acceptor, $[A]_{1/2}$, at which the fluorescence of the donor is reduced to half its original value (i.e., the value it had in the absence of acceptor). The value of $[A]_{1/2}$ follows from the equality

$$k_{ET}[D^*][A] = k_d[D^*]$$
$$[A]_{1/2} = \frac{k_d}{k_{ET}} \qquad (5)$$

and the critical distance R_0 is linked to $[A]_{1/2}$ by the relation

$$R_0(\text{Å}) = \frac{7.35}{\sqrt[3]{[A]_{1/2}}} \qquad (6)$$

Table 3-1 lists the critical transfer distances for three donor–acceptor pairs that have favorable spectral overlap. They show that the dipole resonance mechanism allows for energy transfer over distances corresponding to several molecular diameters.

TABLE 3-1 Förster Transfer between Donors and Acceptors with Favorable Spectral Overlap [1]

Donor	Acceptor	R_0 Theory (Å)	R_0 Experimental (Å)
Bis(hydroxyethyl)-2,6-naphthalene dicarboxylate	Sevron Yellow	26	27 ± 1
Pyrene	Sevron Yellow	39	42 ± 2
Pyrene	Perylene	36	36 ± 2

The transfer rate for a given molecular separation R can be written in the form

$$k_{ET}(R) = \frac{1}{\tau_D}\left(\frac{R_0}{R}\right)^6 \quad (7)$$

This relation is plotted in logarithmic form in Fig. 3-7.

Förster transfer does not require diffusional encounters between molecules, it is not therefore too dependent on solution viscosity. This is illustrated in the data of Table 3-2, which refer to energy transfer between pyrene and perylene in solutions of widely varying viscosity.

Exchange Transfer

While Förster transfer depends on the oscillator strength (intensity) of the donor fluorescence and acceptor absorptance, the second coupling mechanism is free of these constraints. It is based on electron exchange between two molecules as indicated in the Hückel diagram of Fig. 3-8.

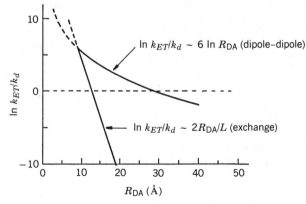

Figure 3-7 The ratio of the rate of energy transfer, k_{ET} to that of spontaneous deactivation, k_d, as a function of transfer distance. (Schematic). Upper curve: dipole resonance transfer; lower curve; electron exchange transfer. (Reproduced from N. J. Turro [1]. © Benjamin-Cummings, 1978.)

TABLE 3-2 Energy Transfer between Pyrene and Perylene as a Function of Solvent Viscosity [13]

Viscosity (cP)	k_{ET} ($M^{-1} s^{-1}$) ($\times 10^{10}$)
0.88	1.33
2.88	0.67
8.38	0.46
71.1	0.20
256	0.11

Dexter [15] has derived the distance dependence for this process in the form of Eq. (8).

$$k_{ET}(\text{exchange}) = KJ \exp(-2R_{DA}/L) \tag{8}$$

Here K is a constant for a given donor–acceptor pair, J is a fully normalized spectral overlap integral, and L is the sum of the van der Waals radii of donor and acceptor. The rate of exchange transfer falls off even more rapidly with increasing molecular separation than the rate of dipole resonance transfer, as can be seen in Fig. 3-7.

The exchange mechanism requires orbital overlap between donor and acceptor and works therefore only over "collisional" distances, of the order of 10 Å. While it also depends on spectral overlap, it does not depend on the intensities of the radiative transitions and is therefore a more general energy transfer mechanism. Exchange transfer is rather nonspecific and does occur on almost every encounter, provided the process is exothermic (i.e., $\{E_{D^*} - E_{A^*}\} < 0$). This is illustrated by the values of k_{ET} shown in Table 3-3, which are of the order of diffusional encounter rates.

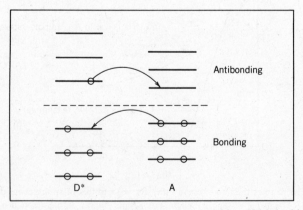

Figure 3-8 Energy transfer by electron exchange as represented in a Hückel orbital diagram.

TABLE 3-3 Energy Transfer Rate Constants and Exothermicity in Exchange Transfer [16]

Donor	Acceptor	k_{ET} ($M^{-1} s^{-1}$)	($E_{D^*} - E_{A^*}$) (kcal)
Triphenyl	Naphthalene	2×10^9	-6
Naphthalene	Biacetyl	9×10^9	-5
Benzophenone	Naphthalene	1×10^{10}	-8
Acetophenone	Naphthalene	1×10^{10}	-12
Acetone	Dibromonaphthalene	3×10^9	-18
Acetone	Biacetyl	5×10^9	-23

Exchange transfer is also less sensitive to changes in multiplicity than resonance transfer. It is therefore the mechanism responsible for the triplet-to-singlet energy transfer in the spectral sensitization processes of organic photochemistry [17].

Elegant experimental proof for various types of energy transfer, that is, between singlet states, triplet states, and between singlet and triplet states, can be found in five short but significant papers of Bennet, Kellog, and co-workers at Du Pont [18–22]. A lucid exposition of the thermodynamic fundamentals of energy transfer is available in an article by Balzani et al. [23].

The Perrin Formula. A simplified view of energy transfer without reference to a specific coupling mechanism was proposed early on by Perrin [24] and was later extensively used by Terenin [25] and his school [26] in describing energy transfer (quenching) in solid systems. It is based on the idea that energy transfer occurs between two molecules if they find themselves together within a sphere of influence or "quenching sphere," as indicated in Fig. 3-9.

The assumption that energy transfer is certain within the sphere, and impossible outside the sphere, leads to the Perrin formula, Eq. (9).

$$\frac{\phi_F}{\phi_F^0} = \exp(VN[A]) \qquad (9)$$

Here ϕ_F is the fluorescence quantum yield in the presence of the quencher concentration [A], ϕ_F^0 is the fluorescence quantum yield in the absence of quencher, V is the volume of the quenching sphere, and N is Avogadro's number.

Energy Migration

So far we have considered energy transfer between unlike molecules, namely, an energy donor and an acceptor, and it was noted that overlap between the emission spectrum of the donor and the absorption spectrum of the acceptor is

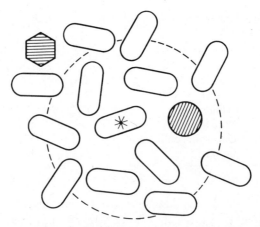

Figure 3-9 The Perrin quenching sphere. The dark particles represent the quencher, the excited molecule is marked with an asterisk. The Perrin model assumes that quenching is completely efficient inside the sphere, inefficient outside.

a condition for efficient energy transfer. It has been shown, however, that even the small overlap between the absorption and the emission spectrum of a molecule at or near the fundamental 0,0-transition is enough to make energy transfer possible. This transfer possibility is the basis of energy migration in solid polymeric systems. As an example, Fig. 3-10 shows the absorption and emission spectra of toluene [27], which may serve as a model for the phenyl groups in, for example, polystyrene.

The migration of excitation energy was first discovered through the emission characteristics of organic crystals, (Bowen et al. 1949) [28]. For example, the blue fluorescence of anthracene is easily quenched by trace amounts of an impurity and, if the impurity is capable of fluorescing, is replaced by the fluorescence of that impurity. Less than 0.01% of tetracene is enough to change the blue emission of an anthracene crystal to the green fluorescence of tetracene [29]. Since the great majority of photons is absorbed by anthracene molecules, the fact that only the tetracene guests finally emit must mean that the excitation is mobile in the system and that virtually every quantum in its lifetime encounters a tetracene center.

The migration process can be treated as a random walk in the lattice [30] and its range estimated from the effect of quencher (tetracene) concentration on the intensity of the anthracene fluorescence. Some actual data calculated by Rosenstock [31] from the experimental results of Schulman et al. [32] will put the phenomenon into perspective. In crystalline anthracene the average number of individual transfer steps is

$$n = 2.25 \times 10^4$$

the average number of different anthracene molecules visited during the

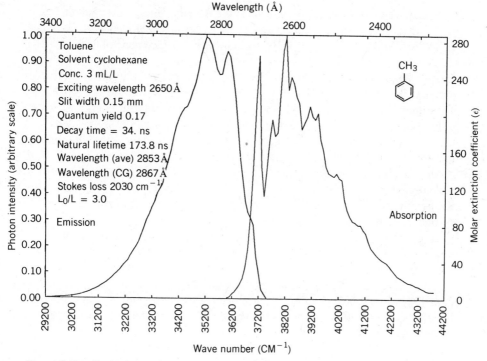

Figure 3-10 Absorption and emission spectra of toluene in solution. Note the mirror image relation between the absorption and the emission spectrum and the small spectral overlap at the frequency of the fundamental vibronic transition. (Reproduced with permission from I. B. Berlman [31]. © Academic Press, New York, 1965.)

random walk excluding multiple visits to the same site, is

$$(n)^{1/2} = 150$$

This corresponds to a mean displacement of 900 Å, the distance a between lattice points being 6 Å. By dividing the fluorescence lifetime ($\tau_F = 3 \times 10^{-8}$ s) by the number of transfers, one obtains the average residence time at a lattice site,

$$\tau_{res} = 1.3 \times 10^{-12}$$

The mean velocity of the quantum along its path is then given by

$$V = \frac{n \cdot a}{\tau_F} = 4.6 \times 10^4 \, \text{cm} \cdot \text{s}^{-1} = 460 \, \text{m} \cdot \text{s}^{-1}$$

the mean velocity of displacement in any particular direction is 300 cm · s^{-1}. In a latter study on crystalline N-isopropyl carbazole doped with perylene, Klöpffer [33] found $n = 4.1 \times 10^4$ and a mean displacement of 1350 Å · s^{-1}.

Energy Migration in Polymers. The physics of energy migration were formulated by Frenkel in 1931, [34], long before they could be tested by experiment. Frenkel introduced the term "exciton" for the energy quantum mobile in an ensemble of identical molecules. Davydov (1951) [35] described the spectral characteristics of such systems in more detail and illuminated thereby a number of unexplained phenomena in solid state physics. Both predicted the possibility of energy migration not only in crystals but also in partially ordered arrays and even in amorphous systems. A few years later Kallmann and co-workers [36] observed long-range energy transfer in polystyrene and in some other polymers.

It was thought that in these systems the exciton was traveling along the phenyl groups in the side chains of the polymer, in a process termed "downchain migration."

This view was confirmed through the work of Fox et al. [37] and David et al. [38] who studied the fluorescence behavior of copolymers and of polymer blends. For example, Cozzens and Fox [39] found that in copolymers of styrene and 1-vinylnaphthalene containing only 1% naphthalene and 99% of benzene units, naphthalene phosphorescence was observed almost exclusively.

In contrast, the blending of 1% poly(1-vinylnaphthalene) with 99% of polystyrene produced mainly polystyrene phosphorescence. It was concluded that energy migration must occur intramolecularly via the manifold of phenyl groups attached to the polyvinyl backbone. The same conclusion was reached from the observation, at 77 K, of delayed fluorescence in highly dilute solid solutions of poly(1-vinylnaphthalene) in organic glasses. Delayed fluorescence is here the result of the encounter of two triplet excited states and therefore evidence for triplet migration along the chain. On encounter, the two triplet states disproportionate to form an excited singlet state (which emits fluorescence) and a singlet ground state.

$$T_1^* + T_1^* \longrightarrow S_1^* + S_0 \tag{10}$$

If the sequence of naphthalene units is interrupted, for example, by copolymerization with methyl methacrylate, the delayed fluorescence is diminished and is observed only as long as significant blocks of naphthyl groups still remain. For the same reasons, delayed fluorescence, for example in poly(2-vinylnaphthalene), depends on molecular weight, that is, on chain length [40].

Down-chain energy migration has since been investigated rather extensively and is now known to be a common phenomenon, which plays an important role in the degradation and stabilization of polymers. The polymer chain acts here as an "antenna" conducting energy either to a reactive site or away from the point of incidence to a quencher where it can be harmlessly dissipated. The range of energy migration in some of these materials can be considerable; in poly(N-vinylcarbazole) Klöpffer [41] found an average range of $n = 10^3$. Early work by Geuskens and co-workers in Bruxelles [42] and more recent studies by Guillet and co-workers in Toronto [43–45] and Webber and co-workers at Austin [46, 47] may serve as entries into the extensive literature on this subject (see also References 48–51).

Side chain interaction in polystyrene and in similar polymers does not only promote energy migration but also the formation of excimers that occurs in solution mainly as a result of conformational transitions, which bring neighboring fluorophores into juxtaposition. In bulk polymers excimer emission is much more pronounced since energy migration also occurs *between* chains. Excimer emission occurs now from monomer pairs that have the required conformation already in the ground state.

Hirayama [52] has investigated the geometric constraints of excimer formation on small molecules. He found that excimers are formed most efficiently when the two interacting chromophores are separated by three carbon atoms, as in 1,3-diphenylpropane. Excimer formation is less facile in 1,4-diphenylbutane, even less so in 1,5-diphenylpentane, and does not occur at all in 1,2-diphenylethane. Vinyl polymers conform to Hirayama's rule and excimer formation in these systems is the rule rather than the exception.

Excimers have a lower excitation energy than the isolated monomer and excimer sites in a polymer matrix act therefore as exciton traps. It is for this reason that, for example, polystyrene films emit exclusively excimer fluorescence although the concentration of excimer sites is low, of the order of 1 mol% [53] or, as in the case of poly(N-vinylcarbazole) 0.1% [41]. In these systems the excimer sites are supplied with excitation energy by exciton migration. In fact the concentration of excimer sites limits the exciton migration range in these systems. Thus, as a result of the presence of 0.1% of exciton trap sites in poly(N-vinylcarbazole) the exciton migration range in the polymer is about $n = 10^3$. In the crystals of the model compound N-isopropylcarbazole where there are no excimer traps, the migration range is $n = 4 \times 10^4$ [33].

In dilute solid solutions of polymers the excitons are restricted to an intramolecular path along the polymer backbone, in solid films of the neat material they can migrate intermolecularly from one chain to another. An example of this is provided by the photopolymer PPDA based on the *p*-phen-

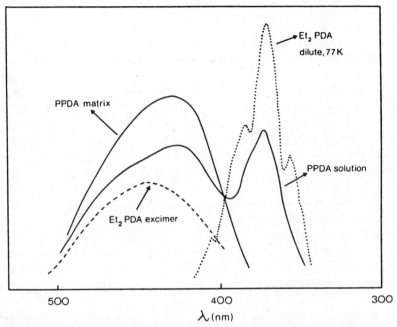

Figure 3-11 Fluorescence spectra of *p*-phenylenediacrylic acid diethylester (Et$_2$PDA) in dilute and in very concentrated solution, and of a polyester of *p*-phenylenediacrylic acid (PPDA) in solution and as a solid film. (Reproduced with permission from P. L. Egerton et al. [54].)

ylenediacrylate chromophore described in Chapter 2.

$$\left[\text{OOCCH}=\text{CH}-\langle\text{O}\rangle-\text{CH}=\text{CHCOOCH}_2\text{CH}_2\text{O}-\langle\text{O}\rangle-\text{OCH}_2\text{CH}_2\right]_n$$

PPDA

Here the chromophores are part of the polymer backbone, and since they are not adjacent to each other, down-chain energy migration is not possible. Interestingly, a certain amount of excimer fluorescence is observed along with the predominant monomer fluorescence, when a dilute solution of the polymer is irradiated (see Fig. 3-11). This indicates that the polymer chain folds upon itself and that a few chromophores are able to assume a contact configuration that produces excimer emission [55]. When solid films of the same polymer are irradiated, no monomer fluorescence is seen and only excimer fluorescence is observed [54] (see Fig. 3-11). The excitation produced at the monomer sites must be freely mobile in the solid matrix to reach an excimer site within its lifetime.

Energy Migration in Photoreactive Polymers. Energy migration plays an important role in some photopolymers by channeling energy to reactive sites. In the photopolymer PCP, based on the chromophore diphenylcyclopropene (see Chapter 2, p. 31), less than 2% of chromophores are located at reactive sites, yet the initial quantum yield of the material is 85% of the theoretical maximum. The high quantum yield means that in the early stages of irradiation almost every quantum finds a reactive site. This is possible only if the excitation is mobile in the polymer matrix and if the reactive sites are also energy traps [56].

Farid et al. [57] have determined the range of triplet energy migration in the following group of generic photopolymers.

Number	Photopolymer	E_T (kcal/mole)	Reference
I	⊢OOC—CH=CH—⟨O⟩	54.8	60
II	⊢OOC—CH=CH—⟨O⟩⟨O⟩ (naphthyl)	48.5	60
III	⊢OOC—CH=CH—⟨O⟩—CH=CH—COOR	49.3	60
IV	⊢OOC—◁(Ph)(Ph) **PCP**	51.0	61
V	—OOC—CH=CH—⟨O⟩—CH=CH—COO— ·(CH₂)₂O—⟨O⟩—O(CH₂)₂· **PPDA**	49.3	60

The polymers were doped with a fixed quantity of a triplet sensitizer having a triplet energy somewhat higher than that of the reactants (thioxhanthone, $E_T = 56$ kcal/mole). Variable quantities of a triplet quencher (1-pyrenebutyric acid methyl ester), were added. Its triplet level, $E_T = 48$ kcal/mole, is below that of the reactants. The effect of increasing the quencher concentration on the photographic sensitivity of the material was monitored, which means that the crosslinking photoreaction itself was used to indicate the arrival of an exciton at a reactive site.

A random walk analysis of this situation leads to the following expression; φ stands here either for a quantum yield of the reaction or for a fluorescence

TABLE 3-4 Migration Range of Triplet Excitons in a Group of Photopolymers [57]

Polymer	n	$\phi_r(T)$
I	6	0.045
II	45	0.13
III	33	0.10
IV	125	1.70
V	40	0.035

yield.

$$\frac{\phi_0}{\phi} = \left(1 + \frac{[Q]}{[A]}\right)^{\ell+1}\left(1 + \ell' n \frac{[Q]}{[A]}\right) \quad (11)$$

In this equation [A] is the concentration of reactant (acceptor) molecules in the solid matrix and [Q] is the concentration of quencher. In the solid matrix the surroundings of the reactive site and of the quencher cannot be treated as a continuum, instead, they are characterized by average coordination numbers in analogy with the description of a reactant in a crystal lattice. The average coordination number of a reactive site, ℓ, is the number of molecules, A or Q, in contact with one molecule of A located at a reactive site; the coordination number of the quencher, ℓ', is the number of A molecules surrounding the quencher in the matrix. The number of energy transfers during the lifetime of the excitation is indicated by the "jump number" n. Taking $\ell = 3$ and $\ell' = 4$, a jump number of 5.7 is obtained for poly(vinyl cinnamate). Experimental results for this and other polymers are listed in Table 3-4.

SPECTRAL SENSITIZATION

Spectral Mismatch between Reactant and Illuminant. The spectral sensitization of a photoreactive system is a means of utilizing radiative energy that is not directly absorbed by the reactants. The most common source of UV radiation is the mercury lamp, which emits in several quite narrow wavebands. If the photoreactive system does not absorb in one of these regions, most of the radiative energy falling onto the system is wasted. Poly(vinyl cinnamate) is a classical example. Figure 3-12 shows the absorption spectrum of poly(vinyl cinnamate) together with the emission spectrum of a medium pressure mercury arc. The spectral mismatch between emitter and absorber is the principal cause of the low photographic sensitivity of unsensitized poly(vinyl cinnamate).

Energy transfer makes it possible to improve the light collecting ability of the system by adding a component that absorbs the available energy efficiently and transmits it to the reactants. This process, which extends the action

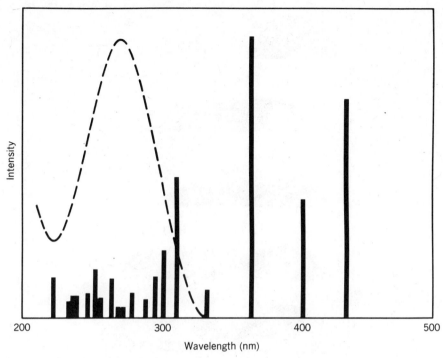

Figure 3-12 Spectral mismatch between the absorption of poly(vinyl cinnamate) and the emission of a medium pressure Hg arc. The absorption spectrum of poly(vinyl cinnamate) misses most of the strong emission lines.

TABLE 3-5 Sensitization of Poly(vinyl cinnamate) [58] (Irradiated with a Medium Pressure Hg Arc)

Sensitizer	Relative Sensitivity
Unsensitized	1
Chrysene	9
Benzophenone	10
p-Dinitrobenzene	19
p-Dinitroaniline	55
Picramide	200
1,2-Benzanthraquinone	250
Michler's ketone	320
1,3-Diazo-3-methyl-1,9-benzanthrone	550

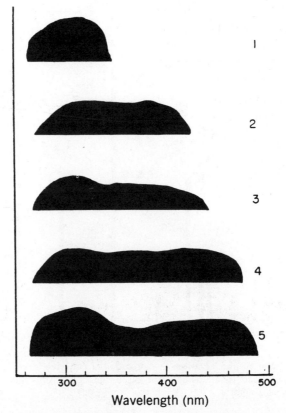

Figure 3-13 Action spectra of crosslinking in poly(vinyl cinnamate) sensitized with various triplet sensitizers. The spectra were obtained by exposing the resist to the continuous spectrum at the back of a spectrograph and developing the resist film after exposure. (1) unsensitized, (2) *p*-nitroaniline, (3) 2,4-dinitroaniline, (4) picramide, (5) 3-methyl-1,3-diaza-1,9-benzanthrone. (Reproduced with permission from E. M. Robertson et al. [58].)

spectrum of the system, is termed "spectral sensitization" or "photosensitization." The effect of sensitizers on the performance of a system can be dramatic; this is illustrated in Table 3-5 in the case of poly(vinyl cinnamate) irradiated with a medium pressure mercury arc [58].

In practice, spectral sensitization always extends the absorption range of the system to *longer* wavelengths, corresponding to lower excitation energies (see Fig. 3-13).

To achieve this by singlet energy transfer with a sensitizer that absorbs at longer wavelength than the reactant would require endothermic energy transfer, which is not a practical proposition. It is often possible, however, to find a sensitizer whose singlet excited state lies *below* that of the reactant, but whose triplet level lies *above* the triplet level of the reactant. The situation is

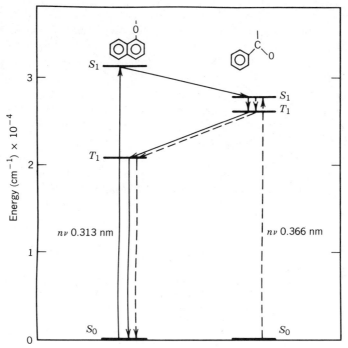

Figure 3-14 Energy level diagram typical for triplet sensitization. It refers to polymer bound naphthyl groups and acetophenone. (Reproduced with permission from L. Merle-Aubry et al. [59].)

schematically indicated in Fig. 3.14, which refers to the pair naphthalene(reactant)–acetophenone(sensitizer) [59].

Sensitization by Energy Transfer

Triplet Sensitization. In triplet sensitization, energy is transferred exothermically from the triplet excited state of the sensitizer to the ground state of the reactant. The great majority of sensitization processes in organic photochemistry is based on the following reaction.

$$S_0(\text{reactant}) + T^*(\text{sensitizer}) \rightarrow T^*(\text{reactant}) + S_0(\text{sensitizer}) \quad (12)$$

Since the overall spin of the system is conserved, the transfer process is spin allowed and occurs efficiently, provided only that transfer is exothermic, that is, $E_T(\text{sensitizer}) > E_T(\text{reactant})$. The operation of this principle is illustrated in Fig. 3-15, which refers to the sensitization of poly(vinyl cinnamate) by a group of triplet sensitizers. The ratio $\phi(\text{sensitized})/\phi(\text{unsensitized})$ is plotted here as a function of the triplet energy of the sensitizers and it can be seen that sensitizers with E_T lower than 53 kcal/mole are ineffective, while all sensitizers with triplet energies higher than 58 kcal/mole are about equally effective. Evidently the triplet energy of the cinnamoyl group must lie between 53 and 58 kcal/mole. Herkstroeter and Farid [60, 61], by a more careful calibration, established the triplet level of the cinnamoyl group at 54.8 kcal/mole.

In triplet sensitization only the triplet state of the reactants is populated. As a result, the photoreaction occurs exclusively from the triplet state and may proceed by a different route than the singlet state reaction. Sometimes this difference can be important. For example, in the singlet excited state, cycloaddition proceeds in a single step; in the triplet state it occurs by a two-step mechanism, where a (triplet) biradical is formed in the first step, but where the cyclobutane ring can close, in a second step, only after one of the spins of the biradical has inverted.

$$
\begin{array}{c}
\overset{\uparrow}{C\cdot} \quad C \\
|\uparrow \; + \; \| \\
C\cdot \quad C
\end{array}
\longrightarrow
\begin{array}{c}
\overset{\uparrow}{C\cdot} \quad \overset{\uparrow}{\cdot C} \\
| \quad \quad | \\
C\text{——}C
\end{array}
\xrightarrow{\text{spin flip}}
\begin{array}{c}
\overset{\downarrow}{C\cdot} \quad \overset{\uparrow}{\cdot C} \\
| \quad \quad | \\
C\text{——}C
\end{array}
$$

$$
\begin{array}{c}
{}^1C \quad C^4 \\
\| \; + \; \| \\
{}^2C \quad C^3
\end{array}
\qquad\qquad
\begin{array}{c}
C\text{——}C \\
| \quad \quad | \\
C\text{——}C
\end{array}
$$

For this reason the quantum yield of cycloaddition in triplet sensitized films of poly(vinyl cinnamate) [$\phi(T) = 0.045$] is lower [63] than the quantum yield of the same process in the unsensitized material [$\phi(S) = 0.11$]. This interpretation is supported by the effect of temperature on the two processes. Singlet cycloaddition in poly(vinyl cinnamate) is virtually independent of temperature; the triplet process has an activation energy of about 1 kcal/mole [62] (see Fig. 3-16). This is a reasonable value for the energy barrier set up by the repulsion of two parallel spins in the (planar) biradical.

All this does not change the fact that the sensitized resist has a much higher photographic speed when irradiated with the full mercury arc. The increase in photographic speed is caused by better light utilization, in spite of the lower efficiency of the photochemical event.

Candidates for triplet sensitizers must at least partially fulfill the following conditions:

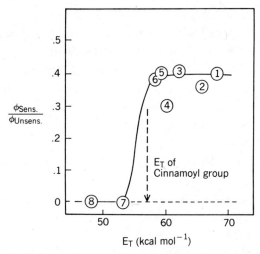

Figure 3-15 Quantum efficiency of crosslink formation in poly(vinyl cinnamate) doped with various triplet sensitizers plotted against their triplet energy values. (1) benzophenone, (2) thioxanthone, (3) Michler's ketone, (4) *p*-nitroaniline, (5) 2-acetonaphthone, (6) 1-acetonaphthone, (7) 9-fluorenone, (8) pyrene. (Reproduced with permission from A. Reiser and P. L. Egerton [62].)

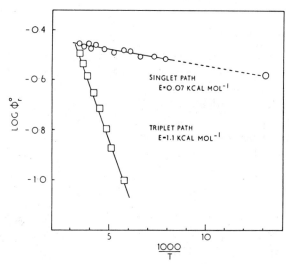

Figure 3-16 The effect of temperature on the quantum yield of cycloaddition in poly(vinyl cinnamate): on direct irradiation, (upper curve) and on sensitization with Michler's ketone. (Reproduced with permission from A. Reiser and P. L. Egerton [62].)

1. They must have a high rate of intersystem crossing from S_1 to T_1 and consequently a high quantum yield, ϕ_T, of triplet formation.
2. The energy difference between the singlet excited state S_1 and the triplet excited state T_1 (termed the singlet–triplet splitting) must be small.
3. The triplet lifetime should be long to increase the probability of energy transfer between sensitizer and reactant.
4. They should absorb strongly the available radiation in a spectral region where the reactant does not absorb.
5. They must be soluble in the reaction medium (solvent or polymer matrix).

In general, these conditions are met in the aromatic ketones where the $n\pi^*$ character of the excited states (triplet and singlet) favors intersystem crossing, high values of ϕ_T as well as a small singlet–triplet splitting. In fact, the majority of common sensitizers are either ketones or contain the carbonyl group somewhere in their structure. Table 3-6 gives some data on a range of molecules that are used as triplet sensitizers.

Table 3-7 lists some of the ketocoumarin sensitizers developed by Farid and co-workers [64] at Eastman Kodak. Here the triplet energies of the donors can be fine tuned by varying the substitution on the basic skeleton. For a more complete listing see Reference 64.

Sensitization by Electron Transfer

Sensitization occurs in some cases not by energy transfer but by the transfer of an electron. The classical example is the spectral sensitization of silver halide emulsions in photography [65]. There the sensitizing dyes inject on irradiation an electron into the conduction band of the silver halide crystal. A similar process may occur on a molecular level. The role of electron transfer in organic photochemistry has been reviewed by Mattes and Farid [66].

Often it is not easy to decide whether sensitization occurs by energy transfer or by electron transfer; in silver halide photography the energy transfer versus electron transfer controversy has lasted for years. Also, one process does not preclude the other and both may occur simultaneously (with different rates) in the same system. One of the criteria by which a decision as to the sensitization mechanism can sometimes be made is that the onset of energy transfer, and its efficiency, correlate with the energy difference between donor and acceptor, while in electron transfer the correlation is with the redox potentials of the components. An example with relevance to photoresists is the sensitization of azide photolysis by aromatic hydrocarbons [66a]. This proceeds by the coupled reactions indicated in Eq. (13).

$$A^* + D \rightleftharpoons D^+ + A^-$$
$$D^+ \longrightarrow \text{decomposition} \qquad (13)$$

TABLE 3-6 Triplet Energies of Organic Molecules That May Be Used as Photosensitizers

Sensitizer	E_T (kcal/mole)	Sensitizer	E_T (kcal/mole)
Benzene	84	Flavone	62
Phenol	82	Michler's ketone	61
Acetone	78	Naphthalene	61
Benzoic acid	78	Nitrobenzene	60
Benzonitrile	77	2-Acetonaphthone	59
Aniline	77	Acridene Yellow	58
Xanthone	74	Chrysene	57
Acetophenone	74	1-Acetonaphthone	57
Diisopropyl ketone	74	Biacetyl	55
Diphenyl sulfide	74	Benzil	54
Diphenylamine	72	Fluorenone	53
Benzaldehyde	72	Pyrene	49
4-CF_3-Acetophenone	71	Anthracene	47
Carbazole	70	1,2-Benzanthracene	47
Triphenylamine	70	1,12-Benzperylene	46
Hexachlorobenzene	70	Eosin	43
Benzophenone	69	3,4-Benzpyrene	42
Thiophene	69	Thiobenzophenone	40
Fluorene	68	9,10-Dichloranthracene	40
Triphenylene	67	Crystal Violet	39
Thioxhanthone	65	Perylene	35
Biphenyl	62	Naphthacene	29
Anthraquinone	62	Oxygen	23
Quinoline	62		
Phenanthrene	62		

Here the hydrocarbon sensitizer is the electron acceptor and the azide functions as the electron donor. When the transfer rate is plotted against the known sensitizer energies, the correlation is not convincing. The correlation with the free energy of charge transfer, ΔG_{CT}, based on redox potentials, however, is excellent (see Fig. 3-17).

The change in the free energy on charge transfer between two molecular species may be calculated by an expression derived by Weller [67, 68].

$$\Delta G_{CT} = E(D/D^+) - E(A^-/A) - \Delta E_{00}^* - e^2/\kappa a \qquad (14)$$

Here $E(D/D^+)$ is the oxidation potential of the donor, $E(A^-/A)$ is the reduction potential of the acceptor (sensitizer), ΔE_{00}^* is the excitation energy of the sensitizer, e is the electronic charge, κ is the dielectric constant of the

TABLE 3-7 Absorption Data, Triplet Energies, and Triplet Yields of Alkoxy-Substituted Carbonyl-bis(coumarins) [64]

R_1	R_2	R_3	R_4	λ_{max} (nm)	$\varepsilon\,(10^3)$	E_T	ϕ_{isc}
H	H	H	H	350	24	57.9	0.90
H	MeO	H	MeO	375	41	57.1	0.65
MeO	MeO	H	H	377	33	56.0	0.46
MeO	MeO	H	MeO	383	43	56.3	0.65
MeO	MeO	MeO	MeO	392	43	56.2	0.70

medium, and a is the distance between the molecular centers at the moment of transfer.

PHOTOCHEMISTRY IN AMORPHOUS SOLIDS

Reactant Sites. In most of their practical applications, photopolymers are used in the form of amorphous solid films. The solid environment affects in

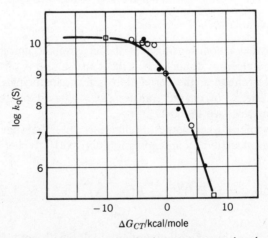

Figure 3-17 Fluorescence quenching of aromatic electron acceptors by phenyl azide (○) and naphthyl azide (●). The quenching rate constant k_q is correlated with the free energy of charge transfer ΔG_{CT}. The solid curve and the points (□) are taken from the work of Rehm and Weller. (Reproduced with permission from L. J. Leyshon and A. Reiser [66a].)

principle all physical and chemical processes, but the effect is most prominent in bimolecular reactions, which require the cooperation of two reactants. Poly(vinyl cinnamate) [69] (see Chapter 2) is a good example of such a system.

In poly(vinyl cinnamate), crosslinks are formed by the photodimerization of two cinnamoyl groups.

$$\text{Ph-CH=CH-C(=O)-} + \text{Ph-CH=CH-C(=O)-} \longrightarrow \text{[cyclobutane dimer]} \quad (15)$$

When the quantum yield of the reaction in the solid film is plotted as a function of the degree of conversion of cinnamoyl groups, curve B in Fig. 3-18 is obtained., It can be seen that the reaction efficiency falls dramatically during the photoprocess and that the reaction comes to a standstill ($\phi = 0$) when about half of all cinnamoyl groups are still intact. This lack of reactivity in some groups must be attributed to the absence of reaction partners in the vicinity. These groups may be said to be located at unreactive sites. In general the reactivity of the whole system is conveniently described in terms of reactant sites.

A reactant site is defined, rather imprecisely, as the reactant taken together with its immediate surroundings.[3] It is characterized by its reactivity, by its geometric configuration, by optical and dielectric properties, and so on. The macroscopic system of the amorphous matrix can be viewed as an ensemble of sites and quantitatively described by the distribution of site properties. For example, the macroscopic quantum yield of the system is the average of the site reactivities, taken over all sites of the system. The decrease in the quantum yield during irradiation, as it is shown in Fig. 3-18, can then be understood, if it is assumed that there is an uneven distribution of site reactivities in the system and that the more reactive sites, by definition, are consumed earlier than the less reactive ones.

---

[3] The sites so defined resemble the "reaction cavity" introduced by Cohen and Schmidt [70] into the photochemistry of organic crystals. However, while all reaction cavities in the crystal are identical, the amorphous matrix contains a range of sites differing in geometry. During the photoprocess the average reactivity of the ensemble of sites in the amorphous solid changes, that of the crystal remains the same. That is the reason why the quantum yield of dimerization in crystals of cinnamic acid (curve A in Fig. 3-18) is constant, while that of the poly(vinyl cinnamate) film decreases.

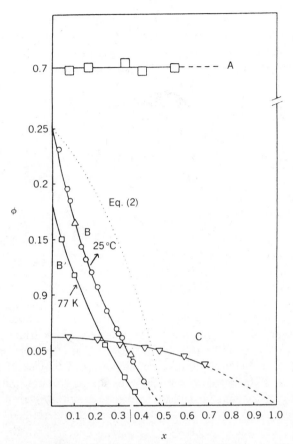

Figure 3-18 Quantum yield of cycloaddition between cinnamoyl groups as a function of reactant conversion: Curve A crystalline cinnamic acid, α-form; Curve B solid films of poly(vinyl cinnamate); Curve C solutions of poly(vinyl cinnamate) in dichloroethane. (Reproduced with permission from P. L. Egerton et al. [69].)

Reactivity Distributions. If the previous interpretation is correct, the site ensembles that correspond to consecutive stages of the photoprocess are truncated versions of the initial ensemble. The shape of the quantum yield function $\phi(x)$ contains, therefore, information on the *original* distribution of site reactivities.

The distribution of site reactivities can be elicited from the experimental data by an iterative procedure, which is described in Pitts and Reiser [71]. The quantum yield function $\phi(x)$ is calculated for various trial distributions, and the distribution that reproduces the experimental $\phi(x)$ data most closely is taken to be the true distribution of site reactivities. Figure 3-19 shows the result of this procedure for the case of poly(vinyl cinnamate). The trial distributions are shown in the upper-right-hand part of the figure in the form

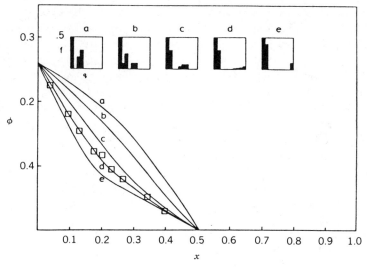

Figure 3-19 Quantum yield versus reactant conversion in poly(vinyl cinnamate). The individual curves correspond to the site reactivity distributions shown as histograms in the upper half of the figure. Curve "d" closely fits the experimental points, histogram "d" represents the reactivity distribution in poly(vinyl cinnamate) film. (Reproduced with permission from E. Pitts and A. Reiser [71].)

of histograms. For the purpose of this analysis the reactant sites have been collected into 10 groups with reaction probabilities, q, in the ranges from 0–0.1 to 0.9–1.0. In the histograms the frequency of occurrence of a given reaction probability is plotted against the value of q. Of the several curves calculated from the histograms curve "d" fits the experimental points best, histogram "d" is therefore closest to the distribution of site reactivities in the solid material.

It can be seen from histogram "d" in Fig. 3-19 that in poly(vinyl cinnamate) a small group of chromophores is responsible for the photosensitivity of the polymer and that the majority of cinnamoyl groups is located at unreactive or almost unreactive sites. This is by no means a general finding. For example, in the photopolymer PPDA based on p-phenylenediacrylic acid (p. 28) where the photoreactive chromophore is part of the polymer backbone, the reactivity distribution is described by histogram "d" in Fig. 3-20. There are no highly reactive sites here, but a large fraction of sites with moderate reactivity. Finally, in the photopolymer PCP based on diphenylcyclopropene (p. 30) we find that there are less than 2% of reactive sites, yet the average quantum-efficiency of the system is almost unity.

Reactivity analysis provides considerable insight into the functional mechanism of photoreactive polymers. It is of practical interest because it identifies the sources of inefficiency in the system and points to the direction in which improvements may be possible. The three polymers discussed in the foregoing paragraph are a good example. All three are based on the same crosslinking reaction, [2 + 2] photocycloaddition, yet while photosensitivity in poly(vinyl cinnamate) is limited by a low concentration of reactive sites, in PPDA the limitation is the low site reactivity. The high quantum efficiency of the diphenylcyclopropene resist PCP is the result of energy migration and trapping. In summary: The performance of a photoreactive system in the solid state depends not only on the inherent reactivity of its functional groups, but also on their spatial distribution in the matrix, on the mobility of the excitation, and on the presence or absence of energy traps. In this last respect some resist films resemble true solid state devices.

Reactivity distributions play a role even in monomolecular reactions, such as the cis–trans isomerization of stilbenes and azobenzenes and the reversible ring-opening and ring-closing reactions of spirobenzopyrans [72]. For the case of the spirobenzopyran shown in Eq. (16)

$$\text{Spirobenzopyran (colorless)} \underset{h\nu', \Delta t}{\overset{h\nu}{\rightleftarrows}} \text{Merocyanine (colored)} \tag{16}$$

where X is S, O, CMe$_2$ the kinetics of the thermal recovery process are strictly first order in fluid solution; in a solid matrix, the log D versus time curve has the form shown in Fig. 3-21.

The faster response in the early stages is caused by an uneven distribution of reactivities in the amorphous solid. Since isomerization is accompanied by a change in the volume requirements of the molecule, differences in site reactivity are caused here by the microscopically nonuniform distribution of free volume [73]. Lawrie and North [74] have used the thermal return reaction of a merocyanine to the corresponding spiropyran as a molecular probe to derive the free volume distribution in a polymer matrix; others have used cis–trans photoisomerization for that purpose [75–77]. More recently Sung and co-workers [78, 79] and Victor and Torkelson [80] have studied free volume distributions in glassy polymers with a number of photochromic probes.

The considerations apply in general to reactions in the glassy state. Above the glass transition temperature reaction rates are usually much higher and follow the same kinetics as in liquid solution. Figure 3-22 shows the effect of temperature on the reaction rate (or rather the change in conversion for a constant radiation dose) for the forward (ring-opening) reaction of a

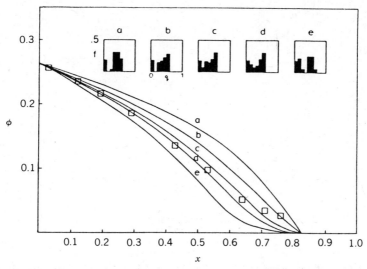

Figure 3-20 Quantum yield versus reactant conversion in a polyester of *p*-phenylenediacrylic acid. The individual curves correspond to the site reactivity distributions shown as histograms. Curve "d" is the best fit to the experimental points. (Reproduced with permission from E. Pitts and A. Reiser [71].)

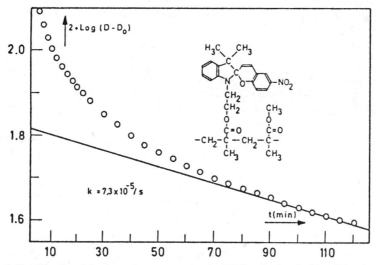

Figure 3-21 Kinetics of the thermal return reaction (at 27°C), after irradiation, of the photochromic polymer indicated in the figure. The process was monitored by following the decay of the absorption maximum at 585 nm. (Reproduced with permission from G. Smets [72].)

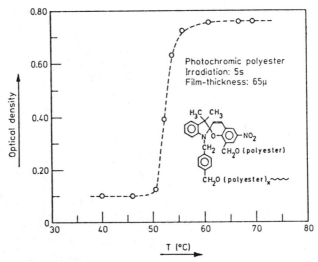

Figure 3-22 Dependance of the photoresponse of a photochromic polyester on temperature. (Reproduced with permission from G. Smets [72].)

spiropyran; the rate increases abruptly as the system goes through the glass transition [72].

Even with respect to purely photophysical processes the concept of chromophore sites is appropriate. The absorption and the emission spectra of small molecules in organic glasses show broad bands that are caused by the superposition of the line spectra of individual molecules placed in sites of different properties, which impose slightly different constraints on the central molecule. Shpolskij [81] has shown that on rare occasions the molecular volume of the guest almost coincides with that of the host crystal and in that case all sites become identical and line spectra, or at least very narrow band spectra, are observed in absorption and in emission. More recently, through the advent of lasers with extremely narrow emission lines, it has become possible to excite a particular group of sites within a broad absorption band (site selection spectroscopy) and elicit narrow band emission from molecules dissolved in amorphous organic glasses [82] (see Fig. 3-23). This is convincing evidence that the concept of reactant sites corresponds to an experimental reality.

Site Geometries and Reaction Products. The primary attribute of a site is its geometric configuration from which other site properties are derived. In a crystalline solid information on site geometry may be gained from diffraction experiments. Cohen and Schmidt [70, 83] used this technique on a large group of crystalline derivatives of cinnamic acid and established a direct correlation between the geometric configuration of the unit cell of the crystal and the stereochemistry of the reaction products. They formulated a law of "minimum displacement," saying in effect that a given geometry of the unit cell produces

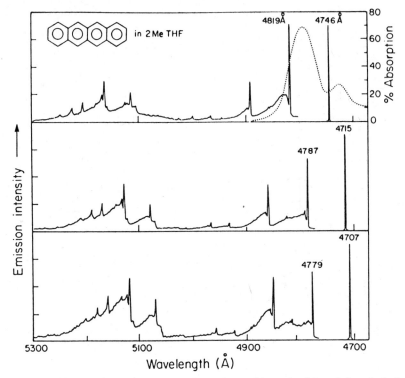

Figure 3-23 Site selection spectroscopy of tetracene in a solid matrix of 2-methyltetrahydrofuran at 4 K. The broken curve in the first figure indicates the absorption spectrum of tetracene. The wavelength of the exciting laser radiation is indicated in angstroms. On the left are the discontinuous (line) emissions of the sites, selectively activated by the monochromatic laser. (Reproduced with permission from W. C. McColgin et al. [82].)

that reaction product which resembles it most closely [83]. A similar correlation between site geometry and product structure may be assumed in amorphous solids and may here be used to estimate the distribution of site geometries.

Egerton et al. [84] have tested this assumption on the photodimerization of ethyl cinnamate. They irradiated the pure liquid and passed the photolyzed sample through a gas chromatography column to isolate the various cyclic products. Up to seven (out of eleven possible) cyclobutane isomers could be determined in this way, the most important being δ-truxinate, β-truxinate, and α-truxillate.

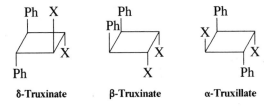

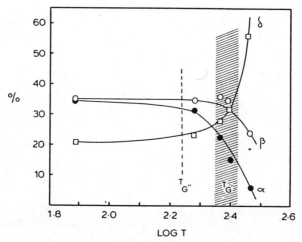

Figure 3-24 The contributions of δ-truxinate, β-truxinate, and α-truxillate to the photoproducts in irradiated ethyl cinnamate as a function of the temperature at which the system was irradiated. (Reproduced with permission from P. L. Egerton et al. [84].)

Irradiation experiments were carried out at room temperature where ethyl cinnamate is a liquid, as well as at lower temperatures down to 77 K where the material is frozen to a hard glass. Figure 3-24 shows the change in product composition that occurs when the system goes through the glass transition. It corresponds to the change from a reactivity-controlled process in the liquid state to the site-geometry-controlled process in the solid matrix. In the liquid state, δ-truxinate, the isomer with the least steric hindrance between the substituents, is the principal product. The reaction is regiospecific and strongly favors head-to-head cycloaddition, hence the low abundance of α-truxillic acid in the products. In the solid system, reactivity is of little concern and the product distribution reflects the distribution of site geometries. As the density of the glass increases, those geometries that allow the closest packing will prevail; they are the compact site geometries that correspond to α-truxillate and β-truxinate. These results illustrate the decisive role of molecular morphology in solid state photochemistry and point to the advantages to be gained if this morphology is controlled by appropriate structural design.

REFERENCES

1. N. J. Turro, *Modern Molecular Photochemistry*, Benjamin-Cummings, Menlo Park, CA, 1978.
2. J. A. Barltrop and J. D. Coyle, *Elements of Organic Photochemistry*, Wiley, New York, 1975.
3. J. B. Birks, *The Photophysics of Aromatic Molecules*, Chap. 7, Wiley, London, 1970.

4. T. Förster and K. Kasper, *Z. Phys. Chem.* (N.F.), **1**, 274 (1954); *Z. Elektrochem.*, **59**, 976 (1955).
5. J. A. Ferreira and G. Porter, *J. Chem. Soc. Faraday II*, **73**, 340 (1977).
6. H. Leonhardt and A. Weller, *Z. Elektrochem.*, **67**, 791 (1963).
7. M. Gordon and W. R. Ware, Eds., *The Exciplex*, Academic, New York, 1975.
8. H. Knibbe and A. Weller, *Z. Phys. Chem.* (N.F.), **56**, 99 (1967).
9. R. F. Bartholomew, R. S. Davidson, P. F. Lambeth, J. F. McKellar, and P. H. Turner, *J. Chem. Soc. Perkin 2*, p. 577 (1972).
10. K. Sandros and H. L. J. Bäckstrom, *Acta Chem. Scand.*, **16**, 958 (1962); **18**, 2355 (1964).
11. T. Förster, *Ann. Phys.*, **2**, 55 (1953); *Discuss. Faraday Soc.*, **27**, 7 (1959).
12. T. Förster, *Die Fluoreszenz Organischer Verbindungen*, Vanderhoek & Ruprecht, Gottingen, 1951.
13. M. Z. Maksimov and I. B. Rotman, *Opt. Spectrosc.* **12**, 237 (1962).
14. A. A. Lamola, in *Energy Transfer and Organic Photochemistry*, Interscience, New York, 1969.
15. D. L. Dexter, *J. Chem. Phys.*, **21**, 836 (1948).
16. Reference 1, Chap. 9.
17. P. J. Wagner, *Creat. Detect. Excited State*, **1A**, 174 (1977).
18. R. G. Bennet, *J. Chem. Phys.*, **41**, 3037 (1964).
19. R. G. Bennet, R. P. Schwenker, and R. E. Kellog, *J. Chem. Phys.*, **41**, 3040 (1964).
20. R. E. Kellog and R. G. Bennet, *J. Chem. Phys.*, **41**, 3042 (1964).
21. R. E. Kellog, *J. Chem. Phys.*, **41**, 3046 (1964).
22. R. G. Bennet, *J. Chem. Phys.*, **41**, 3048 (1964).
23. V. Balzani, F. Bolletta, and F. Scandola, *J. Am. Chem. Soc.*, **102**, 2152 (1980).
24. J. Perrin, *Compt. Rend.*, **184**, 1097 (1927).
25. A. Terenin and V. Ermolaev, *Trans. Faraday Soc.*, **52**, 1042 (1956).
26. V. L. Ermolaev, *Sov. Phys. Usp.* (*Engl. Transl.*), **80**, 333 (1963).
27. I. B. Berlman, *Handbook of Fluorescence Spectra of Aromatic Molecules*, Academic, New York, 1965.
28. E. J. Bowen, E. Mikiewicz, and F. W. Smith, *Proc. Phys. Soc. London, Sect. A*, **62**, 26 (1949).
29. G. T. Wright, *Proc. Phys. Soc. London Sect. A*, **66**, 777 (1953).
30. J. Jortner, S. A. Rice, S. Choi, and J. Katz, *J. Chem. Phys.*, **41**, 309 (1965).
31. H. B. Rosenstock, *J. Chem. Phys.*, **48**, 532 (1968).
32. J. H. Schulman, H. W. Etzel, and J. G. Allard, *J. Appl. Phys.*, **28**, 792 (1957).
33. W. Klöpffer, *J. Chem. Phys.*, **50**, 1689 (1969).
34. J. I. Frenkel, *Phys. Rev.*, **37**, 1276 (1931).
35. A. S. Davydov, *Izv. Akad. Nauk SSSR, Ser. Fiz.*, **15**, 605 (1951); *Theory of Molecular Excitons*, McGraw-Hill, New York, 1962.
36. F. H. Brown, M. Furst, and H. Kallmann, *Discuss. Faraday Soc.*, **27**, 43 (1959).
37. R. B. Fox, T. R. Price, R. F. Cozzens and J. R. McDonald, *J. Chem. Phys.*, **57**, 534 (1972).

38. C. David, W. Demarteau, and G. Geuskens, *Eur. Polym. J.*, **6**, 1397 (1972).
39. R. F. Cozzens and R. B. Fox, *J. Chem. Phys.*, **50**, 1532 (1969).
40. N. Kim and S. E. Webber, *Macromolecules*, **13**, 1233 (1980).
41. W. Klöpffer, *J. Chem. Phys.*, **50**, 2337 (1969).
42. C. David, V. Naegelen, W. Piret, and G. Geuskens, *Eur. Polym. J.*, **11**, 596 (1975).
43. J. Guillet, in *Degradation and Stabilization of Polymers* (G. Geuskens, Ed.), p. 181, Applied Science, London, 1975.
44. J. E. Guillet and W. A. Rendall, *Macromolecules*, **19**, 224 (1986).
45. J. E. Guillet, J. Wang, and L. Gu, *Macromolecules*, **14**, 105 (1986).
46. S. E. Webber, P. E. Avots-Avotins, and M. Deumie, *Macromolecules*, **14**, 105 (1981).
47. F. Bai, C. H. Chang, and S. E. Webber, *Macromolecules*, **19**, 2798 (1986).
48. A. M. North and J. L. Ross, *J. Polym. Sci., Polym. Symp.*, No. 55, 259 (1976).
49. J. L. R. Williams and R. C. Daly, *Prog. Polym. Sci.*, **5**, 61 (1977).
50. J. L. R. Williams, S. Farid, J. C. Doty, R. C. Daly, D. P. Specht, R. Searle, D. G. Borden, H. J. Chang, and P. A. Martic, *Pure Appl. Chem.*, **49**, 523 (1977).
51. S. Farid, P. A. Martic, R. C. Daly, D. R. Thompson, D. P. Specht, S. E. Hartman, and J. L. R. Williams, *Pure Appl. Chem.*, **51**, 241 (1979).
52. F. Hirayama, *J. Chem. Phys.*, **42**, 3163 (1965).
53. W. Frank and L. A. Harrah, *J. Chem. Phys.*, **61**, 1526 (1974).
54. P. L. Egerton, J. Trigg, E. M. Hyde, and A. Reiser, *Macromolecules*, **14**, 100 (1981).
55. M. Graley, A. Reiser, A. J. Roberts, and D. Phillips, *Macromolecules*, **14**, 1752 (1981).
56. M. V. Mijovic, P. B. Beynon, T. J. Shaw, K. Petrak, and A. Reiser, *Macromolecules*, **15**, 1464 (1982).
57. S. Farid et al., private communication, 1987.
58. E. M. Robertson, W. P. Van Deusen, and L. M. Minsk, *J. Appl. Polym. Sci.*, **2**, 308 (1959).
59. L. Merle-Aubry, D. A. Holden, and J. E. Guillet, *Macromolecules*, **13**, 1138 (1980).
60. W. G. Herkstroeter and S. Farid, *J. Photochem.*, **35**, 71 (1986).
61. W. G. Herkstroeter, D. P. Specht, and S. Farid, *J. Photochem.*, **21**, 325 (1983).
62. A. Reiser and P. L. Egerton, *Photogr. Sci. Eng.*, **23**, 144 (1979).
63. T. A. Shankoff and A. M. Trozzolo, *Photogr. Sci. Eng.*, **19**, 173 (1975).
64. D. P. Specht, P. A. Martic, and S. Farid, *Tetrahedron*, **38**, 1203 (1982).
65. W. West and P. B. Gilman, in *The Theory of the Photographic Process* (T. H. James, Ed.), p. 251, Macmillan, New York, 1977.
66. S. L. Mattes and S. Farid, in *Organic Photochemistry* (A. Padva, Ed.), Dekker, Vol. 6, p. 233, New York, 1983.
66a. L. J. Leyshon and A. Reiser, *Trans. Faraday Soc.*, **68**, 1918 (1972).
67. D. Rehm and A. Weller, *Ber. Bunsenges. Phys. Chem.*, **73**, 834 (1969).

68. A. Weller, *Z. Phys. Chem.* (N.F.), **133**, 93 (1982).
69. P. L. Egerton, E. Pitts, and A. Reiser, *Macromolecules*, **14**, 95 (1981).
70. M. D. Cohen and G. M. J. Schmidt, *J. Chem. Soc.*, p. 1996 (1964), and papers following in the series. See also G. M. J. Schmidt, in *Reactivity of the Photoexcited Organic Molecule*, p. 227, Interscience, New York, 1967.
71. E. Pitts and A. Reiser, *J. Am. Chem. Soc.*, **105**, 5590 (1983).
72. G. Smets, *Adv. Polym. Sci.*, **50**, 18 (1983).
73. G. Smets, J. Thoen, and A. Aerts, *J. Polym. Sci., Polym. Symp.* No. 51, 119 (1975).
74. N. C. Lawrie and A. M. North, *Eur. Polym. J.*, **9**, 348 (1973).
75. C. S. Paik and H. Morawetz, *Macromolecules*, **5**, 171 (1972).
76. C. Eisenbach, *Makromol. Chem.*, **179**, 2498 (1978).
77. C. Eisenbach, *Polym. Bull.*, **7**, 348 (1973).
78. C. S. P. Sung, I. R. Gould, and N. J. Turro, *Macromolecules*, **17**, 1447 (1984).
79. W.-C. Yu, C. S. P. Sung, and R. E. Robertson, *Macromolecules*, **21**, 355 (1988).
80. J. G. Victor and J. M. Torkelson, *Macromolecules*, **20**, 2241 (1987).
81. E. V. Shpolskij, *Pure Appl. Chem.*, (1974).
82. W. C. McColgin, A. P. Marchetti, and J. H. Eberly, *J. Am. Chem. Soc.*, **100**, 5622 (1978).
83. M. D. Cohen, *Angew. Chem., Int. Ed. Engl.*, **14**, 386 (1975).
84. P. L. Egerton, E. M. Hyde, J. Trigg, A. Payne, P. Beynon, M. V. Mijovic, and A. Reiser, *J. Am. Chem. Soc.*, **103**, 3859 (1981).

4 PHOTOINITIATED POLYMERIZATION

In crosslinking photoresists, changes in solubility are brought about by the formation of a three-dimensional network from an ensemble of linear chains. A similar result can be achieved by polymerizing a monomer in the presence of polyfunctional components. If polymerization can be initiated by a photoreaction, such a process may be used as an imaging system. Image amplification in crosslinking is brought about through the immobilization of large molecules by a small number of crosslinks. In photopolymerization systems the source of amplification is the chain reaction of the polymerization process.

Of the early reports on the use of photoinitiated polymerization for the purpose of image reproduction we mention those of Eggert et al. [1] (1938) and of Morton [2] (1949), as well as that of the Osters [3] (1957). The pioneering work of Plambeck [4] in designing a viable photopolymer printing plate was patented in 1956 (assigned to Du Pont), and was followed in rapid succession by the patents of Leekley and Sorensen [5] (1956, assigned to Time Inc.) and of Hoerner and Olsen [6] (1958, assigned to BASF, Nyloprint). All the early work, as well as almost all systems currently in use, are concerned with free radical addition polymerization. More recently, cationic ring-opening polymerization has been developed to the point where it offers some important advantages. Finally, the photoinitiated condensation polymerization between thiols and olefinic double bonds has to be considered.

PHOTOINITIATED RADICAL POLYMERIZATION

Photoinitiated radical polymerization can be described by the following reaction sequence, where I stands for the initiator, M for monomer.

$$I \xrightarrow{h\nu} R_2^{\cdot} + R_2^{\cdot} \tag{1}$$

$$R^{\cdot} + M \longrightarrow RM^{\cdot} \tag{2}$$

$$RM^{\cdot} + M \longrightarrow RM_2^{\cdot} \tag{3}$$

$$\vdots \qquad \vdots$$

$$RM_{n-1}^{\cdot} + M \longrightarrow RM_n^{\cdot} \tag{4}$$

$$RM_m^{\cdot} + RM_n^{\cdot} \longrightarrow RM_{m+n}R \tag{5}$$

The first step is the generation of free radicals $R_1^{\cdot}$ and $R_2^{\cdot}$ by a photoreaction, reaction (1). Next, the radicals interact with the monomer and start the reaction chain, reaction (2). Reactions (1) and (2) together constitute the process of initiation. Reactions (3) and (4) represent the propagation of the radical function in the monomer, that is, the growth of the polymer chain. Reaction (5) stands for the many possible termination steps that end the reaction and set a limit to chain growth. We will now consider each of these steps in turn.

PHOTOGENERATION OF RADICALS

The first step in photoinitiation is the formation of free radicals. Molecules or molecular systems that are capable of forming radicals on irradiation are termed photoinitiators. An extensive literature on photoinitiators and photoinitiation is available and some useful references are to be found at the end of this chapter [7–13]. Most photoinitiators in current use are based on one of two mechanisms: photofragmentation and photoinduced hydrogen abstraction.

In photofragmentation, radicals are produced by the scission of a covalent bond,

$$I \xrightarrow{h\nu} R_1^{\cdot} + R_2^{\cdot} \tag{6}$$

In photoinduced hydrogen abstraction the reaction of the excited initiator with a hydrogen donor (coinitiator) produces free radicals.

$$\begin{aligned} I &\xrightarrow{h\nu} I^* \\ I^* + HD &\longrightarrow IH^{\cdot} + D^{\cdot} \end{aligned} \tag{7}$$

For other initiating mechanisms see page 125.

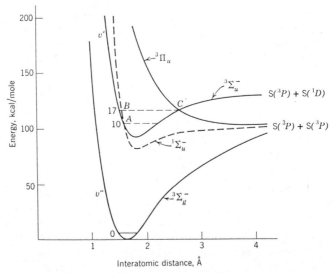

Figure 4-1 Potential energy curves for the ground state and some excited states of the S_2 molecule. (Reproduced with permission from E. J. Bowen, *Chemical Aspects of Light*, 2nd Ed., Clarendon, Oxford, 1946.)

Photofragmentation

The absorption of a photon by a diatomic molecule corresponds to the promotion of an electron from a bonding into an antibonding molecular orbital. This often leads to a dissociative or predissociative state (see Fig. 4-1) and to the formation of free atoms or radicals. For example, in the case of molecular sulfur, the reaction

$$S_2 \xrightarrow{h\nu} 2S \qquad (8)$$

proceeds quantitatively, the difference between the photon energy ($h\nu$) and the bond dissociation energy (D_0) appearing as the kinetic energy of the two sulfur atoms [14].

In a polyatomic molecule the absorption of a photon also promotes one electron into an antibonding orbital, but this may now extend over a larger group of atoms, and excitation does not necessarily lead to fragmentation [15]. If fragmentation occurs in the excited state, it is usually the weakest bond that is broken. Thus, irradiation of diphenylsulfide yields phenylsulfide radicals, the S—S bond being the weakest link in the system.

$$\text{Ph-S-S-Ph} \xrightarrow{h\nu} 2\,\text{Ph-S}\cdot \qquad (9)$$

The photon energy absorbed by the initiator ($h\nu = 110$ kcal/mole) is here considerably higher than the energy of the S—S bond (70 kcal/mole). In these circumstances almost every absorption event leads to bond scission.

In liquid solution, effective radical yields of photodissociation are substantially lower because of *cage recombination*. The solvent surrounding the initiator molecule at the moment of scission forms a molecular "cage," which must rearrange before the initiator fragments can separate. The cage keeps the fragments together for some time during which they may recombine. Cage recombination competes with escape from the cage, and as a result only a fraction of the radicals produced in the scission event becomes available for reaction outside the solvent cage.

The notion of a "cage effect" was introduced into solution kinetics by Franck and Rabinowitch (1934) [15]. Its importance for radical chain polymerization was recognized by Noyes [16] and by others who have contributed to the experimental investigation of the effect [17]. A good review of some of this work is given by Roffey [10]. Noyes in particular has shown that even for a pair of radicals that have escaped from the primary cage, the probability of recombination in its vicinity remains high [18]. This "secondary cage effect" determines the level of initiation efficiency in most practical systems (see also reference 19).

Noyes [18] has calculated the quantum yield of initiation as a function of monomer concentration for an assumed highly reactive monomer. His results are summarized in Fig. 4-2. At low concentrations of monomer most radicals recombine with other radicals and the initiation efficiency is proportional to monomer concentration. At higher values of [M] the efficiency reaches a limit

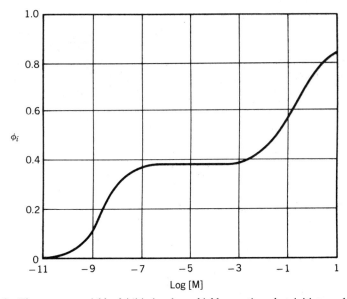

Figure 4-2 The quantum yield of initiation by a highly reactive photoinitiator, plotted as a function of monomer concentration. (Reproduced with permission from L. F. Meadows and R. M. Noyes, [17].)

where the monomer is catching all radicals that have escaped primary and secondary recombination. This is shown in the efficiency plateau between $\log[M] = -7$ and $\log[M] = -2$. If the monomer is very reactive it may compete with secondary recombination at concentrations higher than 0.01 mole/L and the initiation efficiency may increase further. Finally, if the monomer reacts with radicals almost at every encounter then, in pure monomer, the initiation reaction may even compete with primary recombination.

The cage effect is responsible for the strong viscosity (and temperature) dependence of the initiation efficiency. An increase in viscosity slows down the rate of escape from the cage more than it affects the rate of cage recombination. As a result, photogeneration of free radicals by bond scission becomes inefficient in concentrated polymer solutions or in polymer matrices.

The net production of free radicals is the outcome of a competition between a number of processes following the promotion of the molecule into the excited singlet state (see the following diagram).

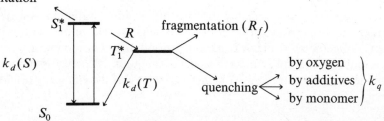

In most cases, fragmentation occurs from the excited triplet state; the triplet yield, ϕ_T, (that is, the fraction of excited states that change from the singlet excited state to the triplet state) is, therefore, an important parameter. The probability of radical recombination in the solvent cage depends also on the multiplicity of the excited state in which bond scission occurs [15]. An excited singlet state dissociates into radicals with pairwise correlated spins; their recombination is spin allowed and may occur within the time of a single nuclear vibration. Radicals formed from an excited triplet have parallel spins and can recombine only after spin inversion, which requires a time of the order of 10^{-7} s. The situation is summarized in the following reaction scheme (10).

$$^1I \longrightarrow {}^1I^* \longrightarrow {}^3I^* \\ \searrow \quad \downarrow \quad \downarrow \\ R_1^\uparrow + R_2^\downarrow \longrightarrow R_1^\uparrow + R_2^\uparrow \tag{10}$$

In the presence of air, fragmentation from the triplet state competes primarily with triplet quenching by oxygen. Molecular oxygen has a triplet ground state, $^3O_2$, which reacts eagerly with excited triplets, leading to the formation of singlet oxygen and the ground state of the initiator.

$$^3I^* + {}^3O_2 \longrightarrow {}^1O_2^* + {}^1I \qquad (11)$$

The overall success of the fragmentation process is measured by the effective quantum yield of radical formation, ϕ_i, which is defined as the number of separated radical pairs (in principle available for initiation), divided by the number of photons absorbed.

$$\phi_i = \frac{\text{separated radical pairs}}{\text{quanta absorbed}} \qquad (12)$$

There are no reliable physical methods for measuring absolute radical concentrations. However, since radical recombination and the quenching of the excited triplet by monomer or by solvent do return the system to its ground state, the effective quantum yield, ϕ_i, can be measured by the consumption of initiator during irradiation, in the absence of oxygen or of other chemical quenchers.

Initiators Based on Photofragmentation

Peroxides. For example, benzoyl peroxide, BPO

$$\text{Ph-C(=O)-O-O-C(=O)-Ph} \xrightarrow{h\nu} 2\ \text{Ph-C(=O)-O}\cdot \qquad (13)$$

BPO

On excitation [20], benzoyl peroxide dissociates into two benzoyl radicals. The excitation energy is of the order of 100 kcal/mole, much higher than the energy of the —O—O— bond (30 kcal/mole) and primary scission is therefore quantitative. Nonetheless, peroxides are now seldom used; they have poor thermal stability and they absorb below 300 nm. Most of all, the benzoyl radical is not very reactive.

Azo Compounds. For example, 2,2′-azobis(butyronitrile), AIBN

$$N\equiv C-\underset{\underset{CH_3}{|}}{\overset{\overset{CH_3}{|}}{C}}-N=N-\underset{\underset{CH_3}{|}}{\overset{\overset{CH_3}{|}}{C}}-N\equiv C \xrightarrow{h\nu} 2\,N\equiv C-\underset{\underset{CH_3}{|}}{\overset{\overset{CH_3}{|}}{C}}\cdot\ +\ N_2 \qquad (14)$$

The 2-cyanoisopropyl radical formed in the fragmentation of AIBN is much more reactive than the benzoyl radical and therefore more effective in initiating polymerization. Furthermore, the fragmentation yield of AIBN is higher: A molecule of nitrogen is eliminated in the photolytic process and the two active radicals are formed at a distance of some 3 Å from each other, which increases their chances of escape from the primary cage.

2,2'-Azobis(butyronitrile) is not thermally stable, indeed it is a popular thermal initiator. Its major drawback, however, is the fact that it absorbs only below 300 nm.

Benzoin Derivatives. The search for a molecule that absorbs at 365 nm (the principal emission of the Hg arc in the UV), and that dissociates on excitation, started with acetophenone. Ketones are known to undergo α cleavage (the Norrish-type I reaction), opening a C—C bond adjacent to the carbonyl group. In 2-phenylacetophenone, for example, this leads to the simultaneous formation of a benzoyl and a benzyl radical.

$$\text{Ph-CO-CH}_2\text{-Ph} \xrightarrow{h\nu} \text{Ph-CO}\cdot + \cdot\text{CH}_2\text{-Ph} \quad (15)$$

Fragmentation in carbonyl compounds occurs from the excited triplet state [15], and the rate of the process depends strongly on substituents at the α carbon. Thus, while reaction (15) is only of minor importance in 2-phenylacetophenone, it is the principal deactivation route in benzoin and its derivatives [21].

$$\text{Ph-CO-CH(OH)-Ph} \xrightarrow{h\nu} \text{Ph-CO}\cdot + \cdot\text{CH(OH)-Ph} \quad (16)$$

Photoscission is particularly efficient in the alkyl ethers of benzoin and these have now become the most widely used photoinitiators throughout the UV curing industry. The rate constants for α cleavage in the group of 2-phenylacetophenone derivatives in Table 4-1 illustrate the increase in fragmentation efficiency that can be achieved by substitution.

The benzoin alkyl ethers have absorption maxima between 330 and 350 nm, their molar extinction coefficients in this region are of the order of 10^2 L/mole·cm [22]. Both the benzoyl radical and the alkoxybenzyl radicals participate in the initiation process [23]. Figure 4-3 shows the absorption spectra of some generic structures.

A drawback of the generic benzoin ethers is their poor shelf life, which is linked to the presence of a labile hydrogen on the ether carbon [22]. This

TABLE 4-1 Rate Constants for α Cleavage in Derivatives of 2-Phenylacetophenone [21]

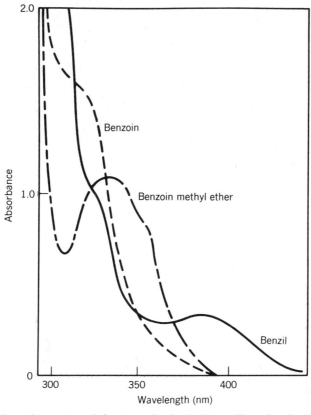

X	Y	k_0 (s^{-1})
H	H	2×10^6
H	CH$_3$	3×10^7
CH$_3$	CH$_3$	1×10^8
H	OH	1.2×10^9
H	OCH$_3$	$> 10^{10}$
H	OC$_4$H$_9$	$> 10^{10}$
H	CN	$< 10^5$

Figure 4-3 Absorption spectra of three generic photoinitiators (Reproduced with permission from C. G. Roffey [10].)

deficiency is removed in the benzil ketals, such as 2,2'-dimethoxy-2-phenylacetophenone (DMPA).

$$\underset{\text{DMPA}}{Ph-\underset{\underset{}{\overset{O}{\|}}}{C}-\underset{\underset{OCH_3}{|}}{\overset{\overset{OCH_3}{|}}{C}}-Ph} \xrightarrow{h\nu} Ph-\overset{O}{\overset{\|}{C}}\cdot + \cdot\underset{\underset{OCH_3}{|}}{\overset{\overset{OCH_3}{|}}{C}}-Ph$$

$$\downarrow$$

$$Ph-\overset{O}{\overset{\|}{C}}-OCH_3 + \cdot CH_3 \quad (17)$$

2,2'-Dimethoxy-2-phenylacetophenone is thermally stable, it has a convenient spectral absorption, its fragmentation yield in alcohol at 25°C is $\phi_i = 0.37$ [24] (that of the isopropyl derivative is $\phi_i = 0.47$) [21]. It dissociates in two steps and finally produces the reactive methyl radical, which accounts for the high initiation efficiency of DMPA (Irgacure 651) [25].

Many other benzoin derivatives have been proposed. One of these is an initiator that has surfactant properties [26].

$$Ph-\underset{\underset{\underset{SO_2^- \ Na^+}{|}}{\underset{O}{|}}}{\underset{\underset{CH_2}{|}}{\overset{\overset{O}{\overset{\|}{}}}{C}}-\underset{\underset{}{|}}{\overset{\overset{H}{|}}{C}}-Ph}$$

In the curing of liquid monomer compositions in air, this initiator accumulates at the air–liquid interface and, on irradiation, provides a high concentration of free radicals at this site that counteracts the inhibiting effect of aerial oxygen and produces good surface cure (see also Reference 27).

Derivatives of Acetophenone. The photoreactive derivatives of acetophenone are exemplified by 2,2-diethoxyacetophenone (DEAP), which fragments in the usual way [28].

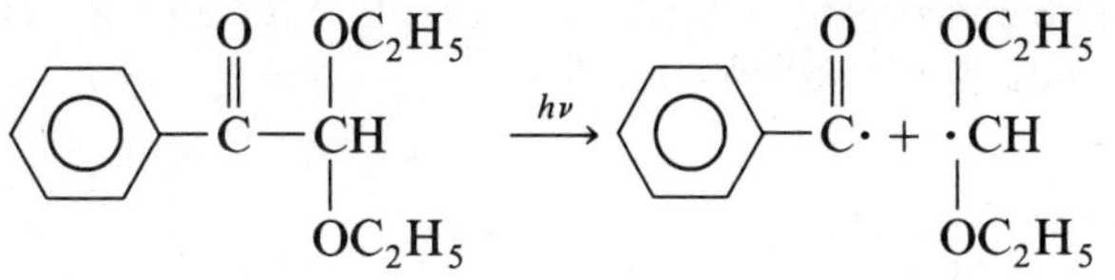

$$Ph-CO-CH(OC_2H_5)_2 \xrightarrow{h\nu} Ph-CO\cdot + \cdot CH(OC_2H_5)_2$$

DEAP

$$\downarrow$$

$$C_2H_5CHO + \cdot CH_2CH_3$$

(18)

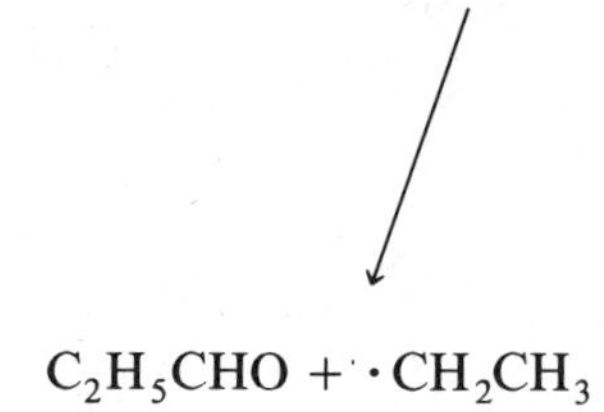

Although some of the radiation quanta are wasted here in a (Norrish type II) cyclization process, DEAP, just as DMPA, is much more effective than simple benzoin ethers. This may be due to the secondary fragmentation step, which favors the escape of the radicals from the primary cage.

Two other acetophenone derivatives deserve mention: 1-benzoylcyclohexanol (Irgacure 184)

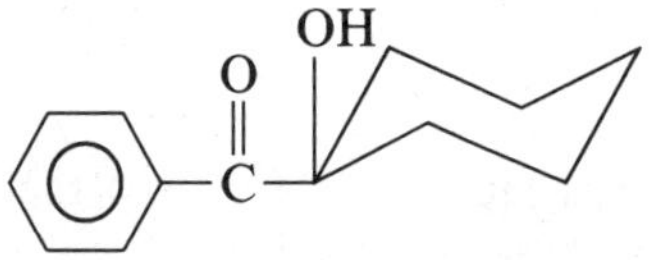

which does not produce the yellow stain typical of many other photoinitiators and can therefore be used for white coatings, and 4-methylmercapto-α,α-dimethyl-morpholino acetophenone,

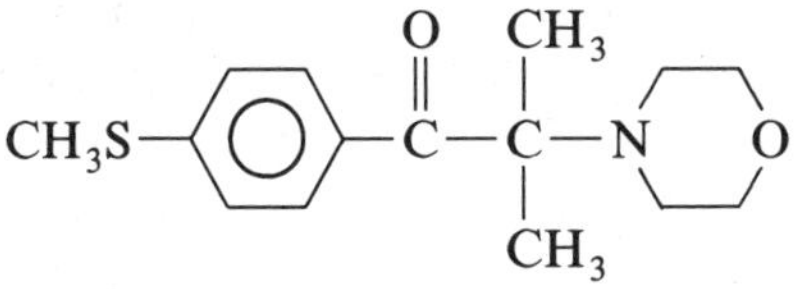

which has a much stronger absorption at 365 nm than other benzoin ethers ($\varepsilon \sim 10^4$ L/mole · cm) and is therefore suitable as a photoinitiator for highly pigmented coatings [28].

Ketoxime Esters of Benzoin. The idea of secondary fragmentation, which counteracts cage recombination and leads eventually to small, highly reactive radicals, is taken one step further in the ketoxime esters first introduced by Delzenne and co-workers [29]. For example, in the photolysis of 1-phenyl-1,2-propanedione-2-*O*-benzoyloxime (PPO), *two* volatile fragments are eliminated after primary scission of the N—O bond.

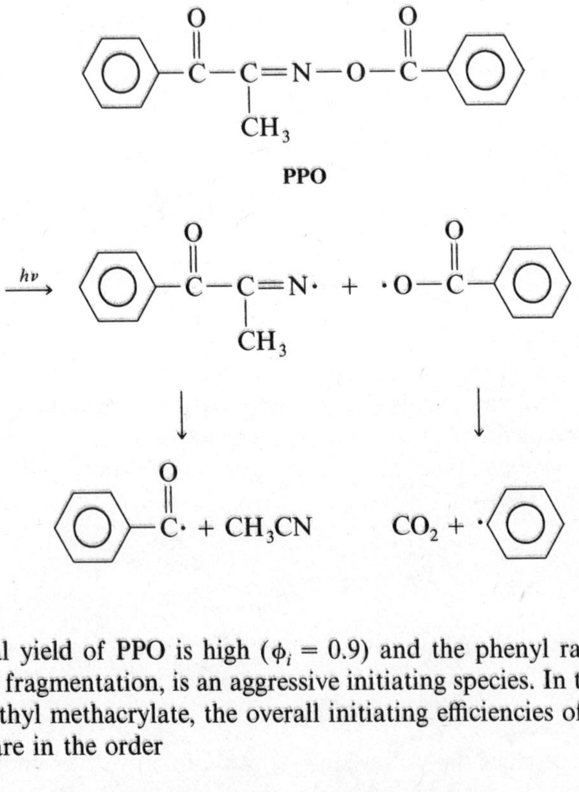

$$Ph-\overset{O}{\underset{\|}{C}}-\underset{\underset{CH_3}{|}}{C}=N-O-\overset{O}{\underset{\|}{C}}-Ph$$

PPO

$$\xrightarrow{h\nu} Ph-\overset{O}{\underset{\|}{C}}-\underset{\underset{CH_3}{|}}{C}=N\cdot \; + \; \cdot O-\overset{O}{\underset{\|}{C}}-Ph \tag{19}$$

$$\downarrow \qquad\qquad\qquad \downarrow$$

$$Ph-\overset{O}{\underset{\|}{C}}\cdot \; + \; CH_3CN \qquad CO_2 + \cdot Ph$$

The radical yield of PPO is high ($\phi_i = 0.9$) and the phenyl radical, the final product of fragmentation, is an aggressive initiating species. In the polymerization of methyl methacrylate, the overall initiating efficiencies of some of these initiators are in the order

PPO > DEAP > benzoin ethers

Triazines. Another system where multiple fragmentation is brought about by α scission are the symmetrical triazines, which on excitation dissociate into three substituted nitriles [30].

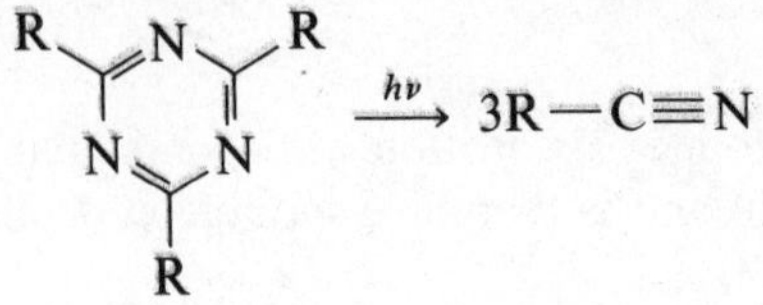

$$\underset{R}{\underset{|}{\underset{N}{\overset{R\diagdown}{\diagup}}}}\underset{N}{\overset{N}{\diagdown}}\underset{N}{\overset{\diagup R}{\diagdown}} \xrightarrow{h\nu} 3R-C\equiv N \tag{20}$$

Here the primary photoproducts are not radicals, but they dissociate into radicals in a secondary thermal step.

$$Cl_3C-C\equiv N \longrightarrow Cl\cdot + \cdot Cl_2C-C\equiv N \qquad (21)$$

Biimidazoles. A novel type of initiator based on the idea of incorporating a particularly weak bond into a molecule are the heavily substituted aryl-biimidazoles [31, 32].

$$(22)$$

They are obtained from the corresponding imidazoles by treatment with mild oxidants. The two moieties of the biimidazole are held together by a weak bond between nitrogen atoms and dissociate on excitation with almost unit efficiency into two imidazolyl radicals that are highly stabilized by resonance and have consequently long radical lifetimes. They do not initiate polymerization, but they will react with hydrogen donors, for example with tertiary amines, and produce in this way active radicals capable of initiating polymerization.

$$(23)$$

Radicals Generated by Hydrogen Abstraction

Hydrogen abstraction by the excited triplet states of ketones is one of the classic reactions of organic photochemistry. Irradiation of benzophenone, for example, in the hydrogen donor solvent isopropanol leads to the following processes:

$$Ph_2CO \xrightarrow{h\nu} {}^1Ph_2CO^* \longrightarrow {}^3Ph_2CO^*$$

$$
{}^3Ph_2CO^* + H-\underset{\underset{CH_3}{|}}{\overset{\overset{CH_3}{|}}{C}}-OH \longrightarrow Ph_2\dot{C}O-H + \cdot\underset{\underset{CH_3}{|}}{\overset{\overset{CH_3}{|}}{C}}-OH \quad (24)
$$

mixed dimers

$$Ph_2\underset{\underset{OH}{|}}{\overset{\overset{OH}{|}}{C}}-CPh_2$$

$$HO-\underset{\underset{H_3C}{|}}{\overset{\overset{H_3C}{|}}{C}}-\underset{\underset{CH_3}{|}}{\overset{\overset{CH_3}{|}}{C}}-OH$$

In the absence of any other reactive species the radicals formed by hydrogen abstraction combine into dimers, but in the presence of a reactive monomer they initiate chain polymerization. The species that absorbs radiation and is promoted to an excited state is termed the initiator, the hydrogen donor is the coinitiator.

The triplet excited states of common (aromatic) ketones are known to have $n\pi^*$ configurations, that is, they are formed by promoting an electron from a nonbonding n orbital on oxygen into an antibonding π^* orbital that is delocalized over the aromatic π system. Charge delocalization produces a partial positive charge on oxygen that may interact with the electron rich C—H bond and lower the energy of the transition state for hydrogen abstraction.

$$
\left[Ph_2C \overset{\delta-}{=} O \cdots \overset{\delta+}{H} - \underset{\underset{CH_3}{|}}{\overset{\overset{CH_3}{|}}{C}} - OH \right]^*
$$

Ketones with $\pi\pi^*$ configurations in the triplet state are less polar and hence less reactive. While $n\pi^*$ triplets abstract hydrogen from alcohols and even from tertiary C—H bonds of hydrocarbons, $\pi\pi^*$ triplets are slow to react with alcohols and will only react efficiently with tertiary amines, the strongest hydrogen donors [33].

In general, the success of the hydrogen abstraction step

$$R_2CO^* + H-D \longrightarrow R_2\dot{C}O-H + \cdot D \quad (25)$$

depends on the bond dissociation energies involved and on polar effects in the transition state [34]. Rate constants for the reaction of benzophenone triplets with a range of hydrogen donors are listed in Table 4-2.

TABLE 4-2 Rate Constants for the Reaction of Benzophenone Triplets with Various Hydrogen Donors [12]

Donor	k_q
Benzene	10^4
Cyclohexane	4×10^4
Methanol	3×10^5
Ethanol	6×10^5
Isopropanol	1×10^6
THF[a]	4.5×10^6
n-Nonylmercaptan	1×10^7
Thiophenol	3×10^8
Triethylamine	2×10^9
(Oxygen)	(2×10^9)
Phenyl dimethylamine	3×10^9

[a] Tetrahydrofuran is abbreviated as THF.

Practical Initiator Systems Based on H Abstraction [35]

Anthraquinone and Tetrahydrofuran

$$\text{Anthraquinone} + \text{THF} \xrightarrow{h\nu} \text{product} \quad (26)$$

Here, the tetrahydrofuryl radical is the principal initiating species [36]. For better solubility, 2-ethylanthraquinone or 2-*tert*-butylanthraquinone are recommended, or in aqueous systems, anthraquinone sulfonate.

Benzophenone and Tertiary Amines. Benzophenone and tertiary alcohols had been used for some time as initiators when Sander et al. [37] discovered the dramatic enhancement of initiating efficiency brought about in these systems by small additions of tertiary amines. It was found that the effect was caused not only by the good hydrogen donor properties of the alkyl groups in the α position to nitrogen, but also by the ability of the amines to act as electron donors and form *exciplexes* (excited complexes, see Chapter 3) with the triplet excited state of benzophenone and other aromatic ketones [38–40]. In fact, interaction between tertiary amines and ketones may not be confined to the excited state and weak donor–acceptor interactions may cause preassociation in the ground state. On excitation, when the two partners have been brought together in the exciplex, they may either transfer a hydrogen atom [41, 42], or an electron [43], or the exciplex may emit or be deactivated nonradiatively [40].

$$Ph_2C=O^* + N(CH_3)_3 \longrightarrow \left[Ph_2C \overset{\delta-}{=} O \cdots \overset{\delta+}{N} \begin{array}{c} CH_3 \\ CH_2 \\ H-CH_2 \end{array} \right]^*$$

(27)

$$
\begin{array}{ccc}
Ph_2\dot{C}-OH & Ph_2\dot{C}-O & Ph_2C=O \\
+ & + & + \\
(CH_3)_2N\dot{C}H_2 & \dot{N}(CH_3)_3 & N(CH_3)_3 \\
\text{H transfer} & \text{Electron transfer} & \text{Deactivation}
\end{array}
$$

The improvement in initiating efficiency on adding amines to the benzophenone-type initiator is so large that tertiary amines have become by far the most important coinitiators. They produce highly reactive carbon-centered radicals that are efficient initiating species. In addition, they have high chain transfer activity for peroxide radicals and for acrylates [44]. The following amines are in frequent use.

$(C_2H_5)_3N$	Triethylamine
$CH_3-N\begin{array}{c}C_2H_4OH\\C_2H_4OH\end{array}$	N-Methyldiethanolamine
$H_2N-\bigcirc-COOCH_3$	4-(Aminomethyl)benzoate
$(CH_3)_2N-\bigcirc-COOCH_3$	4-(Dimethylamino)methylbenzoate
$(CH_3)_2N-\bigcirc-CHO$	4-(Dimethylamino)benzaldehyde

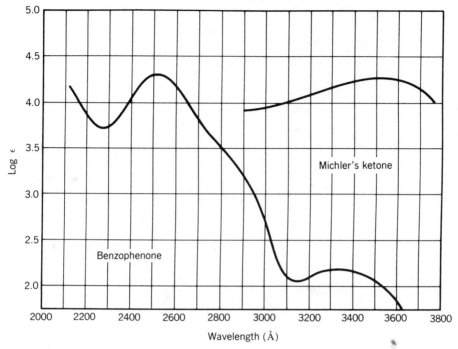

Figure 4-4 Absorption spectra of benzophenone and of Michler's ketone. (After R. A. Friedel and M. O. Orchin, *Ultraviolet Spectra of Aromatic Compounds*, Wiley, New York, 1952.)

Michler's Ketone

$$(CH_3)_2N-\text{C}_6H_4-\overset{O}{\underset{\|}{C}}-\text{C}_6H_4-N(CH_3)_2$$

4,4'-Bis(dimethylamino)benzophenone also called Michler's ketone, is an aromatic ketone as well as a tertiary amine and should on its own be a complete initiating system. In fact, H-abstraction does occur on irradiation and Michler's ketone is a useful photoinitiator [45–47]. It has a much stronger absorption at 365 nm than benzophenone (see Fig. 4-4). Michler's ketone is often used as the coinitiator together with benzophenone, and in that case it acts as the hydrogen donor. A disadvantage of Michler's ketone is the yellowish coloration of the coatings, which for some purposes is not acceptable. This deficiency is absent in two more recent systems, the thioxanthones and the ketocoumarins.

Thioxanthones

The thioxanthones behave like typical aromatic ketones [48]. They are practically colorless, and their triplet lifetime is relatively long, which makes the H-abstraction process efficient. Popular derivatives are 2-chlorothioxanthone and 2-isopropylthioxanthone. They are used in conjunction with various tertiary amines, for example, with the following two compounds.

$$\text{C}_6\text{H}_5\text{—COO—C}_2\text{H}_4\text{—N(CH}_3\text{)}_2$$

$$(\text{CH}_3)_2\text{N—C}_6\text{H}_4\text{—COO—C}_2\text{H}_4\text{—O—C}_4\text{H}_9$$

3-Ketocoumarins

Ketocoumarins were originally developed as triplet sensitizers [49]. They are equally useful as initiators and function here by abstracting hydrogen from suitable donors. By varying the substituents (R_1, R_2, R_3) it is possible to fine tune the absorption spectrum (and the triplet level) of the initiator (see Fig. 4-5). A wide range of structures is now available [50]. Tertiary amines are coinitiators.

Dye Sensitized Initiation

Initiators based on the scission of covalent bonds are limited to the UV. With hydrogen transfer or electron transfer, however, it is possible to generate radicals from molecules with low excitation energies. In this way photoinitiation can be extended into the visible part of the spectrum. Bamford and Dewar [51] observed as early as 1949 that some vat dyes could sensitize the photopolymerization of styrene. In 1954 Gerald Oster [52] found that the polymerization of acrylonitrile and of acrylamide could be photoinitiated by

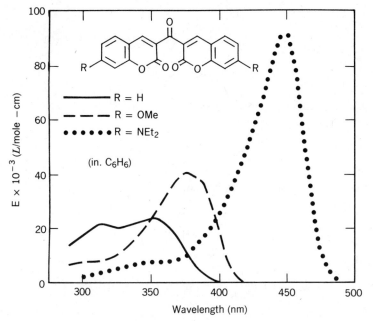

Figure 4-5 Absorption spectra of substituted bis(ketocoumarins). (Reproduced with permission from D. P. Specht et al. [50]. © Pergamon Press, 1982.)

fluorescein, Rose Bengal, and similar dyes, in the presence of reducing agents (phenylhydrazine, ascorbic acid), and oxygen. The amount of dye required was very small, 0.1% of the weight of monomer, and the quantum yield of monomer consumption was in excess of 4000 monomer units per photon. Oster's observations did not only spark interest in dye sensitized initiation, but resulted in much industrial activity in the general field of photopolymerization.

The following reaction sequence describes the mechanism of dye sensitized photoinitiation. Here Q stands for the dye (a quinonoid structure) and HD is the hydrogen or electron donor.

$$Q \xrightarrow{h\nu} {}^1Q^* \longrightarrow {}^3Q^* \tag{28}$$
$${}^3Q^* + HD \longrightarrow HQ\cdot + D\cdot$$

The donor radical $D\cdot$ is the initiating species. The semiquinone radical $HQ\cdot$ is usually not reactive enough to act as initiator, but it may disproportionate

$$2HQ\cdot \rightarrow Q + H_2Q \tag{29}$$

regenerate the quinone Q and form the fully reduced hydroquinone H_2Q. The quinone is also regenerated by the reaction of the semiquinone or the hydroquinone with aerial oxygen [53].

$$HQ\cdot + O_2 \rightarrow Q + HOO\cdot \qquad (30)$$

$$H_2Q + O_2 \rightarrow Q + H_2O_2 \qquad (31)$$

Finally, in a parallel process, oxygen may interact with the triplet state of the dye to produce singlet oxygen.

$$^3Q^* + {}^3O_2 \rightarrow Q + {}^1O_2^* \qquad (32)$$

In the case of fluorescein, in a medium of pH = 7, reactions (28) and (29) take the form

[Fluorescein (quinone form)] $\xrightarrow{HD}$ [Semiquinone radical] (33)

$$2 \text{ [Semiquinone radical]} \rightarrow$$

[Fluorescein quinone form] + [Hydroquinone] (34)

Several classes of photoreducible dyes have been identified [54]; the xanthenes (fluorescein, Rose Bengal), the thiazines (methylene blue, thionine), the acridinium dyes (acriflavin), and some natural products such as riboflavin.

Photoreducible Dye Classes	Dye	λ_{max}
Acridinium	Acriflavine	460
Xanthene	Rose Bengal	565
Thiazine	Thionine	645

A wide range of electron and hydrogen donors have been found to work in these systems [55, 56], and Eaton [54] has given a comprehensive list. Tertiary amines are the most important [57]. They transfer an electron from the lone pair of nitrogen to the acceptor, and this step is followed immediately by the dissociation of a proton from the acidic α C—H bond to yield a resonance stabilized radical as shown in reaction (35).

$$Ph_2\ddot{N}-CH_2-COO^- \xrightarrow{-1e^-} Ph_2\dot{N}^+-CH_2-COO^-$$
$$\downarrow \qquad (35)$$
$$Ph_2\ddot{N}-\dot{C}H-COO^- \longleftrightarrow Ph_2\dot{N}^+-\ddot{C}H-COO^- + H^+$$

TABLE 4-3 Limiting Quantum Yields and Rate Constants for the Reduction of Methylene Blue by Amines [57]

	ϕ^a	k_q^b ($\times 10^8 M^{-1} s^{-1}$)	IP^c (eV)
Tributylamine	0.68	2.00	7.22
Tripropylamine	0.61	2.02	7.23
Triethylamine	0.41	0.468	7.50
Benzylamine	0.77	0.532	7.56
Quinuclidine	0.09	0.697	7.72
Piperidine	0.27	0.111	7.85
Pyrrolidine	0.56	0.0713	7.98
Diethylamine	0.54	0.0710	8.01
Triethanolamine	0.54	0.778	
Aniline	0.84	111	7.70
N-Methylaniline	0.98	147	7.30
N,N'-Dimethylaniline	0.99	156	7.10

^{a}Limiting yield at infinite donor concentration.
^{b}Solvent, methanol.
^{c}Ionization potential.

Other types of donors are arylsulfinates [58], for example, $ArSO_2Na$, ethers like isobenzofurane and even suitably substituted aromatic hydrocarbons [59], enolates of diketones such as dimedone [60], carboxylates (ascorbic acid) [61], and finally organometallics such as benzyltrimethylstannane, $PhCH_2Sn(CH_3)_3$ [62].

There is broad correlation between the ionization potentials and the electron transfer efficiencies of electron donors. This is illustrated in Table 4-3 where the quantum efficiency of the reduction of triplet methylene blue to the semiquinone, together with the rate constant of this process are listed for a number of amines [57].

Applications. At the time of its discovery, dye sensitized photopolymerization promised to develop into a new photography that would in some applications rival silver halide photography [63]. The initiators were sensitive to visible light and the quantum yield of the polymerization process itself was high; values of 10^5 to 10^6 monomer units per photon were observed [3]. Furthermore, the coatings could be sensitized to a single color, which would allow the making of color separations without filters, directly from a color negative. The polymerizable coatings could also be made panchromatic by using a mixture of photoreducible dyes.

In the end, the early promise of dye sensitized photopolymerization was not fulfilled: As graphic arts intermediates the systems were not reliable enough to replace silver halides and in the manufacture of printing plates optimum physical properties were more important than sensitivity to visible light.

Recently, however, the situation has changed. Interest in dye sensitized polymerization has revived with the advent of lasers and laser scanners. The argon ion laser and the helium–neon laser, which are now the preferred radiation sources emit in the visible range (green and red, respectively). To use them for signal recording at a reasonable scan rate a material is required that is highly sensitive to visible light.

Du Pont have developed a visible laser-sensitive resist (Riston LV) intended for the direct laser imaging of printed wire boards (PWBs) [64]. The system is based on the principles of dye-sensitized polymerization. With an argon laser it has an exposure requirement of 10 to 15 mJ/cm^2.

The specifications for laser scanning in the graphic arts are more exacting and sensitivities of the order of 0.1 mJ/cm^2 are considered desirable. Shimizu has designed a high speed photopolymer system that is sensitive at 488 nm, the wavelength of emission of the Ar$^+$ ion laser, and which to a degree approaches the stated speed requirements [65]. The key feature here is the presence of an efficient transfer agent, a heteroaromatic mercaptan. The sensitizers are cyanine-type dyes that transfer energy to the initiator [tris(trichloromethyl-s-triazine) or a biimidazole], which dissociates into radical fragments. These abstract hydrogen from the mercaptan, and the S-centered radical formed in this way initiates polymerization. With a triacrylate monomer and an alkali-

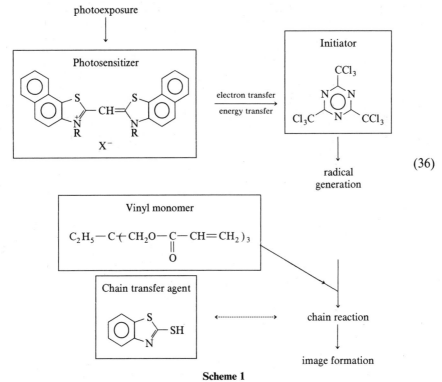

(36)

Scheme 1
Reaction mechanism of Shimizu's laser exposable photopolymer system [65]

soluble binder, exposure doses of 1 mJ/cm² were found to be sufficient for imaging. The reaction mechanism of the system is indicated in reaction Scheme 1 (36).

Dye Sensitization with Photochemical Pumping. High photosensitivity was recently achieved with a dye-sensitized photopolymerization system that incorporates a photochemical pumping mechanism. In this system a brief imagewise exposure to visible light produces a UV absorbing sensitizer capable of transferring excitation energy to a radical initiator. The latent image formed in this first step is photodeveloped by a blanket exposure to UV radiation, absorbed by the UV sensitizer.

In the composition described by Mitchell et al. [66] the initiator is a biimidazole (biIm) (see p. 113), which on excitation through energy transfer from the sensitizer dissociates into imidazole radicals (Im·).

$$\left(\text{biIm} \right)_2 \xrightarrow{h\nu} 2 \text{ Im·} \tag{37}$$

The rather special UV sensitizer (S_{UV}) shown in reaction (38) is generated from its precursor by an oxidative reaction step with the imidazole radical.

$$\text{precursor} + 2 \text{ Im·} \rightarrow S_{UV} + 2 \text{ ImH} \tag{38}$$

In the photodevelopment step the imidazole radical has a double function: It initiates the polymerization of the monomer (M), but it also produces more

UV sensitizer, which in its turn generates more imidazole radicals from the biimidazole. There is here a photochemical pumping mechanism by which an initially weak latent image is amplified.

The successive steps of the process are described by the following reaction sequence.

Imaging Step:

$$S_{vis} + h\nu_{vis} \rightarrow S^*_{vis} \tag{39}$$

$$S^*_{vis} + biIm \rightarrow 2Im^\cdot + S_{vis} \tag{39a}$$

$$2Im^\cdot + H_2S_{UV} \rightarrow 2HIm + S_{UV} \tag{39b}$$

Development Step:

Photochemical loop

$$S_{UV} + h\nu_{UV} \rightarrow S^*_{UV} \tag{40}$$

$$S^*_{UV} + biIm \rightarrow S_{UV} + 2Im^\cdot \tag{40a}$$

$$2Im^\cdot + H_2S_{UV} \rightarrow 2Im + S_{UV} \tag{40b}$$

$$Im^\cdot + M \rightarrow ImM^\cdot \tag{40c}$$

$$ImM^\cdot + M \rightarrow polymerization \tag{40d}$$

The plasticizer (or the monomer) content of the coating can be so adjusted that the original (unexposed) film is tacky, the polymerized film is nontacky. The image is made visible by dusting with a colored toner powder that adheres only to the unexposed (tacky) areas. In this configuration, exposure energies as low as 0.1 mJ/cm^2 are sufficient to produce an image.

Holography is another application where laser sensitive materials are needed, both in the recording of holograms as well as in the photofabrication of holographic gratings where the interference of two laser beams is used to generate a periodic pattern of light and shade [67, 68]. The recording of holographic images is dealt with in more detail on page 169.

Radical Formation by Ligand Exchange

Transition Metal Carbonyls. Some transition metal carbonyls, in conjunction with polyhalogen compounds, are efficient thermal initiators of polymerization [68a]. Bamford and Finch [69] discovered (1962) that these systems could also function as photoinitiators. On irradiation, radicals are formed by a two-step

TABLE 4-4 Absorbance and Photoreactivity of Transition Metal Carbonyls [70]

Transition Metal Carbonyl	λ_{max} (nm)	ε^a_{max} (L/mole · cm)	ε^a (at 365 nm) (L/mole · cm)	Relative Sensitivity
$Re(CO)_{10}$	330		1	
$Mo(CO)_6$	291	14,000	125	1.5
	329	2,000		
$Fe_3(CO)_{12}$				2
$Cr(CO)_6$	282	11,000	176	2.5
	324	2,300		
$Mn_2(CO)_{10}$	345	19,000	9340	6
Cyclopentadiene-$Fe(CO)_2I$				11
Cycloheptatriene-$Cr(CO)_3$	280	6,900	4520	20
	345	5,300		
	500	420		
Benzene-$Cr(CO)_3$	256	7,400	2200	50
	316	10,000		

^{a}Molar extinction coefficient.

ligand exchange reaction.

$$Mn_2(CO)_{10} \xrightarrow{h\nu} Mn_2(CO)_9 + CO \quad (41)$$

$$Mn_2(CO)_9 + CCl_4 \longrightarrow Mn_2Cl(CO)_9 + \cdot CCl_3 \quad (42)$$

The trichloromethyl radical is the species that initiates polymerization.

Wagner and Purbrick [70] found that benzene-chromium-tricarbonyl (BCT) was the most efficient photoinitiator in a group of 70 transition metal carbonyls examined (see Table 4-4 for a sample of the data). The overall result of the photoreaction is indicated in the following reaction mechanism (43)

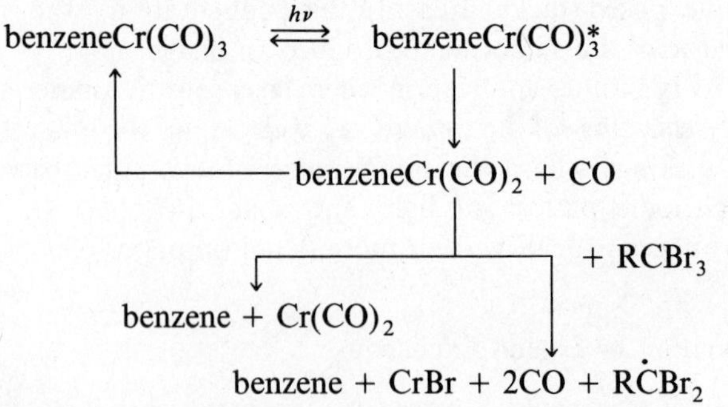

$$\text{benzeneCr(CO)}_3 \underset{}{\overset{h\nu}{\rightleftarrows}} \text{benzeneCr(CO)}_3^* \quad (43)$$

$$\downarrow$$

benzeneCr(CO)$_2$ + CO

+ RCBr$_3$

benzene + Cr(CO)$_2$

benzene + CrBr + 2CO + R$\dot{\text{C}}$Br$_2$

The quantum yield of radical production depends on the halogen content of the solvent. It is 0.23 in dichlorethane, 0.27 in bromoform, and 0.41 in liquid methacrylate and CCl_4.

Ferrocenes. Ferrocenes can function in a manner similar to that of the transition metal carbonyls [71]. They form complexes with polyhalogen compounds such as CCl_4.

$$(C_5H_5)_2Fe + CCl_4 \rightarrow (C_5H_5)_2Fe \cdots Cl-CCl_3 \qquad (44)$$

On irradiation of the ground state complex an electron is transferred from iron to the nearest chlorine atom, leading to the formation of a ferrocinium ion and a $\dot{C}Cl_3$ radical.

$$(C_5H_5)_2Fe \cdots Cl-CCl_3 \xrightarrow{h\nu} (C_5H_5)_2Fe^+ + Cl^- + \cdot CCl_3 \qquad (44a)$$

The trichloromethyl radical is the initiating species.

The outstanding feature of all these systems is that in the ligand exchange reaction only one radical is produced and therefore cage recombination of radicals is avoided [70]. Ligand exchange initiation has been used successfully with a variety of acrylates and methacrylates in viscous compositions and in solid polymer matrices, where radical recombination is particularly harmful.

THE INITIATION STEP

It is not enough to generate radicals, they have to react with the monomer in order to initiate polymerization. If the fraction of radicals that succeed in this is f_i, the product $\phi_i f_i = \phi_j$ is the quantum yield of chain initiation. From a practical point of view ϕ_j is the figure of merit of the initiating system.

The initiation step proper, Eq. (2) in our mechanism on page 103 and Eq. (45) below, competes with the quenching (scavenging) of primary radicals by various additives (Ad) (stabilizers, antioxidants) and by oxygen,

$$R\cdot + M \xrightarrow{k_r} RM\cdot \qquad (45)$$

$$R\cdot + Ad \xrightarrow{k_{Ad}} \text{nonreactive products} \qquad (45a)$$

$$R\cdot + O_2 \xrightarrow{k_O} \text{nonreactive products} \qquad (45b)$$

As a result, f_i can be expressed in the general form

$$f_i = \frac{k_r[M]}{k_r[M] + k_{Ad}[Ad] + k_O[O_2]} \qquad (46)$$

The oxygen reaction is the most important and in the presence of oxygen, $k_O[O_2]$ in the denominator of Eq. (46) is by far the largest term. In these circumstances f_i tends to zero and the initiation process is completely inhibited. This is the origin of the induction period that is often observed in polymerization processes.

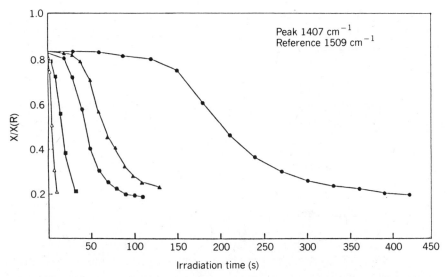

Figure 4-6 Experimental demonstration of the induction period of photopolymerization. Photoconversion of a mixture of 70 wt% of Epocryl, 20% of ethoxyethyl acrylate, and 10% trimethylolpropane triacrylate plotted as a function of irradiation time in the presence of increasing concentrations of the initiator dimethoxy-2-phenylacetophenone (DMPA). Conversion is measured by monitoring an absorption peak of the monomer and an (unchanging) reference peak by Fourier transform infrared spectrometry (FTIR). Dimethoxy-2-phenylacetophenone concentrations: ● 0.5; ▲ 1.0; ◆ 2.0; ■ 4.0; △ 8.0 wt%. (Reproduced with permission from R. D. Small et al. [72].)

Oxygen reacts with most initiating radicals by forming peroxy radicals that do not effectively initiate polymerization.

$$R\cdot + O_2 \rightarrow ROO\cdot \tag{47}$$

In the early stages of irradiation this reaction removes the photogenerated primary radicals. Only when most of the oxygen has been consumed are the primary radicals available for initiation.

Since the rate of generation of primary radicals, at a given light flux, is a function of the concentration of photoinitiator, the induction period depends also on initiator concentration [72]. This is illustrated by the data in Fig. 4-6.

The extent to which oxygen overshadows other common inhibitors can be seen from their transfer constants towards the propagating radicals in the polymerization of styrene and of methyl methacrylate. These are shown in Table 4-5.

In industrial applications a prolonged induction period of polymerization is not acceptable, and for that reason, many photocuring processes are carried out under a nitrogen or carbon dioxide blanket, or in films that are either protected from oxygen by an impermeable coversheet, or exposed in a vacuum frame.

THE INITIATION STEP   129

TABLE 4-5   Transfer Constants for Various Inhibitors and Monomers in the Polymerization of Styrene and Methyl Methacrylate (MMA) [73]

Inhibitor	Styrene^a	MMA^a
Nitrobenzene	0.326^b	
Trinitrobenzene	64.2	
Benzoquinone	518	5.5 (40°C)
Oxygen	14,600	33,000

^{a} The temperature was 50°C except where marked.
^{b} In L/mole s.

In addition, oxygen can be removed from the system by chemical means, the most popular being the reaction with (tertiary) amines. These are often used as hydrogen donors in initiation and form radicals that react with oxygen and are transformed thereby into peroxy radicals. Amines are such effective hydrogen donors that they will react even with these peroxy radicals and will start the following reaction chain [74].

$$R\cdot \ + \ H_3C-N(CH_3)_2 \ \longrightarrow \ RH \ + \ \cdot CH_2-N(CH_3)_2 \quad (48)$$

$$O_2 \ + \ \cdot CH_2-N(CH_3)_2 \ \longrightarrow \ \cdot OOCH_2-N(CH_3)_2 \quad (48a)$$

$$OOCH_2-N(CH_3)_2 \ + \ H_3C-N(CH_3)_2 \ \longrightarrow \quad (48b)$$

$$\longrightarrow \ HOOCH_2-N(CH_3)_2 \ + \ \cdot CH_2-N(CH_3)_2 \quad (48c)$$

The cycle of the mechanism in Eqs. (48) to (48c) is able to consume up to 12 molecules of oxygen by as many molecules of amine, but with the loss of only one radical. That is the reason for the unusually high efficiency of tertiary amines as antioxidants.

Decker [75] has described another interesting way of removing oxygen. In the presence of some sensitizing dyes, such as Rose Bengal (RB), oxygen is transformed into singlet oxygen and can then be scavenged, for example, by diphenylbenzofuran.

$$RB \ \xrightarrow{h\nu} \ ^1RB^* \ \longrightarrow \ ^3RB^* \quad (49)$$

$$^3RB^* \ + \ ^3O_2 \ \longrightarrow \ RB \ + \ ^1O_2^* \quad (50)$$

$$^1O_2^* \ + \ \underset{\phi\ \ \ \ O\ \ \ \ \phi}{\text{(diphenylbenzofuran)}} \ \longrightarrow \ \text{unreactive products} \quad (51)$$

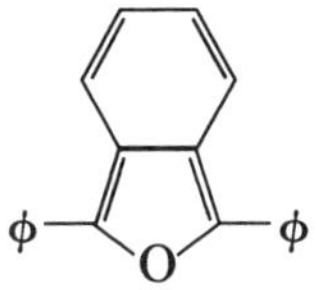

PROPAGATION VERSUS TERMINATION AND THE KINETIC CHAIN LENGTH

The Steady State Treatment

Once a radical chain has been initiated, it propagates spontaneously until it is terminated by encounter with another radical, by disproportionation or in some other way. The average chain length, which ultimately determines the photosensitivity of the system, is linked to the kinetic rate constants of the individual processes. The rates of initiation (i), propagation (p), and termination (t) are described by the following equations [76].

$$R_i = 2f_i\phi_i I_a = 2\phi_j I_a \tag{52}$$

$$R_p = k_p[\text{M}][\text{M·}] \tag{53}$$

$$R_t = 2k_t[\text{M·}]^2 \tag{54}$$

Here I_a is the radiation flux absorbed by the initiator, ϕ_j is the overall initiation efficiency, and k_p and k_t are the rate constants of propagation and termination, respectively. To express the rate of propagation and the average chain length in terms of measurable quantities, the radical concentration must be eliminated from Eqs. (53) and (54). This is readily done for a steady state regime; in the photostationary state the rates of initiation and of termination must be equal, and in these conditions the steady state concentration of M· is given by

$$[\text{M·}] = \left(\frac{\phi_j I_a}{k_t}\right)^{1/2} \tag{55}$$

The rate of propagation can now be expressed in the form

$$R_p = k_p \left(\frac{\phi_j I_a}{k_t}\right)^{1/2} [\text{M}] \tag{56}$$

and the kinetic chain length, which is determined by the competition between propagation and termination, is given by the expression

$$\nu = \frac{R_p}{R_t} = \frac{R_p}{R_i} = \frac{k_p}{2\sqrt{k_t}} (\phi_j I_a)^{-1/2} [\text{M}] \tag{57}$$

The ratio $k_p/\sqrt{k_t}$ is termed the characteristic ratio of the monomer system. It is independent of the mode of initiation and measures the ability of the monomer to support a radical chain reaction. Table 4-6 lists values of the characteristic ratio for a few common monomers.

TABLE 4-6 Values of $k_p/\sqrt{k_t}$ for Some Common Monomers [9]

Monomer	$k_p/\sqrt{k_t}$ (L/mole s)$^{1/2}$
Acrylamide	4.37[a]
Methacrylamide	0.2[a]
Methacrylate	0.21–0.23[a]
Vinyl acetate	0.12–0.18[a]
Methyl methacrylate	0.04–0.08[a]
Styrene	0.006–0.016[a]
1,6-Hexanediol diacrylate	8.6–14[b]
Pentaerythrol tetraacrylate	1.5[b]

[a]At 250°C, aqueous solution [9].
[b]At 40°C, pure monomer [80].

This simple analysis takes no account of the attenuation of light intensity in the system, or of the depletion of initiator, and so on (see, however, Reference 77.) It is further assumed that termination occurs only by bimolecular encounters between macroradicals, leading either to recombination or disproportionation. It implies also that k_p and k_t have indeed constant values. This is true in dilute solutions where the physical properties of the medium do not change appreciably during the reaction. In these conditions the conversion of monomer can be monitored *in situ* by various analytical methods, such as spectrometry [78], refractometry, and dilatometry. For kinetic data on a great number of monomers see Reference 79.

Photopolymerization in Viscous Media

In all imaging applications as well as in radiation curing, polymerization is carried out not in dilute solution, but in the neat monomer or in viscous mixtures of monomer with a polymeric binder. In these concentrated systems the properties of the medium (viscosity, temperature) change during polymerization. In the neat polymer in particular, the viscosity increases as polymerization progresses and this slows down all chemical processes that depend on diffusion. In these conditions, the termination process in which two macroradicals must diffuse towards each other is more strongly affected than propagation, where one of the partners is the mobile free monomer. As a result, the characteristic ratio $k_p/\sqrt{k_t}$ and the polymerization rate increase dramatically in the course of the process, giving rise to a strong *autoacceleration* effect. This is illustrated in Fig. 4-7 for the thermal polymerization of neat methyl methacrylate [81]. The broken line in the diagram corresponds to the idealized case of Eq. (56).

In concentrated, multicomponent systems UV–visible spectroscopy is often not applicable, and *IR spectroscopy* is the analytical method of choice [82, 83].

132   PHOTOINITIATED POLYMERIZATION

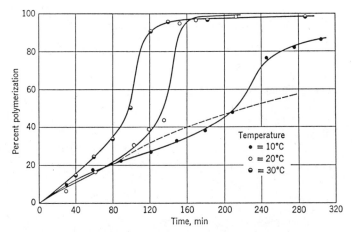

Figure 4-7 Autoacceleration effect in the thermal polymerization of neat methyl methacrylate. [Reproduced with permission from M. A. Naylor, *J. Am. Chem. Soc.*, **75**, 2181 (1953).]

Fast responding FTIR instruments are particularly useful in this context. Infrared analysis, which monitors specific bonds or groups, is a powerful tool in studying the effect of sensitizers, inhibitors, and other addenda [84]. The fraction of monomer, crosslinker, and initiator that has *not* reacted at a particular stage of photopolymerization may be determined by *solvent extraction* [85] and subsequent chromatographic analysis of the extracts. Alternatively, unpolymerized and therefore still volatile components may be estimated conveniently by *thermogravimetric analysis* (TGA) [86].

Fourier transform infrared spectrometry and the other methods just mentioned monitor the degree of monomer conversion as a function of exposure dose. A technique that directly measures the *rate* of polymerization is *photocalorimetry* [87]. In this method the heat evolved by the polymerization reaction is measured in a differential scanning calorimeter. The instrument is fitted with a quartz window to allow irradiation of the monomer sample with UV. Figure 4-8 shows an arrangement used by Tryson and Schultz [88].

The rate of heat evolution (dH/dt) is directly proportional to the rate of polymerization, $R_p = -d[M]/dt$. The proportionality constant is found from an absolute calibration of the instrument in calories per second and from a knowledge of the molar heat of reaction (Q).

$$\frac{-d[M]}{dt} = \left(\frac{-dH}{dt}\right) Q^{-1} \tag{58}$$

Figure 4-9 shows a typical photocalorimetric curve obtained with such an instrument. The degree of monomer conversion at any stage of the polymerization process can be found by stepwise integration under the curve and comparing the area corresponding to time t with the theoretical heat of reaction expected for 100% conversion. The maximum conversion obtainable in the photopolymerization of the material, an important datum in practice, is

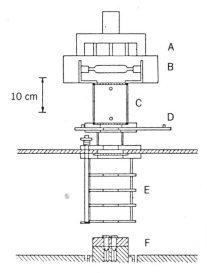

Figure 4-8 Differential scanning photocalorimeter assembly, (modified Perkin-Elmer DSC-2). A: reflector, B: medium pressure mercury vapor lamp, C: circulating water heat filter, D: shutter, E: filters, F: sample cell in calorimeter block. (Reproduced with permission from G. R. Tryson and A. R. Schultz [88].)

found as the ratio of the total heat of photopolymerization [the area under the DSC (differential scanning calorimetry) curve] to the theoretical heat of polymerization calculated for complete monomer conversion from the molar heat of reaction. Table 4-7 gives a few representative data.

From the results of calorimetric experiments and a knowledge of the photon flux absorbed in the sample, the kinetic chain length and the characteristic ratio ($k_p/\sqrt{k_t}$) can be derived from the ratio R_p/R_i. Table 4-8 gives results for the photopolymerization of 1,6-hexanediol diacrylate in the presence of 3% of the initiator DMPA (Irgacure 651).

It should be noted that the ratio $k_p/\sqrt{k_t}$ is characteristic of the material, while the relative polymerization rate $d[M]/[M]\,dt$ depends also on the light flux absorbed by the initiator.

$$\frac{d[M]}{[M]\,dt} = \frac{R_p}{[M]} = \frac{k_p}{k_t}(\phi_j I_a)^{1/2} \tag{59}$$

The data in Table 4-8 show the autoacceleration effect discussed on page 132. The polymerization rate reaches a maximum and finally slows down, long before all monomer is consumed, for lack of radical mobility. The rate of polymerization at the maximum of the photocalorimetric curve is often used as a ready measure of the reactivity of the system. Other data of practical importance are the heat of reaction (usually normalized to the number of monomer units in the system), the maximum conversion achievable in photopolymerization, and the activation energy of propagation. Some of these data are assembled in Table 4-9 for common monomers and crosslinking agents used in photopolymerization.

Monomers and polyfunctional monomers (crosslinkers) are not the only components of practical systems (see p. 154), and polymeric *binders* are added

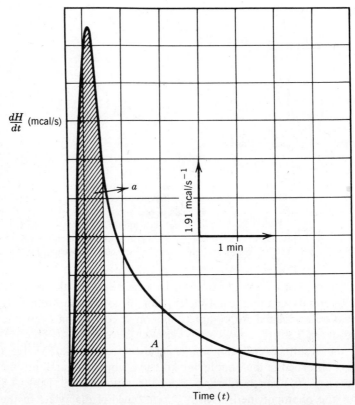

Figure 4-9 Photocalorimetric curve of 1,6-hexanediol diacrylate (HDDA) photopolymerized with 3% dimethoxy-α-phenylacetophenone (DMPA). The rate of heat evolution (dH/dt) is plotted against time (t). The calorimeter was calibrated in millicalories per second (mcal s^{-1}) by the heat of fusion of indium metal. (After G. Wu [80].)

TABLE 4-7 Heat of Polymerization of Some Acrylates [80]

Acrylate	Heat of Polymerization (cal/g)			Maximum Conversion (%)
	Photo	Thermal	Theoretical	
Lauryl acrylate	73	80	80	91
1,6-Hexanediol diacrylate (HDDA)	121	155	170	71
Trimethylolpropane triacrylate (TMPTA)	94	156	195	48

TABLE 4-8 Kinetic Data on the Photopolymerization of 1,6-Hexanediol Diacrylate [80][a]

Conversion (%)	$d[M]/[M]\,dt$[b] (s^{-1})	$k_p/\sqrt{k_t}$[c] (L/mole s)$^{1/2}$	ν[d]
4.9	0.027	5.91	2750
10.5	0.039	8.59	3760
17.4	0.046	9.98	4036
24.4	0.050	10.9	4050
31.6	0.056	12.2	4110
38.7	0.061	13.5	4040
45.5	0.064	14.0	3740
51.3	0.056	12.4	2950
58.2	0.025	5.48	1120
64.1	0.010	2.25	400

[a] At 40°C.
[b] $d[M]/[M]\,dt$ is the relative polymerization rate.
[c] $k_p/\sqrt{k_t}$ is the characteristic ratio.
[d] The kinetic chain length ν, all at a particular conversion.

in many compositions. The function of the binders is to increase the solids content and improve the physical properties of the coatings. It is found that this also improves the photographic sensitivity. In photopolymerization, photographic speed is ultimately determined by the mass of material, which the photoreaction is able to immobilize as a result of unit exposure. The high molecular weight of the binder, caught in the network of the polymerizing monomer, adds substantially to the mass immobilized by a given number of photons.

In many applications it is desirable to use highly viscous, almost solid coatings. These solidify during the early stages of polymerization and a large fraction of monomer remains trapped in the system. To overcome this deficiency "reactive binders" or "reactive prepolymers" are used that contain unsaturated groups. They take part in the reaction and become genuinely locked into the network by a process of "crosslinking polymerization." The physical properties of the cured films are improved and a higher degree of conversion can be achieved.

In the multicomponent, viscous systems that are used in practice, the simple steady state treatment of page 130 is not applicable even as a first approximation. The kinetic problem is too complex and cannot be solved in closed form, but *computer simulation* of comparatively simple models has recently produced interesting results.

Using a percolation model, Kloosterboer and Lippits [89, 90] found that in the early stages of photopolymerization large inhomogeneities are created in the system. Figure 4-10 shows the result of a computer simulation for the case of an acrylate polymerized to 25% conversion: A single polymer chain meanders through several centers of agglomeration (Fig. 4-10a); in Figs. 4-10b and c

TABLE 4-9 Heat of Polymerization, Maximum Conversion, Peak Rate of Polymerization, and Activation Energy [79, 80, 87][a]

Monomer	Abbreviation	H_p^b	x_{max}^c	$R_p[M]^d$	E_a^e
Acrylates					
Methyl acrylate	MA	18.6		1.68	
Ethyl acrylate	EA	18.2		1.8	
Butyl acrylate	BuA	17.3		2.2	
Lauryl acrylate	LA	16.2	78.6	2.51	5.04
2-Ethylhexyl acrylate	EHA	18.8	91.3	1.73	
2-Hydoxyethyl acrylate	HEA	18.2	88.3	13.9	5.98
Ethylene glycol diacrylate	EGDA	12.8	62.1	12.5	5.68
1,3-Butanediol diacrylate	BDDA	12.6	62.1	11.7	7.47
1,6-Hexanediol diacrylate	HDDA	16.4	79.6	14.8	4.38
Neopentyl glycol diacrylate	NPGDA	11.8	57.3	9.6	8.08
Trimethylolpropane triacrylate	TMPTA	8.8	42.7	14.2	1.81
Pentaerythritol tetraacrylate	PET4A	6.7	32.5	10.0	3.62
Methacrylates					
Methyl methacrylate	MMA	9.2	83	0.95	5.0
2-hydroxyethyl MA	HEMA	9.94	73	4.25	4.0
Ethylene glycol DMA	EGDMA	7.16	52.6	2.11	3.93
1,6-Hexanediol DMA	HDDMA	8.79	64.6	3.23	3.27
Neopentyl glycol DMA	NPGDMA	7.03	51.7	2.79	3.03
Trimethylolpropane TMA	TMPTMA	5.29	38.9	3.24	2.50

[a] At 30°C.
[b] Molar heat of reaction per acryloyl group.
[c] Calculated as total heat of reaction to theoretical heat of polymerization.
[d] $d[M]/[M]dt$ in percent.
[e] Derived from the temperature dependence of the peak rate of polymerization.

the kinetic chain length has been artificially limited to 20 and to 2, and the degree of inhomogeneity is decreased. Figure 4-10d corresponds to the step-reaction polymerization of a tetrafunctional acrylate. The formation of microgel particles depicted in the model calculations has been confirmed by experiment. It is in agreement with ESR measurements that show the almost indefinite survival of some radicals (lifetimes of days and weeks) in the dense regions of the system where radical mobility is minimal. These conclusions are supported by DSC: a second polymerization exotherm is observed in cured systems when their temperature is increased above T_g and radical mobility is restored.

Inhomogeneities formed in the early stages of photopolymerization lead to local changes of composition, and eventually to phase separation. For exam-

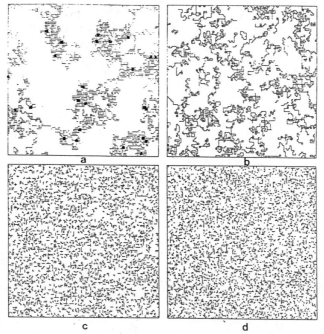

Figure 4-10 Computer simulation of the polymerization in two dimensions of a tetrafunctional monomer. Bond conversion is 25%. (*a*) A single living radical is considered; no limit to chain length; (*b*) kinetic chain length limited to 20 steps; (*c*) kinetic chain length limited to 2 steps; (*d*) step polymerization, kinetic chain length = 1. (Reproduced with permission from H. M. J. Boots [91].)

ple, Maerov [92] recently showed that a mixture of 25% of a triacrylate (TMPTA) and 75% of a binder (50:50 polystyrene: *sec*-butylmaleate) is biphasic with spherical domains rich in monomer (365 Å average domain diameter) and a binder-rich disperse phase. On irradiation, the size of the spherical domains increases (from 365–450 Å to 1000 Å) and finally, after some 40 units of exposure, the phases coalesce (see Fig. 4-11). With a different binder exactly the opposite was observed; an initially homogeneous mixture gradually separated into two distinct phases.

Phase changes of this kind have a large effect on the physical properties of the polymerized image and play an important role in the design and performance of practical systems.

Photopolymerization in Amorphous Solids

In going from a liquid monomer to the highly viscous pastes used, for example, in dry resists, the polymerization behavior and the physical proper-

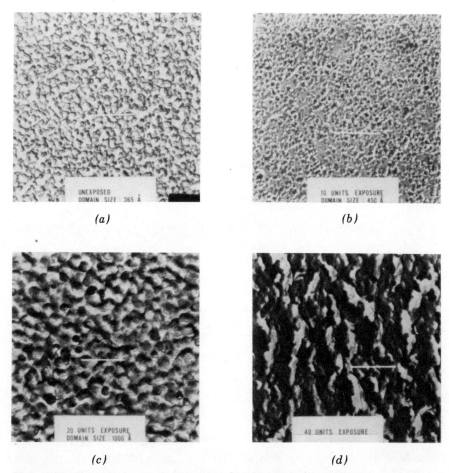

Figure 4-11 Phase changes during photopolymerization. Transmission electron micrographs of the freeze-fracture surfaces of a copolymer of 25% trimethylolpropane triacrylate (TMPTA) and 75% of binder (1:1 polystyrene: *sec*-butylmaleate). Radiation dose applied: (*a*) 0; (*b*) 88; (*c*) 175; (*d*) 350 mJ/cm^2. The marker bar represent 0.5 μm. (Reproduced with permission from S. B. Maerov [92].)

ties of the cured polymers change gradually without any discontinuity. This is not so when an amorphous system is taken from its rubbery state to its glassy state. At the glass transition almost all physical properties of the system change abruptly, and so do its polymerization kinetics. Figure 4-12 shows an example. Here values of the glass transition temperature, T_g, and of the crosslinking quantum yield, Φ, are plotted as a function of plasticizer content. The polymer concerned in this experiment carries free methacryloyl groups and has the following composition [93].

$$\left[-CH_2-\underset{\underset{OCH_3}{\underset{|}{CO}}}{\overset{\overset{CH_3}{|}}{C}}- \right]_m \left[-CH_2-\underset{\underset{CH_3-C=CH_2}{\underset{|}{\underset{CO}{\underset{|}{\underset{O}{\underset{|}{\underset{CH_3-CH}{\underset{|}{\underset{CH_2}{\underset{|}{NH}}}}}}}}}}}{\overset{\overset{CH_3}{|}}{\underset{|}{\overset{|}{C}}}}\underset{}{-} \right]_n$$

$m = 10, n = 1$

Polymer I

Unplasticized, this polymer has a T_g of 117°C. By adding increasing quantities of a plasticizer (dibutylphthalate) T_g is gradually lowered; when T_g reaches room temperature, which is the temperature at which the quantum yield of crosslinking was measured, the system changes from a glass to a thermoplastic polymer and the quantum yield of crosslinking increases steeply.

Many applications require a polymer with a high T_g, well above room temperature. The coatings made from these materials are hard, with a specularly reflecting surface, they have a long shelflife, need no protective cover, and can be used for the preparation of "presensitized" lithoplates [94]. In such systems, exemplified by polymer I, shown on the preceding page, photoinitiated polymerization produces crosslinking. The quantum yield of crosslink formation Φ is here the product of three factors.

$$\Phi = \phi_i f_i \nu \qquad (60)$$

ϕ_i is the quantum yield of radical formation by the initiator, f_i is the efficiency of initiation, and ν is the kinetic chain length of the propagation process. While ϕ_i and f_i cannot exceed unity, the kinetic chain length is not limited in this way and it is the principal factor determining photosensitivity.

Below T_g, chain propagation is slow and the average chain length is usually short. In the absence of oxygen, chain termination occurs to a large extent by radical occlusion, that is, the radicals find themselves trapped at locations where there are no reaction partners available. Table 4-10 gives some photokinetic data for unplasticized polymer I, for polymer I plasticized with 10% of a monoacrylate (II) and plasticized with 10% of a tetraacrylate (III).

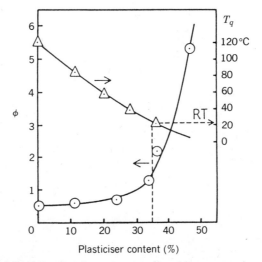

Figure 4-12 Transition from the glassy state to the rubbery state of a photopolymer film. Quantum yield of crosslinking and glass transition temperature as functions of plasticizer content for a derivative of PMMA (polymer I) containing 19% of free methacryloyl groups. (Reproduced with permission from P. L. Egerton et al. [93].)

TABLE 4-10 Kinetic Data on Solid State Photopolymerization [93]

Polymer (see text)	Φ^a	ν^b	k_p (cm³/mole · min) ($\times 10^2$)	k_t (s⁻¹)	f_{occl}^c	D^a (cm² · s⁻¹) ($\times 10^{-16}$)
I	1.6	7	3.5	0.027	0.93	0.16
II	4.1	19	37	0.081	0.87	1.6
III	6.0	27	126	0.35	0.49	5.6
IVe	5.3	24	2700	0.45		120

^{a}Quantum yield of crosslink formation, Φ.
^{b}Kinetic chain length, ν.
^{c}The fraction of chains ending in occlusion, f_{occl}.
^{d}The diffusion coefficient of polymer-bound radicals in the solid matrix, D.
^{e}This system, polymer I plasticized with 47% of dibutylphthalate, is above its T_g.

Although the kinetic chain length in the glassy polymer is rather low, the quantum yield of crosslinking compares very favorably with the values (0.1–0.3) found in conventional crosslinking polymers. Polymer I together with an efficient transition metal carbonyl initiator has been developed into a highly sensitive lithographic plate [94].

Photopolymerization in the Crystalline Phase

In amorphous solids, photoinitiated radical chains are in general short and are often terminated after a few steps when the propagating radical finds itself in a locality where no monomer is within reach. This failure is ultimately caused by the absence of order; in a random system there are always regions that are empty of reactants and where, therefore, the reaction comes to a standstill. To ensure the continual availability of reaction partners, which is the condition for long reaction chains, a degree of order in the solid phase is required. That is the reason why polymerization in crystalline monomers is of interest.

Early work by Morawetz and co-workers on crystalline acrylates [95, 96], has shown that radical chains do in fact propagate in these systems, but the rate of the process is exceedingly slow; in the acrylate crystals the separation of the active double bonds is too large for a facile reaction between radical and monomer. The C—H bonds of the ethylenic groups act here as spacers and determine the dimensions of the unit cell. Only in systems where these spacers are absent is there hope for rapid radical propagation. Such systems are the diacetylenes, rediscovered by Wegner in 1969 [97].

Diacetylenes. In some diacetylene crystals the conjugated triple bonds are so arranged that a 1,4-addition process, once initiated, can propagate spontaneously as indicated in Eq. (61). The reaction can be initiated by heat, UV, high energy radiation, and even by exposure to Cl_2, Br_2, and other reactive gases. For a recent review of the field see Reference 98.

$$\tag{61}$$

In the monomer crystal, the π system of diacetylene is isolated and the monomer absorbs at short wavelength, between 240 and 280 nm. In the polymer, conjugation extends all along the chain, the excitation energy of the system is lowered, and the polymer absorbs in the visible region, between 500 and 600 nm. As a result, the polymerization of crystalline diacetylenes is accompanied by a remarkable color change, the monomer being colorless while the polymers are blue, or in some cases blue-red (see Fig. 4-13). The color change that occurs in diacetylenes when they are kept under ambient laboratory conditions was known long before Wegner discovered the nature of the underlying reaction [98a, b]

Only those diacetylenes are reactive where the geometric change in going from the monomer to the polymer crystal is not too severe. In crystallographic

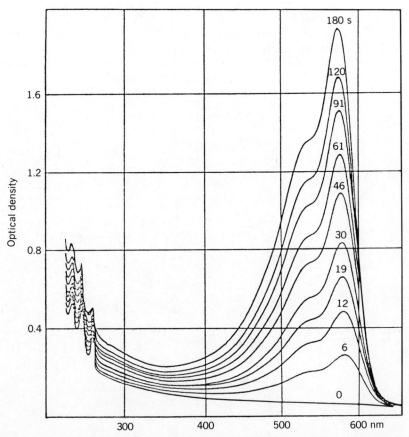

Figure 4-13 Evolution of the UV absorption spectrum of a diacetylene containing block copolymer exposed to 254 nm radiation. The in-growing absorption band at 580 nm is that of the polydiacetylene. The polymer is listed in Table 4-11 with the acronym HDI-PBD-35-1350. (Reproduced with permission from R. C. Liang and A. Reiser [106].)

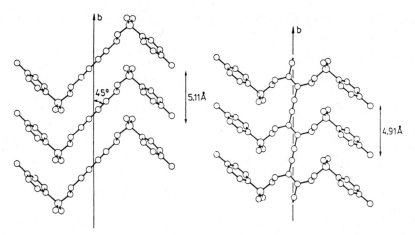

Figure 4-14 Models of diacetylene-bis(*p*-toluenesulfonate) monomer and polymer. Crystal projections onto the plane of the polymer backbone. (Reproduced with permission from V. Enkelmann [98].)

terms this means that the interplanar distance in the monomer crystal must be of the order of 5 Å and that the tilt angle between the principal crystal axis and the long axis of the diacetylene group must fall within a certain range of values. These so-called "topotactic" conditions of diacetylene polymerization are rather strict and even small changes in molecular structure can have a profound effect on crystal reactivity. It is found that the structure of the diacetylene crystal and hence its reactivity are determined largely by the substituents on the diacetylene moiety. Of particular interest are the toluenesulfonic acid esters and the urethanes. The structural change that occurs, for example, in the crystals of a toluene sulfonate is represented in the crystal projections of Fig. 4-14 where it can be seen that the displacements of most of the atom groups in the chemical transformation are quite small.

Quantum yields from 10^{-2} to 200 have been reported for these systems [99, 100]. The high values indicate that diacetylenes are able to support reaction chains of considerable length and could, in principle, be used as imaging materials of high sensitivity. Penzien [101] was the first to find an imaging application for the diacetylenes. He dispersed crystals of a diacetylene toluenesulfonate in a polymeric binder and recorded images in terms of the blue coloration of the polydiacetylene. Sohn et al. [102] used diacetylenes for the recording of x-ray images. More recently, Richter et al. [103] were able to record holograms on the surface of diacetylene monocrystals, making use of the high resolving power (100 Å) of which these systems are capable.

Liang and Reiser [104–106] have used the photoinitiated polymerization of diacetylenes as a crosslinking mechanism, in the hope of exploiting the amplification factor of the chain reaction. The diacetylene function was incorporated into segmented block copolymers of the general structure

$$\left[O-Z-O\overset{O}{\overset{\|}{C}}NH-R-NH\overset{O}{\overset{\|}{C}} \right] \left[OCH_2-C\equiv C-C\equiv C-CH_2O\overset{O}{\overset{\|}{C}}NH-R-NH\overset{O}{\overset{\|}{C}} \right]$$

Soft segment   Hard segment

where R stands either for hexamethylene, introduced via hexamethylene diisocyanate (HDI) or for cyclohexylene (CHDI). Here Z indicates the soft segment component, for example, polybutadiene (PBD). Table 4-11 lists the quantum yield of crosslinking (Φ) and the gel dose (E_G) for a group of these materials [104].

The most serious drawback of the diacetylenes as imaging materials are their absorption characteristics; they absorb in the deep UV, with molar extinction coefficients of the order of 100. There have been attempts to overcome this difficulty by sensitization, but these have not been successful.

More recently, Wegner and co-workers [107] discovered the possibility to polymerize diacetylenes in the form of monomolecular Langmuir–Blodgett films. Here, some sensitizers can be built into the films without violating the topotactic conditions of the polymerization process, and in that case sensitization is successful [108, 109].

PHOTOINITIATED CONDENSATION POLYMERIZATION

The Thiol–Ene System

A unique system of photoinitiated polymerization is based on the reaction of thiols (mercaptans) with olefinic double bonds.

$$B + RSH \xrightarrow{h\nu} RS\cdot + BH \tag{62}$$

$$RS\cdot + C=C \longrightarrow RS-C-C\cdot \tag{62a}$$

$$RS-C-C\cdot + RSH \longrightarrow RS-C-C-H + RS\cdot \tag{62b}$$

This is a self-propagating chain reaction that produces linear polymers if carried out with dithiols and with bifunctional olefins. Marvel and Nowlin [110] have investigated the reaction extensively and found that its kinetics

TABLE 4-11 Reactivity of Some Diacetylene Containing Block Copolymers [104]

Polymer[a]	E_G (mJ/cm^2)	Φ
HDI-PBD-50-1350	1.2	82
HDI-PBD-35-1350	0.5	380
CHDI-PBD-35-1350	6.9	38
HDI-PTMG-50-2000	4.3	11
HDI-PTMG-35-2000	2.9	16
HDI-PTMG-20-1000	12.9	3.9
HDI-PCL-35-1250	4.3	36
HDI-PCL-20-1250	14.3	14

[a] HDI—R in hard segment is hexamethylene; CHDI—R in hard segment is cyclohexylene; PBD—soft segment based on polybutadiene; PTMG—soft segment based on poly(oxytetramethylene)glycol; PCL—soft segment based on poly(caprolactone)diol; -20- or -35- or -50- is the contents (%) of the rigid segment; and 1250, 1350 or 2000 is the molecular weight of the soft segment.

follow those of condensation polymerization. The average degree of polymerization (DP) is not identical with the kinetic chain length of the radical reaction on which it is based, but is governed rather by Caruthers' equation

$$DP = \frac{2}{2 - xf} \quad (63)$$

where x is the degree of conversion and f is the average functionality of thiols and olefins in the system. For dithiols and dienes $f = 2$, and so, for a conversion of 0.95, for example, the degree of polymerization will be 20.

If the average functionality is higher than 2, crosslinking occurs in the system and the reaction produces insoluble networks at comparatively low conversions [111]. The linear polymers that can be formed from dithiols and dienes have not found any particular application, but the polythiols and polyenes make highly photoreactive compositions, which are useful in UV curing and in the manufacture of polymeric printing plates.

The polythiols are accessible by the reaction of polyfunctional alcohols with, for example, mercaptoacetic acid [112].

$$C(CH_2OH)_4 + 4HS-CH_2-COOH \longrightarrow$$

$$C(CH_2-O-CO-CH_2-SH)_4 + H_2O \quad (64)$$

Polyenes can be prepared conveniently by the reaction of epoxy groups with diallylamine and similar reagents.

$$2(CH_2=CH-CH_2)_2NH +$$

$$\underset{CH_2}{\overset{O}{\diagdown}}CH-CH_2-O-\phenyl-\phenyl-O-CH_2-\underset{CH}{\overset{O}{\diagdown}}CH_2$$

↓

$$(CH_2=CH-CH_2)_2N-CH_2-\underset{OH}{CH}-CH_2-O-\phenyl-\phenyl-$$

$$-O-CH_2-\underset{OH}{CH}-CH_2-N(CH_2-CH=CH_2)_2$$

Back et al. [113] showed that the reactivity of double bonds towards thiols decreases with conjugation. In the reaction of n-butylmercaptan with n-heptene the kinetic chain length is 1600, but decreases to 980 for the reaction with styrene and even to 26 for the reaction of this mercaptan with the conjugated diene, isoprene.

Morgan et al. [111] have investigated the reactivity of n-pentylmercaptan with a wider range of olefins. Their results are listed in Table 4-12. It can be seen that the decisive factor is the electron density on the double bond, which is enhanced by electron donating substituents (alkyl ethers) and reduced by conjugation and by electron-withdrawing substituents such as the cyano group in acrylonitrile. On the basis of these results Morgan et al. [111] designed tetrafunctional olefins that were considerably more sensitive than, for example, the more conventional carbamate based systems.

TABLE 4-12 Relative Reactivities of Ene Compounds Towards 1-Pententhiol [111]

Olefin	Relative Rate
n-Butyl vinyl ether	10.7
n-Butyl propenyl ether	6.7
n-Butyl allyl ether	1.6
Heptene-1	1
2-Pentene	0.4
Ethylacrylate	0.45
Styrene	0.14
n-Alkyl methyl carbamate	0.1
Acrylonitrile	0.01

The thiol–ene reaction is initiated by radicals or, more effectively, by aromatic ketones with low lying $n\pi^*$ triplet states such as benzophenone or thioxanthone, which in the excited triplet state readily abstract hydrogen from the thiol and set the chain reaction in motion. Phosphines sensitize also, but are particularly useful in mixtures with ketones where a synergistic effect produces very sensitive systems [114].

Photopolymers based on the thiol–ene reaction are widely used in UV curing. In the graphic arts they are used for the preparation of flexographic printing plates. The W. R. Grace Company has developed a range of successful products on the basis of the thiol–ene process [115].

PHOTOINITIATED CATIONIC POLYMERIZATION

Initiation by Onium Salts

The incentive for the development of photoinitiated cationic polymerization was the desire to make use of the unique properties of epoxy resins in imaging applications and in radiation curing. Ethylene oxide and other epoxy derivatives, for example vinylethers cannot be polymerized by a radical mechanism, but polymerize readily by a cationic route [116].

In these systems chain propagation is based on the attack of a carbocation on the negatively polarized oxygen of the oxirane [117]. Initiation of the chain is either by another cation or by a strong electrophile, for example, a Lewis acid (BF_3) or a protonic Brønsted acid. The process is indicated in reaction mechanism (66, 67).

Initiation

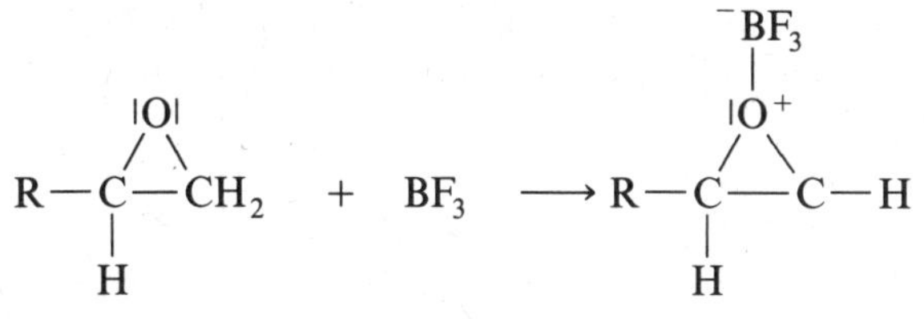

$$R-\underset{H}{\overset{\overset{|\overset{..}{O}|}{\diagup\,\diagdown}}{C}}-CH_2 \;+\; BF_3 \;\longrightarrow\; R-\underset{H}{\overset{\overset{\overset{-BF_3}{|}}{|\overset{..}{O}^+}}{\overset{\diagup\,\diagdown}{C}}}-C-H \qquad (66)$$

Oxirane   Boron trifluoride

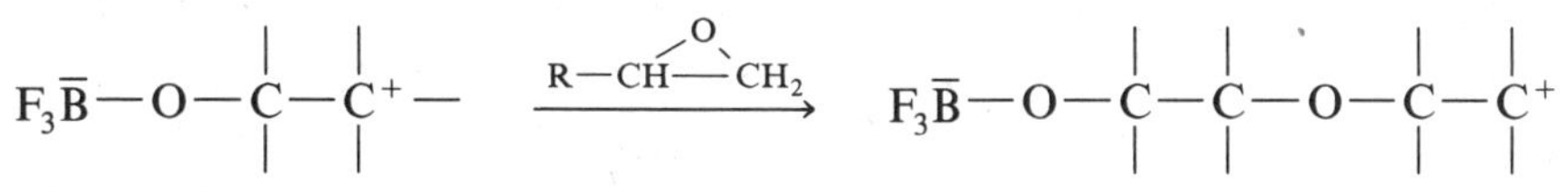

$$F_3\bar{B}-O-\overset{|}{\underset{|}{C}}-\overset{|}{\underset{|}{C}}{}^+ - \quad\xrightarrow{R-CH\overset{O}{\frown}CH_2}\quad F_3\bar{B}-O-\overset{|}{\underset{|}{C}}-\overset{|}{\underset{|}{C}}-O-\overset{|}{\underset{|}{C}}-\overset{|}{\underset{|}{C}}{}^+$$

Carbocation $\hspace{8cm}$ (67)

TABLE 4-13 Heat of Polymerization for Cyclic Ethers Calculated from Ring Strain [118]

Monomer	Ring Size	ΔH (kcal/mol)
Ethylene oxide	3	22.6
Trimethylene oxide	4	19.3
Tetrahydrofuran	5	3.5
Dimethylene formal	5	6.2
Tetrahydropyran	6	−1.3
Trimethylene formal	6	0.0
Tetramethylene formal	7	4.7
Pentamethylene formal	8	12.8

The propagation step is fast; it is helped by the coulombic interaction between the carbocation and the negatively polarized ether oxygen, but the most important factor is the strain in the three-membered ring. This is illustrated by the data in Table 4-13. Termination occurs by the reaction of the carbocation with an adventitious nucleophile (base, e.g.) or, very occasionally, with the anion of the initiator.

$$\cdots O-\overset{|}{\underset{|}{C}}-\overset{|}{\underset{|}{C^+}} + BF_4^- \longrightarrow \cdots O-\overset{|}{\underset{|}{C}}-\overset{|}{\underset{|}{C}}-F + BF_3 \quad (68)$$

In order not to terminate the reaction, and hence inhibit propagation, the counter anion must have very low nucleophilicity. Strong nucleophiles or bases will terminate the reaction immediately. Nevertheless, a small amount of water (1–2%) can be tolerated, which is important for the practical usefulness of the system. Oxygen, which acts as a biradical, has no effect on cationic polymerization, and that too is an important practical advantage. Finally, the epoxy polymers that are the result of the curing process have excellent mechanical properties: They are heat resistant, dimensionally stable, colorless, nontoxic, and chemically inert.

The propagating carbocations [Eq. (67)] do not recombine and as a result, polymers of high molecular weight can be obtained by ionic polymerization. In meticulously purified and dry systems termination can be eliminated altogether and living polymers are observed.

Since cationic polymerization is initiated by Lewis acids or Brønsted acids, photoinitiation requires a photochemical means of producing these species. Aryldiazonium salts were found to produce Lewis acids on photolysis.

$$Ar-N_2^+ BF_4^- \xrightarrow{h\nu} ArF + N_2 + BF_3 \quad (69)$$

Aryldiazonium salt

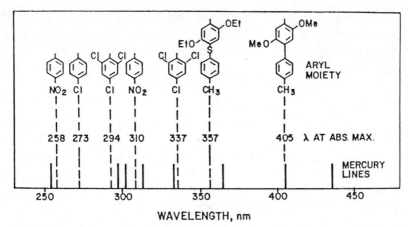

Figure 4-15 Absorption maxima of aromatic diazonium salts with aryl moieties as indicated. The principal mercury lines are marked on the wavelength axis. (Reproduced with permission from S. I. Schlesinger [121].)

Boron trifluoride (BF_3) is a Lewis acid capable of initiating cationic polymerization. Licari and Crepan [119] appear to have been the first to realize the potential of the classical Schiemann reaction (66, 67) in the manufacture of printed circuit boards. Fischer [120a] introduced aryldiazonium salts into the UV curing of epoxy resins. Early patents by Smith deserve mention in this context [120b].

The early initiators absorbed well below 300 nm. Schlesinger [121] prepared diazonium salts with an extended spectral range. These are shown in Fig. 4-15. He was able to polymerize various mono- and bifunctional epoxy monomers, furanes, dioxycyclopentadiene, oxycyclohexene, as well as oligomers (Araldites) and epoxidized novolacs. Exposure requirements of the most sensitive of these systems were of the order of 15 to 30 mJ/cm².

A serious practical drawback of the diazonium salts is their short pot life; coatings have to be made from freshly mixed solutions. Diazonium salts have therefore been displaced by iodonium and sulfonium salts. These are crystalline, stable, and colorless compounds that are soluble in common solvents as well as in many cationically polymerizable monomers. Crivello and Lam have prepared a large number of onium salts of which the following are a selection [122, 123].

Cations	Anions
$\left(R-\langle O \rangle \right)_2 I^+$ $\left(R-\langle O \rangle \right)_3 S^+$	BF_4^- PF_6^- AsF_6^- SbF_6^-
R = H, CH_3, CH_3O, isopropyl, tertbutyl, Cl, Br,	$SnCl_5^+$

The photolysis of the diaryliodonium salts and the triarylsulfonium salts proceeds differently from the photolysis of the diazonium systems. The reaction of the iodonium salts is described by reaction mechanism (70).

$$Ar_2I^+X^- \longrightarrow [Ar_2I^+X^-]^* \longrightarrow Ar\dot{I}^+ + Ar\cdot + X^- \quad (70)$$

solvent, HD

$$ArI^+H + D\cdot \longrightarrow ArI + H^+ + D\cdot \quad (71)$$

Triarylsulfonium salts behave in a similar way.

$$Ar_3S^+X^- + HD \longrightarrow Ar_2S + H^+ + D\cdot + Ar\cdot + X^- \quad (72)$$

There is ample evidence for this mechanism: The aromatic components do not get incorporated into the polymer during polymerization, and if fluorinated anions are used, no fluorinated aromatics are among the photoproducts. That means that the complex anions survive photolysis intact and that no strong Lewis acids are produced. Instead, the corresponding Brønsted acids, for example, $H^+BF_4^-$ or $H^+PF_6^-$, and so on, are generated, which initiate cationic polymerization by the action of the proton on the epoxy monomer.

$$R-\overset{O}{\overset{\diagup\ \diagdown}{CH-CH_2}} + H^+ \longrightarrow HO-\underset{R}{CH}-CH_2^+ \quad (73)$$

Epoxy monomer   **Proton**   **Carbocation**

The complex anions such as AsF_6^- have a dramatic effect on the rate of polymerization, as shown in Fig. 4-16. This is presumably caused by changes in the availability of the protons, which in the nonaqueous medium, exist as ion pairs with the cations. The bonding between anion and cation decreases as their size increases and this makes the proton more active in its catalytic role.

The diaryliodonium and the triarylsulfonium salts have absorption maxima in the vicinity of 250 nm and do not usefully absorb above 300 nm. The iodonium salts can be spectrally sensitized, however, by a number of dyes, such as Acridine Yellow (λ_{max} 411 nm), benzoflavin (λ_{max} 460 nm), Acridine Orange (λ_{max} 539 nm), and others. The effect of sensitization on the decomposition of ditoluyliodonium hexafluoroarsenate in acetonitrile is shown in Fig. 4-17.

Triarylsulfonium salts are not sensitized by acridine derivatives or other common dyes, but they can be sensitized by perylene and by other polynuclear

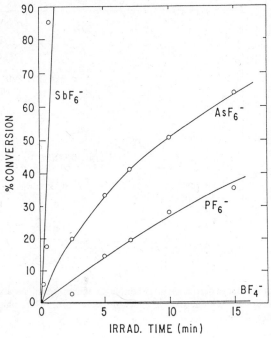

Figure 4-16 Photoinitiated polymerization of cyclohexene oxide, with diphenyliodonium salts (0.02 M), the anions of which are indicated in the figure. (Reproduced with permission from J. V. Crivello and J. H. W. Lam [122].)

aromatic hydrocarbons [116]. More on the sensitization of onium salt photolysis can be found in Chapter 7.

The "living" character of ionic polymerization has already been mentioned. In the absence of adventitious terminators polymerization will go on long after irradiation has ceased, and will often continue until most of the monomer has been exhausted. This is illustrated in Fig. 4-18, which refers to the polymerization of tetrahydrofuran.

Such postirradiative polymerization may be an advantage in UV curing, but it is undesirable in imaging applications where it can be detrimental to resolution and sharpness. Crivello and Lam [123–125] have found a solution to this problem through the development of a new class of cationic initiators, the dialkylphenacylsulfonium salts,

$$\text{Ar}-\overset{\overset{\displaystyle O}{\|}}{C}-CH_2-\overset{+}{S}\!\!\underset{R}{\overset{R}{\diagdown}}\quad X^-\;\underset{\text{dark}}{\overset{h\nu}{\rightleftarrows}}\;\text{Ar}-\overset{\overset{\displaystyle O}{\|}}{C}-C=S\!\!\underset{R}{\overset{R}{\diagdown}}\;+\;H^+X^- \qquad (74)$$

Brønsted acid

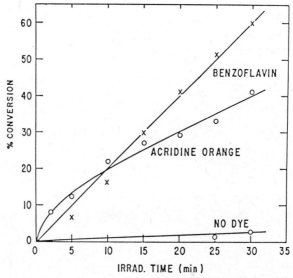

Figure 4-17 Photosensitized decomposition of ditoluyliodonium hexafluoroarsenate in acetonitrile (0.07 M), sensitized with Acridine Orange and with benzoflavine. (Reproduced with permission from J. V. Crivello [116].)

and the dialkyl-4-hydroxyphenyl sulfonium salts.

$$\text{Sulfonium salt} \underset{\text{dark}}{\overset{h\nu}{\rightleftarrows}} \text{Ilyd (keto form)} + H^+X^- \quad (75)$$

      Sulfonium salt   Ilyd (keto form)   Brønsted acid

In contrast to the photolysis of the triarylsulfonium ions who decompose irreversibly on irradiation, the photolysis of the new initiators is reversible. On irradiation, a photostationary state is established, as indicated in Eqs. (74) and (75). As soon as irradiation ceases, the ilyd and the Brønsted acid revert to the original sulfonium salt. The acidity of the medium is needed for the propagation of the ionic chain, and when the acid disappears, the growth of the chain is inhibited and polymerization stops.

Initiation by Iron–Arene Complexes

The polymerization of epoxides can also be initiated by a new class of initiators (Zweifel and co-worker [126, 127]), which are ferrocenelike iron–arene complexes. On irradiation these undergo a ligand exchange reaction where the tridentate arene is replaced eventually by the oxygen atoms of the epoxy groups.

(76)

The initiation step, namely, the opening of the oxirane ring, occurs presumably within the ligand sphere of the iron. Polymerization follows, provided the system is at a high enough temperature. Although the iron–arene complexes are less effective initiators of cationic polymerization than for example, sulfonium salts, they are nonetheless of practical interest because of their stability and the stability of the cured images against degradation.

The iron–arenes have been prepared not only with benzene, but with naphthalene moieties and even with pyrene, which extends the spectral range of the photoreactive compositions well into the visible region. Figure 4-19 shows the absorption spectra of the three complexes.

The iron–arenes are photobleached on irradiation and this allows light to penetrate deeply into the polymerizable layer making it possible to form images with high relief.

While the iron–arenes are capable of initiating the cationic chain, the polymerization process needs thermal activation to proceed. Curing of epoxides with these initiators is a two-step process: photoinitiation at low temperature, followed by thermal polymerization. Epoxides differ in their reactivity towards polymerization; those attached to aliphatic ring compounds are usually more reactive and, once initiated, polymerize at lower temperature than, for example, glycidyl derivatives.

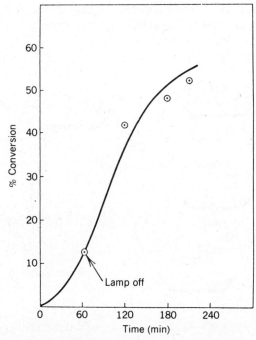

Figure 4-18 Photoinitiated polymerization of tetrahydrofuran, with 0.5×10^{-2} M of triphenylsulfonium hexafluorophosphate. (Reproduced with permission from J. V. Crivello and J. H. W. Lam [123].)

PRACTICAL SYSTEMS

Typical Components

Although initiator and monomer are the only indispensable components of photopolymerization, practical systems are a good deal more complex. They contain polyfunctional monomers (so-called *crosslinkers*), polymeric or oligomeric *binders*, as well as various additives and modifiers: *antioxidants* to stabilize they systems, *radical scavengers* to prevent thermal polymerization before exposure, amines to remove oxygen during irradiation, wetting agents, adhesion promotors, coating aids, and dyes. Table 4-14 lists some typical components [10].

Binders. The monomers and crosslinkers used in photopolymerization were described in an earlier section. The polymeric binders, which are the only other bulk component, are usually chemically inert polymers of moderate molecular weight. They must be at least partially soluble in the monomer, and polymers of the monomer itself are therefore suitable. Their function is to increase the viscosity of the coating mixture and to allow the preparation of thick layers with high solids content.

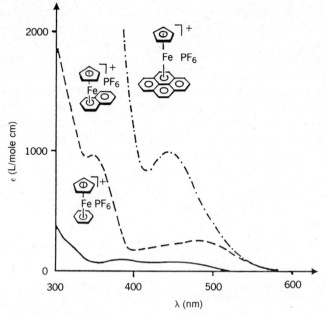

Figure 4-19 Absorption spectra of iron–arene complexes, (Reproduced with permission from K. Meyer and H. Zweifel [126].)

TABLE 4-14 Typical Components of Practical Photopolymerization Systems

Monomers	Acrylates, methacrylates, acrylamide, and styrene
Crosslinkers	1,6-Hexanediol diacrylate, triethyleneglycol diacrylate, N,N'-methylenebis(acrylamide), trimethylolpropanetriacrylate, pentaerythritol triacrylate, and pentaerythritol tetraacrylate
Binders	Polymers of the monomer used, polyesters, polyurethanes, nylons, polycarbonates, and cellulose derivatives
Fillers	Organophilic silicas and clays
Initiators	Benzoin derivatives, anthraquinones plus hydrogen donors, and benzophenones and amines
Stabilizers	p-Methoxyphenol, hydroquinones, naphthols

Since the binder is present in large quantity, it has a decisive influence on the properties of the coating and on those of the cured film. For example, if the binder is alkali soluble, the nonimage areas of the exposed resist layer can be removed with an aqueous developer. Furthermore, some of the additives may be incorporated into the binder; a small proportion of free acrylic acid in a polymethacrylate binder acts as an adhesion promotor. The binder contributes also to photographic speed: The large macromolecule, caught in the network of the polymerizing monomer, enhances the mass-to-photon ratio of the system.

An important component of practical systems are *reactive binders*. These are macromolecules of modest size, (molecular weights between 5000 and 50,000), which carry some unsaturation. Common examples of reactive binders are the unsaturated polyesters obtained by the condensation of maleic and fumaric acid with glycols [128, 129]. On irradiation in the presence of an initiator and some monomer the fumaric acid moieties take part in a process of crosslinking polymerization, as indicated here.

$$\cdots O-\underset{\parallel}{C}-\underset{\mid}{CH}-CH-\underset{\parallel}{C}-O-R-\cdots$$
$$\text{(O)}\quad R \quad \text{(O)}$$

$$\left[\text{(Ph)}-CH-CH_2 \right]_n$$

$$\cdots O-\underset{\parallel}{C}-CH-CH-\underset{\parallel}{C}-O-R-\cdots$$

Because of the high crosslink density that is achieved in this way, the cured coatings have excellent hardness and are heat and water resistant.

The physical properties of the cured films can also be modified by the use of functionalized oligomers. An example are polyfunctional acrylates based on bisphenol A. These are obtained by the reaction of bifunctional epoxides with unsaturated acids [130].

$$H_2C\overset{O}{\diagup}CH-CH_2-O-\text{(Ph)}-\text{(Ph)}-O-CH_2-\underset{H_2}{C}\overset{O}{\diagdown}CH_2$$

$$+ 2CH_2=CH-\underset{\parallel}{C}-OH \xrightarrow{R_3N} \qquad (77)$$

$$CH_2=CH-\underset{\parallel}{C}-O-CH_2-\underset{\mid}{CH}-CH_2-O-\text{(Ph)}-\text{(Ph)}-$$
$$\qquad\qquad O \qquad\qquad OH$$

$$-O-CH_2-\underset{\mid}{CH}-CH_2-O-\underset{\parallel}{C}-CH=CH_2$$
$$\qquad\qquad OH \qquad\qquad O$$

The free OH groups of the prepolymer can be further esterified with the mixed anhydride of acetic and acrylic acid to give a tetrafunctional acrylate [131].

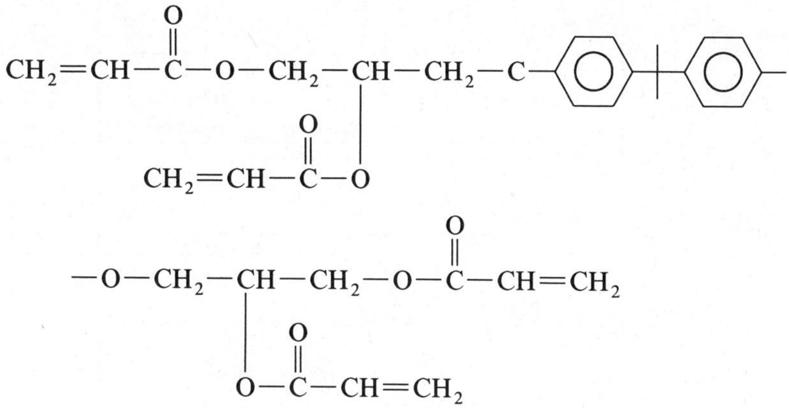

$$CH_2=CH-\overset{O}{\underset{\|}{C}}-O-CH_2-\underset{\underset{CH_2=CH-\overset{O}{\underset{\|}{C}}-O}{|}}{CH}-CH_2-C-\underset{}{\underset{}{\bigcirc}}-\underset{}{\underset{}{\bigcirc}}-$$

$$-O-CH_2-\underset{\underset{\overset{O}{\underset{\|}{O-C-CH=CH_2}}}{|}}{CH}-CH_2-O-\overset{O}{\underset{\|}{C}}-CH=CH_2$$

A combination of epoxy chemistry with the esterification of acrylic acid anhydrides leads to a whole range of polyfunctional reactive prepolymers, such as the following structure:

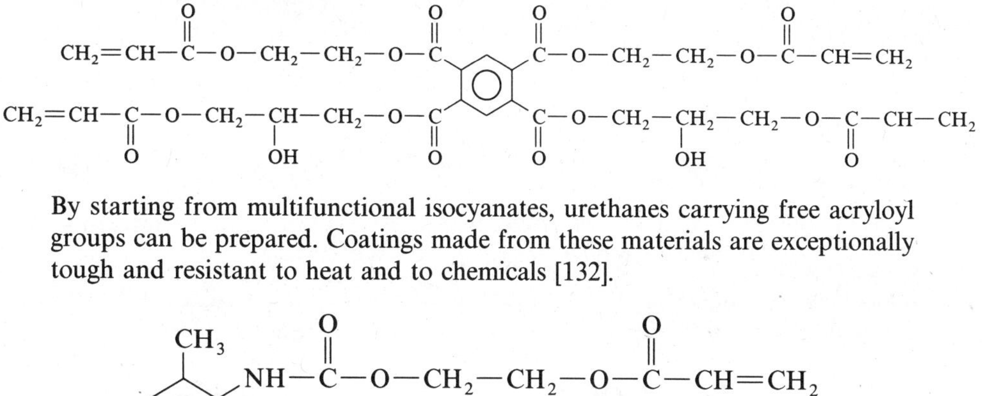

By starting from multifunctional isocyanates, urethanes carrying free acryloyl groups can be prepared. Coatings made from these materials are exceptionally tough and resistant to heat and to chemicals [132].

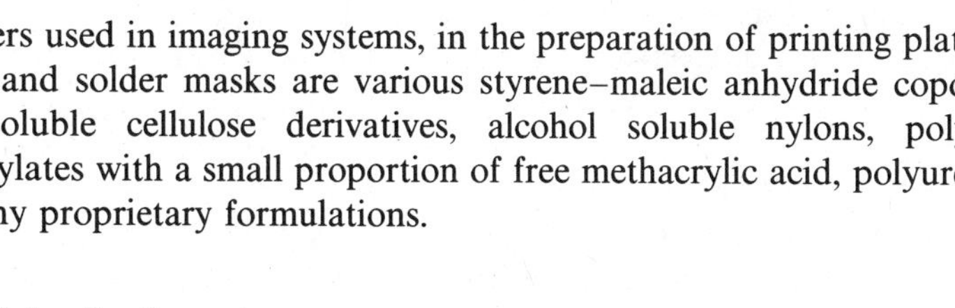

Binders used in imaging systems, in the preparation of printing plates, wire boards, and solder masks are various styrene–maleic anhydride copolymers, alkali soluble cellulose derivatives, alcohol soluble nylons, polymethyl methacrylates with a small proportion of free methacrylic acid, polyurethanes, and many proprietary formulations.

Principal Applications

Photoinitiated polymerization has found its main use in the following three areas.

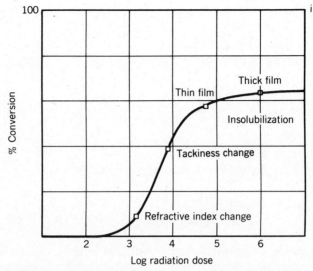

Figure 4-20 Schematic of monomer conversion versus radiation dose for photopolymerizable composition. (After P. Walker [133].)

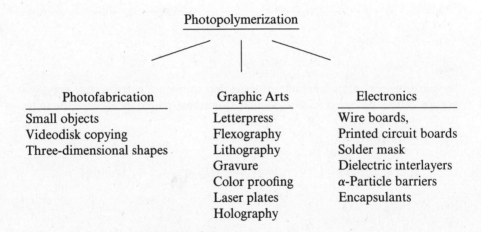

From the chemist's point of view, these applications differ in the material properties that are affected by the action of light. To illustrate this, let us consider the conversion of monomer in a polymerizable coating and plot the degree of conversion as a function of the incident radiation dose. Such a plot, first introduced by Walker [133] is shown in Fig. 4-20.

At the outset, the polymerizable composition is in a liquid or in a semisolid, highly viscous state. In the early stages of polymerization, the primary physical change is an increase in density, which manifests itself as an increase in the index of refraction. As polymerization progresses, the glass transition temperature of the system will increase and the coating will change from a rubbery

PRACTICAL SYSTEMS 159

(and possibly tacky) state to a glass. At this point the coating remains still somewhat permeable to solvent, but gradually as the crosslink density increases, the material becomes solvent resistant; it will not swell significantly and it will be mechanically strong even in the presence of solvent.

All these effects—changes of permeability, changes of tackiness, of solubility, of mechanical wet strength, and of index of refraction, have been exploited for the recording of images. Following Delzenne and co-workers [134] we will present the practical uses of photopolymerization grouped according to the physical properties that have been modulated by the radiation signal.

The Modulation of Solubility

The systems that rely on the modulation of solubility are by far the most numerous. In the graphic arts they comprise letterpress plates, flexographic plates (flexible letterpress plates), and lithographic plates. The composition and chemical mechanism of all three are similar, but they differ in the thickness of the polymer layer used and in the nature of the mechanical support.

Letterpress and flexographic plates are coated to a thickness in the range from 150 to 5000 μm, except for halftone plates, which are thinner (75–150 μm). Lithographic plates have no relief and are prepared with coating thicknesses from 2 to 5 μm. The classical letterpress plate is well described by Plambeck [4] in the first important patent of the field (see Fig. 4-21).

Letterpress Plates. Letterpress plates are relief printing plates (see Chapter 1), which in the printing operation, transfer ink from the raised parts of the letter

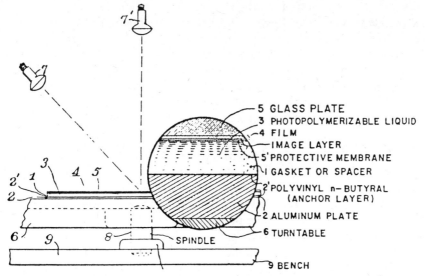

Figure 4-21 Cross section of early polymer relief plate. (Reproduced with permission from L. Plambeck [4].)

Figure 4-22 Finished letterpress polymer plate mounted on printing cylinder. [Reproduced with permission of NAPP Systems (USA), Inc.]

profile onto the receiving sheet. Guttenberg's moving-type wooden plates and typeset metal plates are of this kind. Since 1960, metal plates have been largely replaced by polymer plates.

The composition of the early polymer plates was made of methacrylate–acrylate mixtures with an acrylated cellulose acetate. Other polymer plates that appeared almost at the same time used alcohol soluble polyamides as binders (the polyamide plate of Time Inc.) [5] or various nylon derivatives (Nyloprint and related products of BASF) [6]. The original photoinitiators were benzoin derivatives. Later on, anthraquinone initiators and other systems based on hydrogen abstraction became prevalent. These compositions are coated, sometimes over a primer, onto a metal support, either steel or alumina, or in the case of flexographic plates on rubber or other flexible materials. Figure 4-22 shows a developed letterpress plate in current use [135].

The stages in the preparation of a letterpress plate are the following:

1. A process negative is brought into contact with the working surface of the relief plate.
2. This sandwich is exposed to a bank of UV lamps.
3. The negative is removed and the plate is developed by high pressure sprays of solvent or, more often today, aqueous alkali.
4. The developed and washed plate is dried.
5. Finally, it is flood exposed to crosslink further and harden the developed image.
   It is now ready for use [136].

Polymer plates are used in the production of newspapers, business stationary, forms, folders, file cards, securities, checks, banknotes, postage stamps, and so on. Print runs of 500,000 and more are routinely achieved.

Flexographic plates, which are relief plates prepared on a flexible support, are used for coarser and larger scale work such as the printing on corrugated surfaces, shopping bags, paper towels, milk cartons and various food packages, envelopes, and posters.

Variable Area–Variable Depth Gravure Plates. The depth of the profile achieved in the development of a relief plate depends strongly on the width of the unexposed area. As a result, it is possible to use virtually the same materials for the preparation of polymeric *variable area–variable depth* gravure plates (see Chapter 1). These plates are coated to a thickness of about 0.2 mm (200 μm) on a metal support and protected before use with a polyolefin cover. The plate is first given a short flood exposure that is not sufficient to seriously insolubilize the material, but which slows down dissolution in general. The image exposure is then given through a matted gravure screen positive, which defines the gravure cells. On exposure, the screen frame becomes fully crosslinked, and the highlights of the original image also receive heavy exposures, but the shadow areas will be exposed only lightly and will remain therefore partly soluble. On development the highlight cells remain shallow, and will not carry much ink in the subsequent printing operation, but the shadow cells will develop into deep troughs and will later hold large quantities of ink. This kind of gravure plate, which is currently being launched by BASF, is reasonably competitive with lithography even for shorter runs, while offering some of the quality advantages of variable depth gravure. Figure 4-23 shows a scanning electron microgram of the surface cells of a polymeric gravure plate.

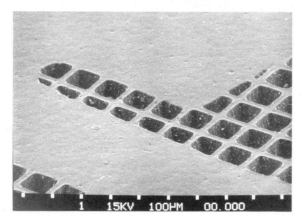

Figure 4-23 Scanning electron micrograph of the cells of a polymer gravure plate. (Reproduced with permission from a BASF original micrograph.)

Lithographic Offset Plates. Lithographic plates formulated from polymerizable components have a composition similar to that of letterpress plates, but with higher initiator and sensitizer content to achieve adequate light absorption in the thinner layers (2–4 μm). Because the layers are thin, oxygen inhibition is more of a problem than in relief plates and exposure is usually in a vacuum frame. Also, tertiary amines are added to the composition to scavenge oxygen chemically during exposure. Image discrimination in lithographic plates is based on ink acceptance and rejection and the composition is almost invariably coated on grained and anodized alumina, which is able to carry an ink repellent water film [137]. Because the coatings are thin, resolution in lithographic plates is good, (better than 6 μm in the best plates) and photographic sensitivity is of the order of 10 mJ/cm^2.

Lithographic plates are now the leading print medium for short and medium long runs, which includes scientific books and textbooks, catalogs, posters, periodicals, and even local daily and weekly papers.

Laser-exposable lithographic plates are a new and interesting application of this technique. In recent years an increasing amount of information is being transmitted and stored in digital form. After retrieval this information is reconstituted in alphanumeric or in pictorial form, for example by way of a laser scanner, where a laser beam scans the surface of a receiving element while being modulated by a suitable electronic device. The pixel density required for adequate reproduction of printed characters or line drawings is 200 to 500 dots per inch (dpi); for pictorial material it is in the range from 1000 to 1500 dpi [138]. At the high scan rates demanded by the industry (at least one standard page per minute for pictorials and 20 pages per minute for text) and with the limited power density of available lasers, a highly sensitive recording material is needed.

At present only the xerographic method is capable of handling such data streams. The sensitivity requirements for such a system in various exposure modes are compared in Fig. 4-24 [65].

Shimizu [65] has recently described a laser exposable lithographic plate, which comes near to these specifications and requires about 1 mJ/cm^2 exposure to the Ar$^+$ laser at 448 nm. Even more recently Plambeck and co-workers [139] have disclosed a hybrid silver halide–photopolymer system that is based on a photographic color coupler chemistry (see Chapter 2) and which reaches the target sensitivity of about 0.1 mJ/cm^2.

Dry Resists

Another important imaging system based on solubility modulation is the "dry resist" used extensively in the manufacture of wire boards and printed circuit boards. Dry resists have evolved from the composition of letterpress plates [140]. This composition can be sandwiched between a polyester base sheet and a polethylene protective cover foil, as shown in Fig. 4-25.

In use, the resist film is separated from its base sheet and laminated onto the work piece, usually a copper-clad insulating support. The resist is exposed

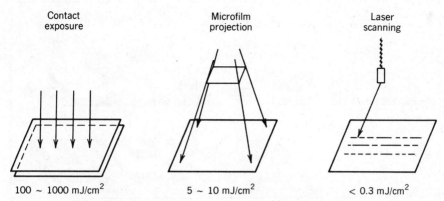

Figure 4-24 Comparison of sensitivity requirements for polymer plate in contact printing mode, in projection mode, and in laser scanning mode. (Reproduced with permission from S. Shimizu [65]. ©TAGA 1986.)

through the polyethylene film to a negative of the desired circuit pattern, the "photo tool." After removing the cover foil, the latent image formed on exposure is spray developed and dried. The metal support may then be etched through the resist mask. This operation is indicated schematically in Fig. 4-26. Alternatively, the pattern is plated up with an etch resistant metal, the resist stripped, and finally, the copper etched away.

Printed Circuit Boards. The classical dry film resist is "Riston," which was introduced in 1970 by Du Pont and has been kept up to date and competitive until this day. Others have followed, and dry resists are today the leading medium in printed circuit board (PCB) and printed wire board (PWB) manufacture. A modern wire board with high density circuitry is shown in Fig. 4-27. (Compare this with the 1975 printed circuit board of Fig. 1-17.)

With the introduction of even denser and more complex printed circuit boards, the circuit pattern is commonly generated by computer. The informa-

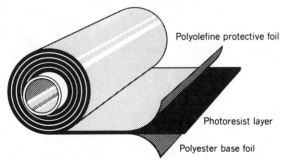

Figure 4-25 Dry resist film of the Riston type. The photoresist layer is sandwiched between a polyester base foil and a polyolefin cover sheet. (Reproduced with permission from a Nylotron prospectus of BASF.)

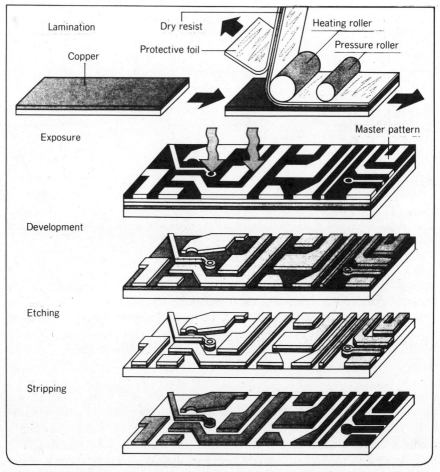

Figure 4-26 Processing steps in the manufacture of a printed circuit board. The individual steps are indicated in the drawing. (Reproduced with permission of the Fonds der chemischen Industrie, Frankfurt, Germany.)

tion is stored on disc and is converted into raster scan image data, which are transferred by a laser scanner onto photographic film. Even more recently, laser scanning technology has improved to the point where *laser direct imaging* onto a laser-sensitive photoresist laminated onto the PCB has become economical. No intermediate photographic negatives are required in this case; the system is extremely flexible and suitable for customization. The number of operations is reduced, which makes for higher accuracy and fewer defects, and the technology is ideal for full automation. The problem is finding a truly suitable imaging material.

Du Pont announced two products that are sensitive to the UV and visible emission of argon lasers, (Riston LUV and Riston LV) [64]. Riston LV is based on the principles of dye sensitized polymerization. Its exposure require-

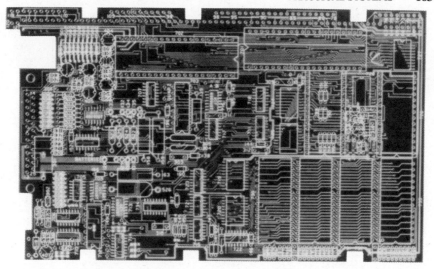

Figure 4-27 High density circuitry on a contemporary wire board. (Reproduced with permission from a Vacrel prospectus of Du Pont.)

ments are in the region of 10 to 15 mJ/cm^2; a standard 18 in. × 24 in. panel can be scanned in 30 s.

The requirements of accuracy and of high resolution, so well publicized in the field of integrated circuit (IC) manufacture, are reflected also in the design trends of printed circuit boards. In 1970, for example, feature sizes were of the order of 20 mil, in 1980 they had shrunk to 8 mil and are projected to reach 3 mil size in 1990. Figure 4-28 [141] compares trends in IC and in printed circuit manufacture.

The increase in the operational speed of electronic devices has a profound impact on wire board design. Signal frequencies are now in the microwave domain and signals have to be routed essentially by microwave transmission lines rather than simply by electron conductors. As a result, dimensional tolerances have shrunk well below 1 mil and the patterning technology used in printed circuit boards will have to follow in the footsteps of IC microlithography [142].

Solder Mask. Another important application of dry resists is their use as solder masks [138]. In the last phases of the manufacture of calculators and processors, the components of the circuit are in place on the circuit board and need to be interconnected by soldering. At that stage a resist is applied to the back and/or to the front of the board where it covers everything, the circuits as well as the contact pads (see Fig. 4-29). It may be applied as a liquid, or with more manufacturing convenience, as a dry resist film. Only the contact pads are subsequently uncovered by photoimaging. A wave of liquid solder is sent across the board; it wets the contact pads and completes the interconnec-

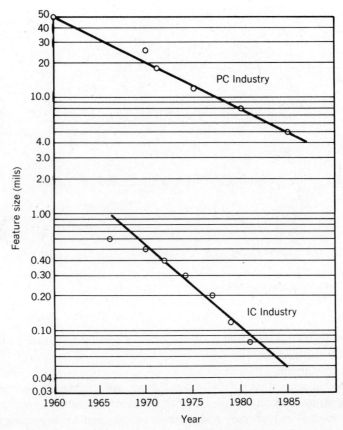

Figure 4-28 Evolution of feature sizes (in mils) in printed circuit board production and in the production of integrated circuits. (Reproduced with permission from R. D. Rust and D. A. Doane [141].)

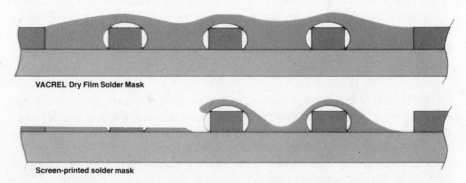

Figure 4-29 Cross section of a dry film solder mask and a screen printed solder mask overcoating the features of a wire board. (Reproduced with permission from a Vacrel prospectus of Du Pont.)

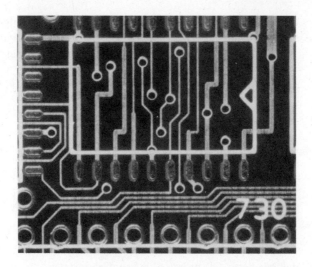

Figure 4-30 Detail of a wire board where the connections were made by the solder mask–solder wave technology. (Reproduced with permission from a Vacrel prospectus of Du Pont.)

tions in a single passage. The resist material must of course be sufficiently heat resistant and mechanically stable. In some instances the solder mask film is not removed after the operation, but serves as a permanent protective encapsulant for the circuitry [143]. Figure 4-30 shows detail of a circuit board interconnected by solder mask technology [144].

The Modulation of Permeability

The permeability of a crosslinked polymer film to a gas or to a liquid depends strongly on the degree of crosslinking. In a system in which crosslinks are produced in response to irradiation, permeability can be controlled by exposure. Gravure films for the preparation of gravure printing plates are based on this effect [145].

Variable Depth–Constant Area Gravure. Gravure films are similar to dry resist films. They are applied in a thickness of about 10 to 15 μm onto a dimensionally stable and transparent polymer base, which is coated at the back with an antihalation layer. The gravure film has been lightly crosslinked either chemically or by an overall preexposure to give it some internal cohesion. In use, it is exposed to a negatively rastered process *positive* where the raster lines are transparent. Through the process positive highlights receive a large radiation dose while shadows receive little exposure.

After exposure the film is transferred, face down, onto the metal printing plate or cylinder and the support sheet is removed. There is no development step. The metal is now etched through the exposed film, which acts as an imagewise etch barrier. The cell walls are fully exposed through the transparent screen and are not etched at all; they form a network that defines the nonprinting plane of the plate over which the ink-removing doctor blade will slide later.

Variable depth-constant area gravure is used for the highest quality color reproduction in art books, glossy magazines, sales brochures, and other publications where quality is more important than price.

The Modulation of Tackiness

Crosslinking increases the glass transition temperature of a polymer. When the T_g of the composition passes through room temperature, the system changes from the rubbery and tacky state to the glassy and nontacky state. The change in tackiness on irradiation has been exploited in a number of well-known off-press proofing materials.

Off-Press Color Proofing. Off-press color proofing is a way to demonstrate and assess the quality of color rendition expected in a particular printing operation [146]. Color reproduction starts with the preparation on photographic film of color separation negatives or positives. Before these are used to cut the cyan, magenta, yellow, and black printing plates, they are first "proofed" by exposing them onto color proofing materials. These are polymer films of essentially the same character as dry resists. They are coated on a dimensionally stable transparent polyester base and covered with a thin protective cover sheet. The composition of the photosensitive polymer is so formulated that its T_g is just below room temperature [147–149].

The color proofing film is laminated onto a receiving sheet that is similar to the surface on which the final print will be made. It is then exposed through the transparent support to, say, the cyan separation positive. On exposure, the material is crosslinked in the highlight and there loses its tackiness, but remains somewhat tacky in the shadow areas. The support of the proofing film is now peeled away and the film dusted with a cyan pigment that is as nearly as possible identical with the ink to be used in production [150]. The pigment adheres in the tacky areas and not in the others (see Fig. 4-31). The whole process is repeated with the magenta, the yellow, and the black printer separations, and since the pigment images are formed in strict register on the same receiving sheet a full color image is produced. Alternatively, the transfer operation can be eliminated by simply overlaying the three pigmented films in register. This second method is now preferred in the profession.

An inverse process has been proposed where the polyester support is first coated with a tonable adhesive layer, this is overcoated with the slightly tacky photopolymer and finally a cover sheet is applied [151]. When this composite is exposed through a process negative (or positive), the exposed areas adhere more strongly to the cover sheet than to the support and are peeled away with the cover sheet after exposure. This uncovers the tonable layer, which can be pigmented to produce an inverse color image (see Figure 4-32).

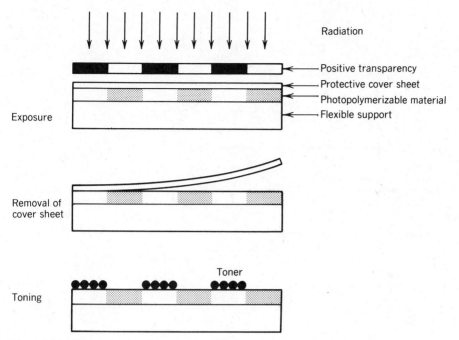

Figure 4-31 Schematic of positive working adhesive color proofing system, Cromalin type. The film is exposed to a process positive, the cover sheet is removed and the unexposed image areas made visible by toning.

Related to the modulation of tackiness is the modulation of adhesion to an adjoining surface. This phenomenon has been used for the preparation of various graphic arts transfers. In these systems a photohardenable composition is coated between two transparent plastic sheets. After exposure the adhesion of the middle layer to the support is greatly decreased so that when the composite is peeled apart the exposed areas are taken away with the cover sheet, the unexposed areas remain with the base. In these "peel-apart" systems a positive and a negative are obtained simultaneously (e.g., in the Crolux engineering film).

The Modulation of Index of Refraction

Holographic Materials. The modulation of refractive index is utilized in the recording of holograms [152–154]. In holography, two beams of coherent radiation (from a laser source) create an interference pattern in the plane of a recording material. The record of this pattern is called a hologram (Fig. 4-33a). When the hologram is illuminated by the reference beam alone, the pattern recreates a virtual image of the object by diffracting the light beam in an appropriate manner (Fig. 4-33b). Since it is the scattering or diffractive

170 PHOTOINITIATED POLYMERIZATION

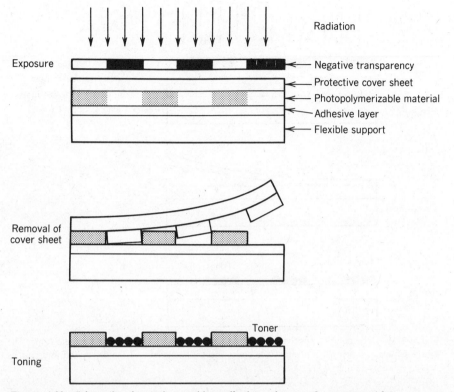

Figure 4-32 Schematic of negative working adhesive color proofing system. After exposure, removal of the cover sheet uncovers the adhesive layer, which is subsequently toned.

power of the pattern that counts in the reconstruction step, imagewise refractive index modulation is here a particularly effective recording method.

Holographic materials must have a high diffraction efficiency, reasonable photographic speed, they must be sensitive to visible light (laser output is usually in the green and in the red part of the spectrum), and the material should be solid and available in precoated plates or sheets.[1] Two practical systems will be mentioned. A material developed by Du Pont is based on pentaerythritol triacrylate and other polyfunctional acrylates as monomers, on cellulose acetate-butyrate as binder and on dye sensitized bisimidazoles as initiators [156, 157]. A system more recently devised by Polaroid is based on Li-acrylate as monomer, N,N'-methylenebisacrylamide as crosslinker, poly(vinylpyrrolidone) as binder. The initiating system is methylene blue

---

[1]Silver halide materials (e.g., the Kodak 649 F plate and the Newport Holographic Plates) are used in holography because of their high sensitivity, but they make only holograms of moderate quality. The highest quality holograms are obtained with dichromated gelatin, but this material has to be freshly coated shortly before use and its photographic speed is poor. See, however, Reference 155. Conventional crosslinking resists produce holograms in terms of a surface relief pattern that can be mechanically replicated.

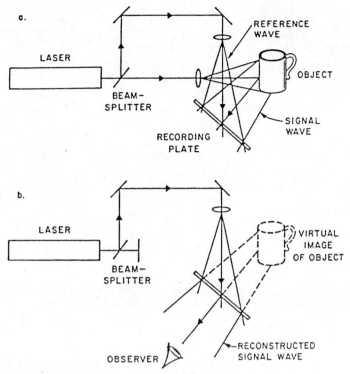

Figure 4-33 Experimental arrangement for the recording and for the viewing of transmission holograms (see text). (Reproduced with permission from W. J. Tomlinson and E. A. Chandross [152].)

together with various branched polyethylene imines, which act as hydrogen donors and simultaneously as reactive binders [158].

With these photopolymerization systems holograms are produced in a three stage operation. First, the film receives a brief (image) exposure to the aerial diffraction pattern; (exposure dose 1 mJ/cm^2). The object beam (see Figure 4-33) is then blocked off, and after an appropriate waiting period the film is given a nonimage flood exposure to the reference beam alone; the exposure dose of this second exposure is of the order of 50 mJ/cm^2. No further development or fixing of the image is required and the hologram is immediately ready for use.

During the first stage of hologram preparation a moderate amount of polymerization occurs in the irradiated areas. As a result, the monomer concentration in these areas is lowered, causing a diffusional flow of monomer from the nonimage into the image areas. This process of internal material transport is the key mechanism in holographic recording with photopolymers. The process continues to some extent during the final flood exposure, which polymerizes all monomer in the film and thereby perpetuates the nonuniform distribution of material that is ultimately the cause for a difference in refractive index [159].

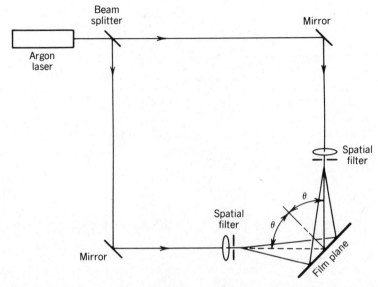

Figure 4-34 Determining the diffraction efficiency of a holographic reproduction material. The interference of the two laser beams in the film plane creates a diffraction grating. The diffraction intensity of this grating is the figure of merit of the holographic material. (Reproduced with permission from W. S. Colburn and K. A. Haines [159].)

The critical property of a holographic film is the diffraction efficiency that can be achieved with it. Holographic materials are evaluated by letting two coherent beams interfere in a film of the material and produce a diffraction grating. Figure 4-34 shows a simple arrangement that can be used for this purpose [159].

The diffraction efficiency (DE) of this grating defined as the ratio of the light intensity (I_1) in the first-order diffraction maximum to the intensity (I_0) of the incident light,

$$DE = I_1/I_0$$

is the figure of merit of the holographic recording material.

REFERENCES

1. J. Eggert et al., U.S. Patent 2,115,198 (1938).
2. T. H. Morton, *J. Soc. Dyers Colour.*, **65**, 597 (1949).
3. G. K. Oster, G. Oster, and G. Prati, *J. Am. Chem. Soc.*, **79**, 595 (1957).
4. L. Plambeck, Jr., assigned to Du Pont, U.S. Patent 2,760,863 (1956).
5. R. M. Leekley and R. L. Sorensen, assigned to Time Inc., Ger. Patent 954,127 (1956).
6. H. Hoerner and B. Olsen, assigned to BASF, Belg. Patent 575,159 (1958).
7. S. P. Pappas and V. D. McGinnis, in *UV Curing: Science and Technology* (S. P. Pappas, Ed.), Vol. I, Chap. 1, Technology Marketing Corp., Norwalk, CT, 1978.

8. G. A. Delzenne, *Adv. Photochem.*, **11**, 1 (1979).
9. V. D. McGinnis, in *Developments in Polymer Photochemistry* (N. S. Allen, Ed.), p. 1, Applied Science, London, 1982.
10. C. G. Roffey, *Photopolymerization of Surface Coatings*, Wiley, New York, 1982.
11. L. R. Gatechair, *J. Radiat. Curing*, **10**(3), 4 (1983).
12. J.-P. Fouassier, *J. Chim. Phys.*, **80**, 339 (1983).
13. J. Calvert and J. Pitts, *Photochemistry*, Wiley, New York, 1965.
14. N. J. Turro, *Modern Molecular Photochemistry*, Benjamin–Cummings, Menlo Park, CA, 1978.
15. J. Franck and E. Rabinowitch, *Trans. Faraday Soc.*, **30**, 120 (1934).
16. R. M. Noyes, *J. Am. Chem. Soc.*, **77**, 2042 (1955).
17. L. F. Meadows and R. M. Noyes, *J. Am. Chem. Soc.*, **82**, 1872 (1960).
18. R. M. Noyes, *Z. Elektrochem.*, **64**, 153 (1960).
19. A. Garton, D. J. Carlsson, and D. M. Wiles, *Makromol. Chem.*, **181**, 1841 (1980).
20. P. K. Sen Gupta and J. C. Bevington, *Polymer*, **14**, 527 (1973).
21. F. D. Lewis, R. T. Lautenbach, H.-G. Heine, W. Hartmann, and H. Rudolph, *J. Am. Chem. Soc.*, **97**, 1519 (1975).
22. H.-G. Heine and H.-J. Traenkner, *Prog. Org. Coat.*, **3**(2), 115 (1975).
23. S. P. Pappas and R. A. Asmus, *J. Polym. Sci., Polym. Chem. Ed.*, **20**, 2643 (1982).
24. G. Wu and A. Reiser, unpublished results, 1983.
25. A. Borer, K. Kirchmayer, and G. Rist, *Helv. Chim. Acta*, **61**, 305 (1978).
26. H. L. Hageman, E. A. Giezen, and L. G. J. Jansen, Eur. Patent Appl. 37,604 (1982).
27. A. Hult and B. Ranby, *ACS Polym. Prepr.*, **25**(1), 379 (1984).
28. C. L. Osborn, *J. Radiat. Curing*, **3**(3), 2 (1976); M. R. Sandner and C. L. Osborn, *Tetrahedron Lett.*, p. 415 (1974).
29. G. A. Delzenne, U. Laridon, and H. Peeters, *Eur. Polym. J.*, **6**, 933 (1970).
30. H. Sakuraji, M. Yoshida, H. Kinoshita, K. Utena, K. Tokumaru, and M. Yoshiro, *Tetrahedron Lett.*, p. 1529 (1978).
31. R. Dessauer and C. Looney, *Photogr. Sci. Eng.*, **13**, 287 (1979).
32. T. E. Dueber, A. G. Anderson and B. M. Monroe, *SPSE 41st Ann. Conf.*, Paper Summaries, p. 78, Arlington, VA, (May 22–26, 1985).
33. P. Suppan and G. Porter, *Trans. Faraday Soc.*, **61**, 1664 (1965).
34. D. Braun and K. H. Becker, *Makromol. Chem.*, **147**, 91 (1971).
35. S. G. Cohen, A. Parola, and G. H. Parsons, Jr., *Chem. Rev.*, **73**, 141 (1973).
36. A. Ledwith, G. Ndaalio, and A. R. Taylor, *Macromolecules*, **8**, 1 (1975).
37. M. R. Sander, C. L. Osborne, and D. J. Trecker, *J. Polym. Sci. Part A-1*, **10**, 3173 (1972).
38. A. Ledwith, in *The Exciplex* (M. Gordon and W. R. Ware, Eds.), p. 209, Academic, New York, 1975.
39. A. Ledwith, J. A. Bosley, and M. D. Purbrick, *J. Oil Colour Chem. Assoc.*, **61**, 95 (1978).
40. A. Ledwith, *Pure Appl. Chem.*, **49**, 431 (1977).

41. A. Ledwith and M. D. Purbrick, *Polymer*, **14**, 521 (1973).
42. R. S. Davidson and S. P. Orten, *J. Chem. Soc. Chem. Commun.*, p. 209 (1974).
43. R. F. Bartholomew, R. S. Davidson, P. F. Lambeth, J. F. McKellar, and P. H. Turner, *J. Chem. Soc. Perkin 2* p. 577 (1972).
44. S. P. Pappas, *J. Radiat. Curing*, **8**(3), 28 (1981).
45. T. H. Koch and A. H. Jones, *J. Am. Chem. Soc.*, **92**, 7503 (1970).
46. D. I. Schuster and M. D. Goldstein, *J. Am. Chem. Soc.*, **95**, 986 (1973).
47. V. D. McGinnis and D. M. Dusek, *ACS Polym. Prepr.*, **15**(1), 480 (1974).
48. M. J. Davis, J. Doherty, A. A. Godfrey, P. N. Green, J. R. A. Yung, and M. A. Parrish, *J. Oil Colour Chem. Assoc.*, **11**, 256 (1978).
49. D. P. Specht, C. G. Houle, and S. Y. Farid, U.S. Patent 4,289,844 (1981).
50. D. P. Specht, P. A. Martic, and S. Farid, *Tetrahedron*, **38**, 1203 (1982).
51. C. H. Bamford and M. J. S. Dewar, *Nature*, **163**, 214 (1949).
52. G. Oster, *Nature*, **173**, 300 (1954).
53. R. S. Davidson and K. R. Tretheway, *J. Chem. Soc. Chem. Commun.* p. 674 (1975).
54. D. F. Eaton, in *Advances in Photochem.*, **13**, 427 (1986); D. F. Easton, *Pure Appl. Chem.*, **56**, 1191 (1984).
55. G. A. Delzenne, H. K. Peeters, and U. L. Laridon, in *Non-Silver Photographic Process*, (R. J. Cox, Ed.), p. 23, Academic, New York, 1975.
56. W. J. Tomlinson and E. A. Chandross, *Adv. Photochem.*, **12**, 201 (1980).
57. R. H. Kayser and R. H. Young, *Photochem. Photobiol.*, **24**, 39, 403 (1976).
58. U. L. Laridon, G. A. Delzenne, and H. K. Peeters, assigned to Afga-Gevaert, U.S. Patent 3,847,610 (1974).
59. U. Steiner, G. Winter, and H. E. A. Kramer, *J. Phys. Chem.*, **81**, 1104 (1977).
60. R. J. Allen and S. Chaberek, assigned to Technical Operations, Inc., U.S. Patent 3,488,769 (1970).
61. G. Delzenne, S. Toppet, and G. Smets, *J. Polym. Sci.*, **48**, 347 (1960).
62. D. F. Eaton, *Photogr. Sci. Eng.*, **23**, 150 (1979).
63. G. Oster, *Photogr. Sci. Eng.*, **4**, 237 (1960).
64. G. Y. Y. Chen, *Tech. Semin.*, PC Fab Expo, Orlando, FL, Jan. 1986.
65. S. Shimizu, *Proc. TAGA* p. 232 (1986).
66. R. D. Mitchell, W. J. Nebe, and W. M. Hardom, *J. Imaging Sci.*, **30**, 215 (1986).
67. S. Sjolander, *J. Imaging Sci.*, **30**, 151 (1986).
68. H. Werlich, G. Sincerbox, and B. Yung, *J. Imaging Technol.*, **10**, 105 (1984).
68a. G. L. Geoffrey and M. S. Wrighton, *Organometallic Photochemistry*, Academic, New York, 1979.
69. C. H. Bamford and C. A. Finch, *Proc. R. Soc. A*, **268**, 553 (1962).
70. H. M. Wagner and M. D. Purbrick, *J. Photogr. Sci.*, **29**, 230 (1981).
71. K. Tsubakiyama and S. Fuyusaki, *J. Polym. Sci., Polym. Chem. Ed.*, **10**, 341 (1972).
72. R. D. Small, J. A. Ors, and B. S. H. Royce, *ACS Symp. Ser.*, **242**, 325 (1984).
73. F. W. Billmeyer, *Textbook of Polymer Science*, 3rd ed., p. 65, Wiley, New York, 1984.

74. R. F. Bartholomew and R. S. Davidson, *J. Chem. Soc. Chem. Commun.*, p. 2347 (1971).
75. C. Decker, *Makromol. Chem.*, **180**, 2027 (1979).
76. M. J. Bowden, in *Macromolecules* (F. A. Bovey and F. H. Winslow, Eds.), Chap. 2, Academic, New York, 1979.
77. A. R. Schultz and M. G. Joshi, *J. Polym. Sci., Polym. Phys. Ed.*, **22**, 1753 (1984).
78. C. H. Bamford and C. F. H. Tippers, in *Comprehensive Chemical Kinetics*, Vol. 14A, *Free Radical Polymerization*, Elsevier, Amsterdam, 1976.
79. J. Brandrup and E. H. Immergut, *Polymer Handbook*, Interscience, New York, 1966.
80. G. Wu, M.S. Thesis, Polytechnic University, Brooklyn, New York, 1984.
81. M. A. Naylor and F. W. Billmeyer, Jr., *J. Am. Chem. Soc.*, **75**, 2181 (1953).
82. J. L. Koenig, *Appl. Spectrosc.*, **29**, 293 (1975).
83. G. L. Collins, D. A. Young, and J. R. Costancza, *J. Org. Coat. Technol.*, **48**(615), 48 (1976).
84. G. L. Collins and I. R. Costanza, *J. Org. Coat. Technol.*, **51**(648), 57 (1979).
85. J. G. Kloosterboer, H. P. M. van Gerrichten, G. M. M. van de Hei, G. M. J. Lippits, and G. P. Melis, in *Characterization of Highly Crosslinked Polymers*, ACS Natl. Meet., 185th, Seattle, WA, Mar. 1983.
86. J. G. Klossterboer, G. M. M. van de Hei, R. G. Gossink, and G. C. M. Dortant, *Polym. Commun.*, **25**, 322 (1984).
87. J. E. Moore, in *UV Curing: Science and Technology* (S. P. Pappas, Ed.), p. 134, Technology Marketing Corp., Norwalk, CT, 1978.
88. G. R. Tryson and A. R. Shultz, *J. Polym. Sci., Polym. Phys. Ed.*, **17**, 2059 (1979).
89. J. G. Kloosterboer and G. J. M. Lippits, *J. Imaging Sci.*, **30**, 177 (1986).
90. J. G. Kloosterboer, G. M. M. van de Hei, and H. M. J. Boots, *Polym. Commun.*, **25**, 354 (1984).
91. H. M. J. Boots, in *Integration of Fundamental Polymer Science and Technology* (L. A. Kleinties and P. J. Lemstra, Eds.), Chap. 3, Applied Science, London, 1985.
92. S. B. Maerov, *J. Imaging Sci.*, **30**, 235 (1986).
93. P. L. Egerton, A. Reiser, W. Shaw, and H. M. Wagner, *J. Polym. Sci., Polym. Chem. Ed.*, **17**, 3315 (1979).
94. H. M. Wagner and J. F. Langford, *Res. Discl.*, **13**, 420 (June 1975); **15**, 328 (June 1977).
95. H. Morawetz, *J. Polym. Sci.*, Part C, **1**, 65 (1963).
96. J. H. O'Donnell, B. McGarvey, and H. Morawetz, *J. Am. Chem. Soc.*, **86**, 2322 (1964).
97. G. Wegner, *Z. Naturforsch. B*, **24**, 824 (1969).
98. V. Enkelmann, in *Polydiacetylenes* (H.-J. Cantow, Ed.), p. 103, Springer, Berlin, 1984.
98a. G. Wegner, *Makromol. Chem.*, **134**, 219 (1970).
98b. G. Wegner, *Makromol. Chem.*, **154**, 35 (1972).
99. A. Prock, M. L. Shand, and R. R. Chance, *Macromolecules*, **15**, 238 (1982).
100. H. Eckhardt, T. Prusik, and R. R. Chance, *Macromolecules*, **16**, 732 (1983).

101. K. Penzien, *Chem. Z.*, **99**, 413 (1975).
102. J. E. Sohn, A. F. Garito, K. D. Desai, R. S. Narang, and M. Kuzyk, *Makromol. Chem.*, **180**, 2975 (1979).
103. K. H. Richter, W. Guttler, and M. Schwoerer, *Appl. Phys. A*, **32**, 1 (1983).
104. R. C. Liang and A. Reiser, *ACS Polym. Prepr.*, **26**(2), 327 (1985).
105. R. C. Liang and A. Reiser, *J. Imaging Sci.*, **30**, 69 (1986).
106. R. C. Liang and A. Reiser, *J. Polym. Sci., Polym. Chem. Ed.*, **25**, 451 (1987).
107. B. Tieke, H. J. Graf, G. Wegner, B. Naegele, H. Ringsdorf, S. Banerjee, D. Day, and J. B. Lando, *Colloid Polym. Sci.*, **225**, 521 (1977).
108. J. P. Fouassier, B. Tieke, and G. Wegner, *Isr. J. Chem.*, **18**, 227 (1979).
109. C. Bubeck, B. Tieke, and G. Wegner, *Ber. Bunsenges. Phys. Chem.*, **86**, 499 (1982).
110. C. S. Marvel and G. Nowlin, *J. Am. Chem. Soc.*, **72**, 5076 (1950).
111. C. B. Morgan, F. Magnotto, and A. D. Ketley, *J. Polym. Sci., Polym. Chem. Ed.*, **15**, 627 (1977).
112. C. L. Kehr and W. R. Wszolek, *ACS Org. Coat. Plast. Chem.*, **33**(2), 295 (1973).
113. R. Back, G. Trick, C. McDonald, and C. Sivertz, *Can. J. Chem.*, **32**, 1078 (1954).
114. C. R. Morgan and A. D. Ketley, *ACS Polym. Prepr.*, **17**(2), 500 (1976).
115. C. R. Morgan, D. R. Kyle, and R. W. Bush, *ACS Coat. Plast.*, p. 373 (1984).
116. J. V. Crivello, in *UV Curing: Science and Technology* (S. P. Pappas, Ed.), pp. 24–75, Technology Marketing Corp., Norwalk, CT, 1980.
117. R. W. Lenz, *Organic Chemistry of Synthetic High Polymers*, p. 247, Interscience, New York, 1967.
118. F. S. Dainton and K. J. Ivin, *Q. Rev. Chem. Soc.*, **12**, 61 (1958).
119. J. J. Licari and P. C. Crepan, U.S. Patent 3,205,157 (1965).
120. (a) E. Fischer, U.S. Patent 3,236,784 (1966). (b) W. H. Smith, assigned to 3M Company, U.S. Patent 3,779,778 (1973); U.S. Patent 4,250,053 (1979).
121. S. I. Schlesinger, *Photogr. Sci. Eng.*, **18**, 387 (1974).
122. J. V. Crivello and J. H. W. Lam, *J. Polym. Sci., Polym. Symp.*, No. 56, 383 (1976).
123. J. V. Crivello and J. H. W. Lam, *J. Polym. Sci., Polym. Chem. Ed.*, **17**, 977 (1979).
124. J. V. Crivello and J. H. W. Lam, *J. Polym. Sci., Polym. Chem. Ed.*, **17**, 2877 (1979); **18**, 1021 (1980).
125. J. V. Crivello and J. H. W. Lam, *Macromolecules*, **14**, 1141 (1981).
126. K. Meyer and H. Zweifel, *J. Imaging Sci.*, **30**, 174 (1986).
127. K. Meyer and H. Zweifel, *SME Tech. Pap., RadCure* p. 417 (1985).
128. S. H. Schroeter, J. E. Moore, and O. V. Orkin, *ACS Org. Coat. Plast. Prepr.*, **34**(10), 751 (1974); *ACS Plast. Prepr.*, **34**(1), 751 (1974).
129. S. H. Schroeter, in *Non-Polluting Coatings and Coating Procedures* (J. L. Gardon and J. W. Prane, Eds.), p. 109, Plenum, New York, 1973.
130. E. G. Shur and R. Dabal, U.S. Patent 3,772,062 (1973).
131. G. F. D'Alelio, U.S. Patent 3,676,398 (1972).

132. Ford Motor Co., Br. Patent 162,722 (1969).
133. P. Walker, Du Pont, private communication, 1986.
134. E. Brinkman, G. Delzenne, A. Poot, and J. Willems, *Unconventional Imaging Processes*, p. 32, Focal, London, 1978.
135. NAPPLATE prospectus, NAPP Systems (USA), Inc., 360 Pacific Street, San Marcos, CA 92069.
136. Nylotron prospectus, BASF AG, Ludwigshafen, Germany.
137. A. C. Schoenthaler, *SPSE Two Day Semin., Applications of Photopolymers, Prepr. SPSE, 1970.*
138. F. Axon et al., *Circuit World*, **4**, 24 (1978).
139. L. Plambeck, A. Cairncross, W. J. Chambers, C. S. Cleaver, D. S. Donald, and D. Eaton, *J. Imaging Sci.*, **30**, 221, 224, 228 (1986).
140. J. R. Celeste, U.S. Patent 3,469,982 (1969).
141. R. D. Rust and D. A. Doane, *Solid State Technol.*, **29**(6), 125 (1986).
142. W. S. Fujitsubo, *Solid State Technol.*, **29**(6), 161 (1986).
143. N. S. Fox, *ACS Symp. Ser.*, **242**, 367 (1984).
144. Vacrel prospectus, E. I. Du Pont de Nemours, Wilmington, DE 18989.
145. J. E. Gervay and P. Walker, assigned to Du Pont, U.S. Patents 3,718,473 (1973); 3,787,213 (1974); 3,879,204 (1975).
146. M. Bruno, *Principles of Color Proofing*, Gama Communications, Salem, NH, 1986.
147. V. F. H. Chu and A. B. Cohen, assigned to Du Pont, U.S. Patent 3,649,268 (1972).
148. R. P. Held, U.S. Patent 3,854,950 (1974).
149. R. N. Fan, U.S. Patent 4,053,313 (1977).
150. R. H. Boyd and V. F. H. Chu, *TAGA Proc.*, p. 9 (1968).
151. A. B. Cohen and R. N. Fan, U.S. Patent 4,174,216 (1979).
152. W. J. Tomlinson and E. A. Chandross, *Adv. Photochem.*, **12**, 201 (1980).
153. L. Solymar and D. J. Cooke, *Volume Holography and Volume Gratings*, Academic, London, 1981.
154. D. H. Close, A. D. Jacobson, J. D. Margerum, R. G. Brault, and F. J. McClung, *Appl. Phys. Lett.*, **14**, 159 (1969).
155. S. Sjoelander, *J. Imaging Sci.*, **30**, 151 (1986).
156. V. C. Chambers, Jr., U.S. Patent 3,497,185 (1969).
157. E. F. Haugh, U.S. Patent 3,658,526 (1972).
158. H. L. Fielding and R. T. Ingwall, U.S. Patent 4,588,664 (1986).
159. W. S. Colburn and K. A. Haines, *Appl. Opt.*, **10**(7), 1636 (1971).

5 POSITIVE RESISTS BASED ON DIAZONAPHTHOQUINONES

This chapter is concerned with positive resists based on diazonaphthoquinones and phenolic resins, because these systems represent the great majority of positive resists used in the semiconductor as well as in the printing industry. More recently, some positive photoresists have been developed on the basis of functionality changes and changes in molecular weight. Some of these will be considered in Chapter 7.

INTRODUCTION

The positive photoresists described in this section are based on two phenomena: the inhibition by diazonaphthoquinones of the dissolution of phenolic resins in aqueous alkali, and the photochemical transformation of the diazonaphthoquinones into indene carboxylic acids.

$$\text{Diazonaphthaquinone} \xrightarrow[H_2O]{h\nu} \text{Indene carboxylic acid} + N_2 \qquad (1)$$

The phenolic resin is usually, but not always, a phenol–formaldehyde polycondensate of low molecular weight known as a novolac. On their own, these resins are readily soluble in dilute aqueous alkali, but when a suitable diazoquinone derivative is added, their rate of dissolution is greatly reduced:

The nonionizable hydrophobic diazoquinone acts here as a dissolution inhibitor. On irradiation, the diazoquinone turns into an acid that is soluble in aqueous alkali and which is no longer a dissolution inhibitor and may even promote the dissolution of the resin. In practice, a mixture of novolac and, say, 15% of a diazonaphthoquinone is coated as a thin film on a substrate, for example a silicon wafer, the film is exposed to a radiation pattern and subsequently treated with dilute alkali. The exposed areas of the resist dissolve much faster than the unexposed ones; if the process is stopped in time, a patterned resist film remains in the unirradiated areas: The material functions as a positive photoresist.

It should be emphasized at this point that in diazoquinone-novolac resists *image discrimination is based on a kinetic effect.* This is shown in Fig. 5-1 where the dissolution rate is plotted logarithmically as a function of diazoquinone

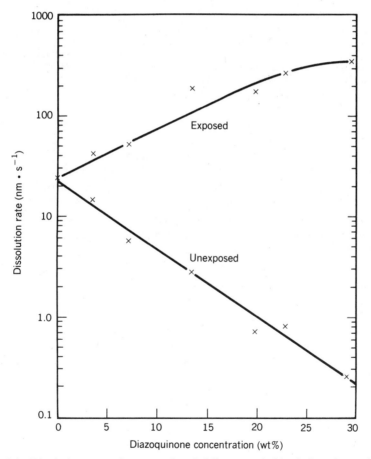

Figure 5-1 Dissolution rates of unexposed and fully exposed (bleached) resist made from Alnavol novolac and a diazoquinone. (Reproduced with permission from D. Meyerhofer [1]. © IEEE, 1980.)

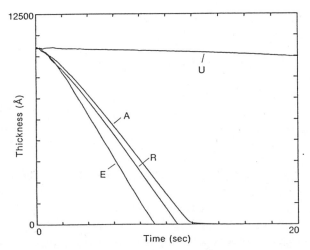

Figure 5-2 Dissolution curves (film thickness versus time in developer) for exposed and unexposed films of a model resist system, pure novolac resin and the indene carboxylic acid–resin mixture. U = unexposed resist, E = exposed resist, R = pure resin, A = acid–resin mixture. The pure resin and the acid–resin mixture dissolve at practically the same rate. (Reproduced with permission from W. D. Hinsberg et al. [3].)

content for irradiated and nonirradiated films of novolac–diazoquinone mixtures. It can be seen that at an inhibitor content of 20% by weight the irradiated films dissolve about 200 times faster than the unirradiated ones [1]. It will be noted that the dissolution rate of the unexposed resist is never zero and, given sufficient time, the resist will dissolve even in the unirradiated areas of the coating. The development process must therefore be stopped as soon as the resist has been completely dissolved in the exposed areas.

It can be seen in Fig. 5-1 that exposure of the resist film not only removes the inhibiting effect of the diazoquinone, but appears to accelerate dissolution above the dissolution rate of pure novolac, indicating that indene carboxylic acid may act as a dissolution promoter. Ouano [2] has investigated this question in some detail and suggested that at least part of the acceleration effect is caused by the evolution of nitrogen in reaction (1). The escaping gas creates additional free volume and facilitates diffusion of the developer into the polymer matrix. To separate the free volume effect from that of the photoacid, Hinsberg et al. [3] and more recently Blum et al. [4] have simulated the exposed resist film by mixing together novolac with indene carboxylic acid. With a rather fast dissolving resin Hinsberg found that such coatings dissolved at essentially the same rate as pure novolac under the same processing conditions (see Fig. 5-2). Blum observed a substantial acceleration effect in a more slowly dissolving resist. Pending a clearer understanding of the mechanism of resist dissolution no general conclusion can be drawn from these experiments.

Diazoquinones have a long history in graphic reproduction. They were used by Koegel and Neuenhaus [5, 6] as the light sensitive component of the Ozalid Process. This was one of the early diazo-type copying systems common in architectural drawing offices. A paper imbibed with diazoquinone and with a phenolic coupler was irradiated through a positive line drawing. The diazoquinone in the exposed areas was destroyed by the photoreaction, and on subsequently treating the exposed sheet with base (ammonia vapor) the diazoquinone remaining in the unexposed areas reacted with the phenolic coupler to produce an azo dye. In this way a positive image of the original drawing was formed.

$$\text{(diazoquinone)} \xrightarrow[\text{Naphthol}]{OH^-} \text{(azo dye product)} \quad (2)$$

At the time, the main advantage of the diazoquinones was their thermal stability, which gave the Ozalid papers a reasonable shelf life. The coupling reaction, however, was slow and the diazoquinones were soon superseded by the more reactive diazonium salts, which are still in use today [7, 8].

Until the 1950s all photomechanical reproduction was based on the dichromated colloids described in Chapters 1 and 2. These negative-working materials require the use of an intermediate photographic negative for the correct reproduction of an original pattern. For a long time the printing trade had been looking for a direct positive printing plate that would eliminate the need for the costly intermediate step. During the second World War the Kalle Laboratories in Wiesbaden, Germany, found that mixtures of phenolic resins with various diazoquinones produced useful positive-working reprographic plates, which were developable in aqueous alkali. German patents [9, 10] on these systems started to appear in 1949, the corresponding American patents were later assigned to the Azoplate Corporation [11] (1956 and following). The positive plates were only a moderate commercial success, but interest in the diazoquinone–novolac systems revived in the 1970s when the resolution requirements of the semiconductor industry outran the capabilities of the negative bis-azide resists. At that time, the diazoquinone resists established themselves as nonswelling, high resolution materials. They now occupy a dominant position in microlithography.

THE CHEMISTRY OF DIAZONAPHTHOQUINONES

The photoreaction of the diazonaphthoquinones was investigated early on (1944) by Suess, [12, 13] who suggested the following mechanism for the overall photoprocess.

POSITIVE RESISTS BASED ON DIAZONAPHTHOQUINONES

[Reaction scheme: Compound I (diazonaphthoquinone) → (hν) → Compound II (carbene) + N_2 → Compound III (ketene) → (H_2O) → Compound IV (indene carboxylic acid)]

(3)

In the first reaction step, nitrogen is eliminated and a carbene (II) is formed. This undergoes a Wolff rearrangement with ring contraction, leading to the ketene (III). In the presence of small amounts of water the ketene is finally hydrated to the carboxylic acid (IV). The crucial feature of the process is the change from a hydrophobic and nonionizable compound to an ionizable hydrophilic species, indene carboxylic acid.

Reaction (3) describes a somewhat idealized process. More recently, Pacansky and others [14–16] investigated the system in more detail. They confirmed the basic accuracy of the Suess mechanism, but found that other, less helpful reactions do also occur. Their results are summarized in Scheme 1.

If the water content of the resist is insufficient, the ketene (III) may react directly with the resin, and may produce crosslinks that are highly undesirable in a positive resist. To suppress this type of crosslinking, ketene scavengers (amines) are sometimes added. They react with the ketene (and prevent crosslinking) but may also decarboxylate the acid on subsequent hydrolysis. The reaction of ketene with the anhydrous base resin has recently been exploited to produce *negative* resist images with strongly undercut profiles. These are required, for example, in various lift-off procedures (see Chapter 10). Itoh et al. [17] have described a system that uses a conventional positive resist and exposes it to a radiation source with a fairly high content of deep-UV radiation (250 nm). The short wavelength radiation is strongly absorbed in the resist and creates a crosslinked crust that is the cause for the overhang profiles shown in Fig. 5-3.

Another crosslink producing side reaction is the formation of a red azo dye as a result of a slow coupling process. This may also occur during development in the unirradiated areas; at that stage, however, crosslinking is no more objectionable.

A third reaction that occurs during postbake is the decarboxylation of indene carboxylic acid, leading to indene and indene dimer [16]. This reaction is of interest because it can be used in the design of a well-controlled *image*

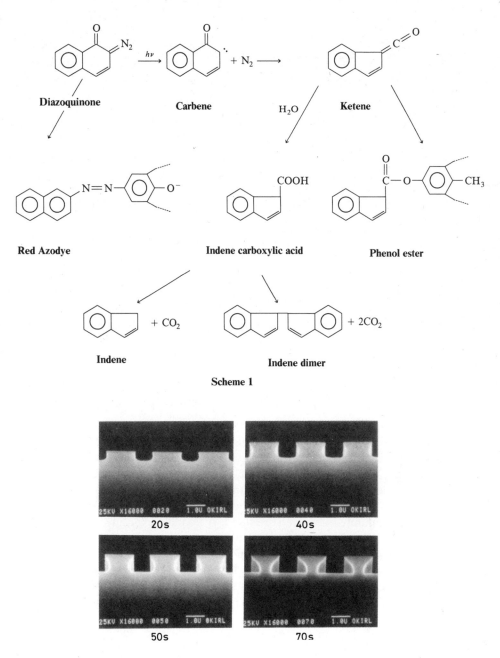

Scheme 1

Figure 5-3 Dependence of resist profiles on development time as a result of isotropic development. (Reproduced with permission from T. Itoh et al. [17].)

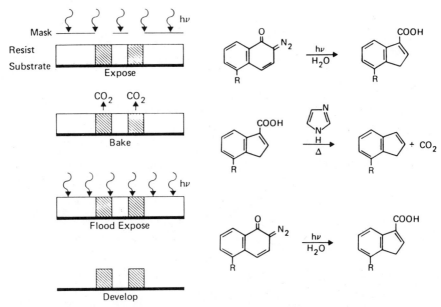

Figure 5-4 Sequence of operations in an image reversal process. The resist is exposed to a pattern and subsequently treated with base to decarboxylate the photogenerated carboxylic acid. A flood exposure of the whole coating results, on development, in a reversal image. (Reproduced with permission from C. G. Willson [20].)

reversal process where a negative image is produced with a conventional positive resist [18]. The decarboxylation process is base catalyzed and can be induced at comparatively low temperatures by the addition of a strong base to the coating solution [19, 20]. Imidazole, monazoline, and triethanolamine give reproducible results and so does treatment of the coating with ammonia vapor [21]. Where the carboxylic acid has been removed the original low dissolution rate of the inhibited novolac is restored. A flooding exposure of the film now produces carboxylic acid (and a high dissolution rate) in the previously unexposed areas. Subsequent development in alkali leads to a negative image of the original mask. This procedure is particularly useful in lift-off processes where a negative and undercut resist image is required. The sequence of operations in the image reversal process is indicated schematically in Fig. 5-4 [22].

Feely et al. [23] have used the acid generated in the photolysis of diazoquinone in yet another way. They added a melamine derivative as a crosslinker to the novolac making the mixture into an "acid hardening" composition. *Acid hardening resins* are usually crosslinked by strong protonic or Lewis acids. Carboxylic acids become effective only at higher temperatures, when the following reaction occurs.

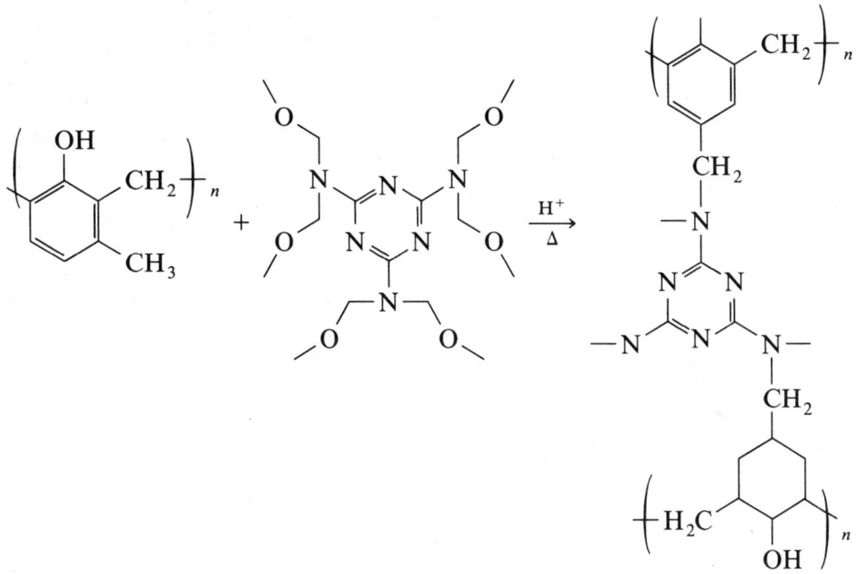

The comparative inactivity of carboxylic acids at room temperature makes it possible to use the system in two different ways. The acid catalyst produced during exposure can either be removed by a basic developer and the system used as a normal positive resist; or the irradiated areas can be crosslinked before development, by heating from 80 to 100°C and subsequently the image areas can be removed by a flood exposure and development in alkali. A negative image of the original is the result.

Feely has shown that in thick films, the depth of light penetration and therefore the spatial extent of the photoprocess can be controlled by a graded (white–gray–black) mask and that quite complex three-dimensional structures can be created, as indicated in Fig. 5-5.

The Photoactive Component

The photoactive components (PACs) most often used in current positive resists are diazonaphthoquinones substituted in the 5-position or in the 4-position with sulfonic acid derivatives.

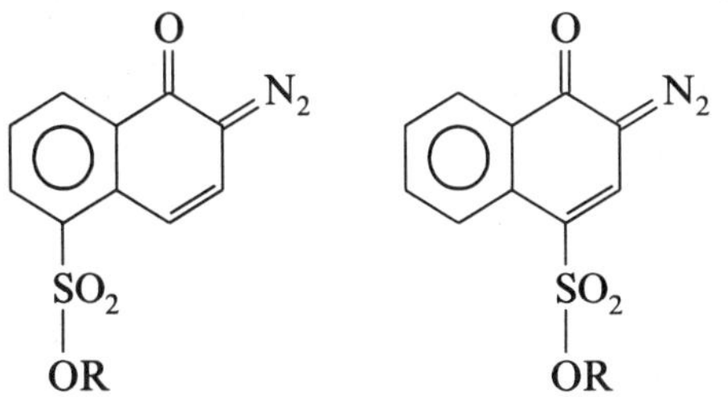

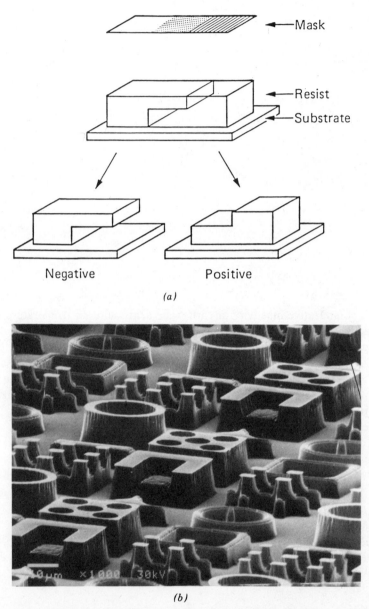

Figure 5-5 Scanning electron micrograph of three-dimensional shapes obtained from a thick acid hardening resist layer processed in the negative mode by heating to 300°C before flood exposure and development. (Reproduced with permission from W. E. Feely et al. [23].)

The starting point of synthesis is usually the corresponding sulfonic acid chloride, which is then esterified with a suitable ballast compound. For resists to be used in the near-UV region good absorption at 365 nm and at 405 and 437 nm is required and aromatic substituents are chosen. The most popular ballast compound is polyhydroxybenzophenone where one to three (four) hydroxyl groups can be esterified with diazoquinone sulfonyl-chloride. An often used inhibitor is a substituted tri-hydroxy benzophenone of the following structure.

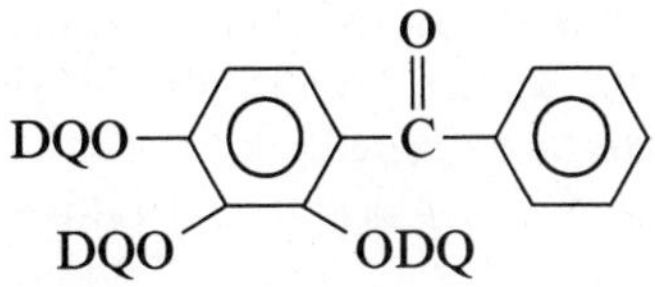

DQ stands here for diazonaphthoquinone. This inhibitor is contained in several commercial resists such as in AZ 1300 and in AZ 4000, in Microposit 1300 and 1400, in Hunt HPR 204 and WX-118, and in the resist OFPR of Tokyo Oka. Another popular diazoquinone derivative is the sulfonyl ester of cumylphenol, which is contained, for example, in Microposit 111 and in the Hunt resist WX-159. Similar systems are used by Polychrome, by Mac Dermid and by Allied [24].

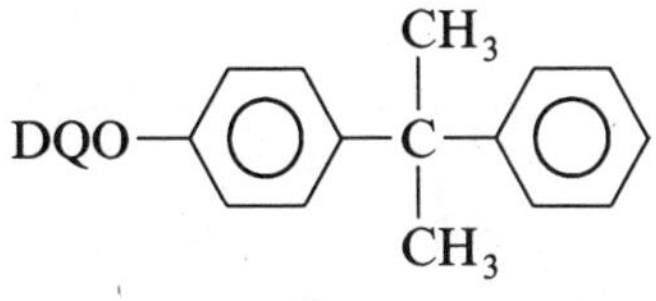

In the Kodak Micropositive Resist 809 the diazonaphthoquinone is directly attached to the phenolic resin, which in this case is a methacrylamide derivative. This resist is particularly suitable for use in multilayer systems where the diffusion of the inhibitor into adjacent resist layers cannot be tolerated.

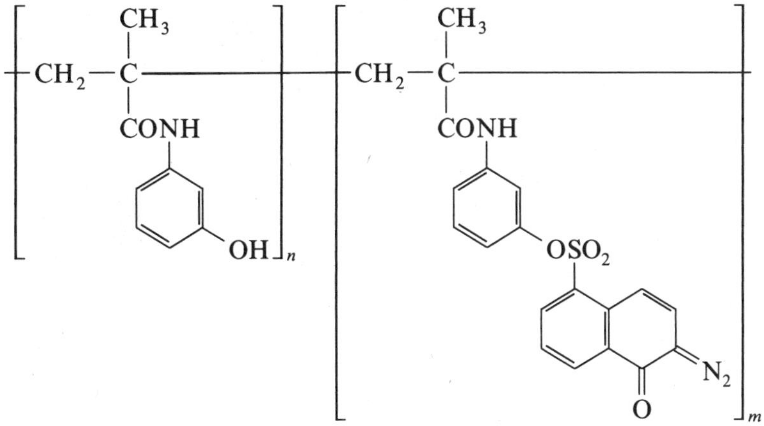

The following physical properties of the photoactive component are important:

1. Good overlap of its absorption spectrum with the emission spectrum of the irradiation source (usually a Hg arc).
2. Bleachability; that is, the absorption at the wavelength of irradiation should disappear ("bleach") during exposure, which means that the photoproduct should not have significant absorption in that region. If that is the case, the exposing radiation and hence the photoreaction will eventually penetrate to the bottom of the resist layer and reach the resist–substrate interface.
3. Compatibility with the base resin. The inhibitor affects the dissolution rate of the resist only if it forms a single phase with the resin.
4. Reasonable thermal stability. The diazoquinone should, for example, not decompose at the prebake temperature.

The absorption spectrum of a representative diazoquinone [25] is shown in Fig. 5-6, together with the spectral emission of the mercury arc. This particular inhibitor shows the absorption bands at 350 and 400 nm that are typical for a diazoquinone substituted in the 5-position of the naphthalene skeleton. The

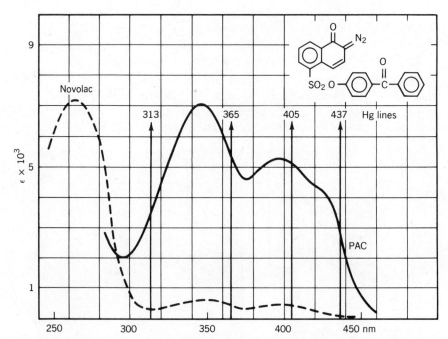

Figure 5-6 Absorption spectrum of a typical diazoquinone inhibitor and novolac base resin shown together with the principal emission lines of the medium pressure Hg arc.

resist AZ 1350 J belongs in this category. In the diazoquinones substituted in the 4-position, such as AZ 2400, the two bands are shifted to 310 and 390 nm (see also Fig. 5-7). The novolac base resin absorbs strongly at about 250 nm but is almost completely transparent at longer wavelengths. The position of the principal lines of the mercury spectrum are indicated in the figure and it can be seen that the resist of Fig. 5-6 can be exposed with the so-called G line at 437 nm. This also means that it is possible to use glass optics in the projection system.

Since the conjugated system of the naphthodiazoquinones is significantly shortened in indene carboxylic acid, most diazoquinones bleach on irradiation and, as mentioned before, that is of great practical importance since it allows the photoreaction to eventually reach the resist–substrate interface. Figures

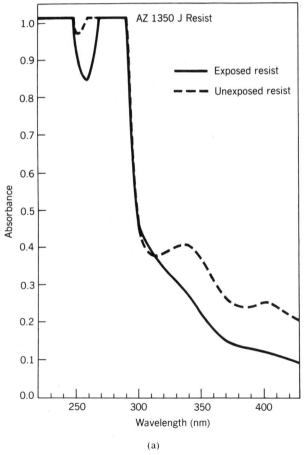

(a)

Figure 5-7a Spectra of unexposed and fully exposed films of the resist AZ 1350 J. (Reproduced with permission from C. G. Willson et al. [26].)

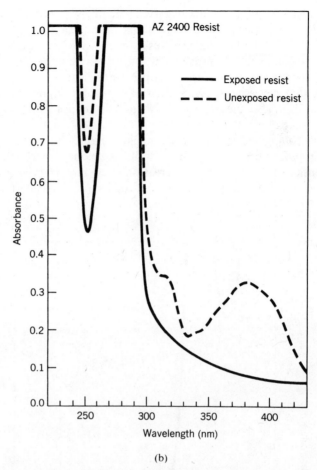

Figure 5-7b Spectra of unexposed and fully exposed films of the resist AZ 2400. (Reproduced with permission from C. G. Willson et al. [26].)

5-7a and 5-7b show the absorption spectra of the classical AZ 1300 J and AZ 2400 resists before and after irradiation [27].

The shape of the diazoquinone spectra can be changed, within limits, by substitution and by the introduction of heteroatoms into the naphthoquinone ring system. Willson et al. [26] have explored these possibilities and have calculated by a semiempirical quantum chemical method (INDO) the expected positions of the principal absorption bands [27]. They found that substitution on the 5-sulfonyl group brought about a red shift of the long wavelength band that decreased in the order

$$N_3 \sim Cl_2 > O\text{-Aryl} > O\text{-Alkyl} > N(CH_3)_2$$

The spectra of some of these compounds are shown in Fig. 5-8.

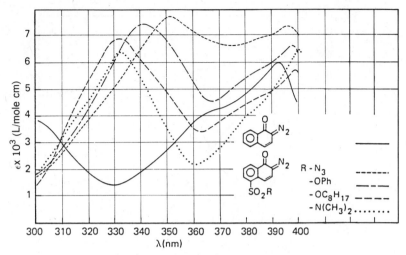

Figure 5-8 Effect of selected 5-sulfonyl substituents on the spectrum of naphthodiazoquinone. (Reproduced with permission from C. G. Willson et al. [26].)

The compatibility of the inhibitor with the base resin is an essential condition, but it does not seem too difficult to achieve. It is one of the virtues of novolac that it mixes rather easily with a wide variety of substances. Diazonaphthoquinones decompose around 120 to 130°C, it is therefore important that the resist film not be exposed to higher temperatures for prolonged periods of time [25].

The Phenolic Base Resin

In most positive diazoquinone resists the phenolic base resin is a novolac, that is, a condensation polymer of various cresoles (or phenol) with formaldehyde. The structure of novolac can be expressed schematically in the following form.

Novolac

The methylene links are not always ortho to the OH groups of the phenols, in particular this is so in *m*-cresol based novolacs where a variety of geometries is possible.

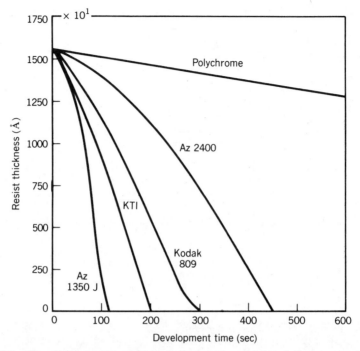

Figure 5-9 Film thickness versus time in developer curves for five similar commercial resists illustrating the variability of dissolution rates in almost identical systems. (Reproduced with permission from J. M. Shaw and M. Hatzakis [29].)

Novolacs and resoles are the oldest synthetic lacquers of the chemical industry. They are also the basis of the first synthetic thermoset resin, Bakelite. Practical synthetic procedures were well established when the resist applications of novolac came along. However, these procedures did not allow for accurate control of the reaction product and, as a result, the performance of positive resists showed great variability [28]. In Fig. 5-9 film thickness is plotted as a function of development time for five nominally similar resist, identically processed in Shipley 2401 developer. Clearly, the Azoplate resist AZ 1350 J and the equivalent Polychrome resist behave very differently [29].

Four material factors that had not been too important in traditional resin manufacture appear to determine resist performance:

1. Molecular weight.
2. The dispersity of the molecular weight distribution.
3. The isomeric composition of the cresoles.
4. The relative position of the methylene linkages.

The molecular weight of novolac resins used in resists is usually between

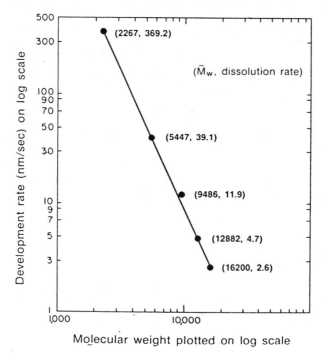

Figure 5-10 Effect of molecular weight on the dissolution rate of a group of otherwise identical resists. (Reproduced with permission from R. A. Arcus [30].)

$M_n = 1000$ to 3000, corresponding to 8 to 25 repeat units in the chain. The dispersity of molecular weight is generally high, although it has been shown that a narrow molecular weight distribution leads to better lithographic performance. Figure 5-10 shows the effect of molecular weight on the dissolution rate of a particular type of novolac resin [30].

Hanabata et al. [31] have shown that also in positive resists a narrow molecular weight distribution, and in particular the absence of low molecular weight fragments is beneficial for good image resolution. They have proposed the so-called "stone wall" model of resist dissolution, where low molecular weight fragments, the small stones in the wall, can be dissolved out from the bulk of the matrix and make the rest of the structure crumble over a wide range of development time and developer concentration. With resins of low dispersity, lines and spaces of 0.5 μm could be resolved by optical (!) lithography.

Chromatography and ^{13}C-NMR have made it possible to determine the isomeric content of the resins, and pure isomers can now be prepared by selective preparation and separation methods. The effect of isomeric composition on the physical and the lithographic properties of novolac resists was recently investigated by Hanabata and co-workers [32, 33]. They characterized the isomeric composition by the *m*-cresol to *p*-cresol ratio and described the

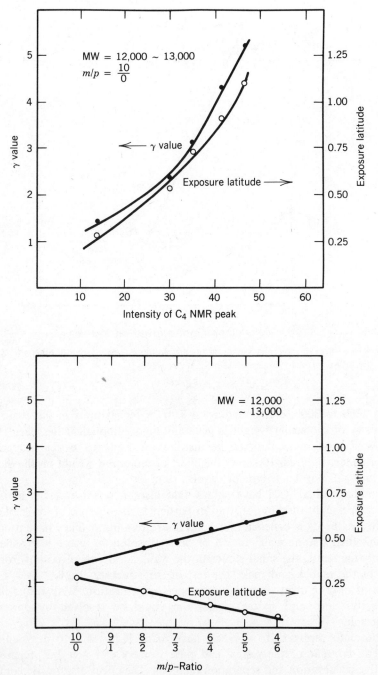

Figure 5-11 Effect of the position of the methylene link (indicated by the intensity of the C_4 NMR peak) and of the *meta-para* cresol isomer ratio on resist contrast in a group of novolac–diazoquinone resists. (Reproduced with permission from M. Hanabata et al. [33].)

TABLE 5-1 Effect of Novolac Structure and Degree of Polymerization on Flow Temperature

$$\left(\underset{R}{\underset{|}{\bigodot}}\text{(OH)}-CH_2\right)_n$$

R	n	T_{flow} (°C)
o-Me	8	85
p-Me	6	119
o-n-Pr	8	73
	15	105
o-sec-Bu	9	69
o-tert-Bu	> 13	> 133
o-Me (70%)	5	85
	7	104
	10	119
p-Me (70%)	10	128

link configuration by the relative intensity of the unsubstituted C_4 peak in the NMR spectrum. The effect of the link configuration on the contrast of the resist is dramatic, that of the m/p ratio is less important (see Fig. 5-11).

The great advantage of novolac resists over the bis-azide/rubber systems is the fact that novolac does not swell during development and is therefore suitable for the reproduction of high resolution pattern. Furthermore, the resin is reasonably resistant to plasma, an important property in present day device

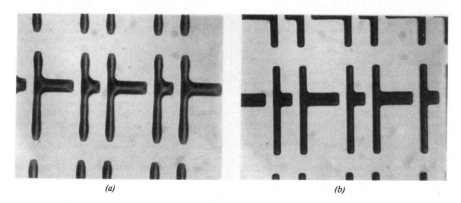

Figure 5-12 Image produced in a novolac–diazoquinone resist. (*a*) Baked at 250°C. (*b*) Cured photomagnetically and then baked at 300°C for 40 s. Note that the untreated resist has flown at 250°C, but that the cured resist retains its shape at an even higher temperature. (Reproduced with permission from P. A. Ruggiero [35].)

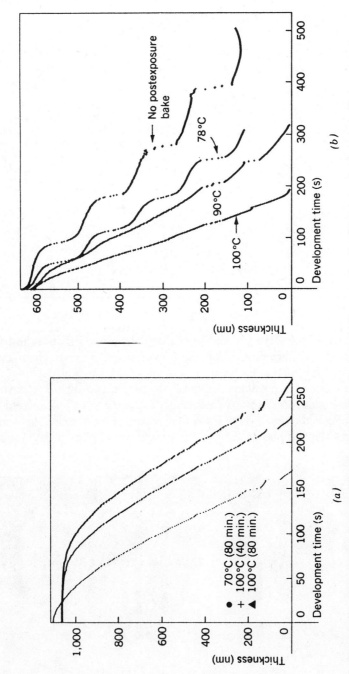

Figure 5-13 The effect of (*a*) preexposure bake and (*b*) postexposure bake on a film of 1350 J AZ resist coated on silicon, exposed to 15 mJ/cm^2 at 405 nm and developed in AZ developer diluted with water 1:1. Preexposure bake at 70 and 100°C; post exposure bake at 78, 90, and 100°C for 20 min. Note the gradual disappearance of the standing wave pattern during postexposure bake. (Reproduced with permission from J. M. Shaw and M. Hatzakis [29].)

fabrication (see Chapter 10). In contrast, the thermal and mechanical properties of novolac are not outstanding and most positive resists have plastic flow temperatures between 120 and 160°C. Even here the effects of isomer composition and of molecular weight play a role as can be seen from the data in Table 5-1.

The image stability of diazoquinone resists can be improved by various radiative treatments that introduce crosslinks and increase the glass transition temperature and also the mechanical strength of the polymer image. Deep-UV exposure [34], as well as so-called pulsed photomagnetic curing, (flash exposure to a high intensity UV arc in the presence of a magnetic field) [35], have been used to prepare positive resist images for high current ion implantation. Figure 5-12a shows a resist pattern on silicon that has been heated well-above the flow point, as indicated by the rounded edges of the pattern. The sample of the same resist in Fig. 5-12b had been previously subjected to photomagnetic curing and did not flow when briefly heated to 300°C.

The behavior of the resist depends not only on material factors, but also on processing conditions, such as the thermal history of the film. Figure 5-13a shows the effect of a preexposure bake on the rate of development of a resist. Figure 5-13b refers to a postexposure thermal treatment that is routinely applied to resist films in order to even out the periodic changes in development rate caused by the interference of the incident and the reflected radiation, the so-called "standing waves" effect [29]. Zampini et al. have recently made a thorough study of the postexposure bake operation [36].

MEASUREMENT OF DISSOLUTION RATES

Image discrimination in positive resists is based on differences in dissolution rates and the differentiation between image and nonimage areas is maximized at the moment when the resist has been just cleared from the irradiated areas. The measurement of dissolution rate is therefore important.

Laser Interferometry. Konnerth and Dill [37, 38] were the first to devise a method of continuously monitoring the thickness of the resist film during development. A simple embodiment of their idea taken from the work of Meyerhofer [1] is shown in Fig. 5-14. A laser beam is directed towards the film that is immersed in the developer. The wavefront reflected from the surface of the film and that reflected from the interface between resist and substrate interfere constructively or destructively [39] depending on the angle of incidence and on the thickness of the film. As the film dissolves in the developer and the film thickness decreases, the photocell registers maxima and minima of light intensity. A record of this signal as a function of time in the developer is shown in Fig. 5-15. The time interval Δt between two extrema on the recorder trace corresponds to a change Δr in film thickness, which is de-

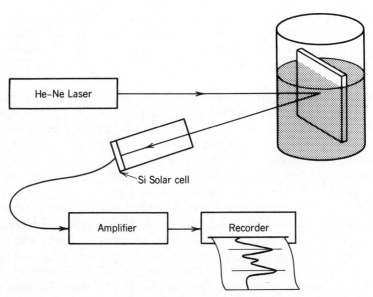

Figure 5-14 Simple arrangement for laser interferometry. The beam from a small He–Ne laser is directed onto the surface of the resist immersed in developer. The beams reflected from the surface of the resist and from the support–resist interface interfere constructly or destructively, depending on the thickness of the resist layer, and the solar cell receives in turn higher and lower light intensities. The corresponding signal is amplified and displayed on a chart recorder. (Reproduced with permission from D. Meyerhofer [1]. © IEEE, 1980.)

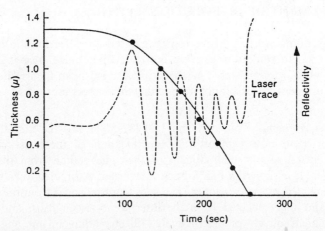

Figure 5-15 Reflectivity versus time in developer monitored by laser interferometry during development. The film thickness versus time in developer curve (full line) can be derived from the recorder trace as indicated in the text. (Reproduced with permission from C. G. Willson [20].)

termined by the Bragg relation

$$2\Delta r = \frac{\lambda \nu}{n} \Delta t \tag{4}$$

where n is the index of refraction of the resist, λ is the wavelength of the laser radiation, and ν is an integer $(1, 2, 3, \ldots)$. The recorder trace can thus be transformed into a film thickness versus development time curve, as it is shown on Fig. 5-15.

Laser interferometers have developed from the simple arrangement shown in Fig. 5-14 into sophisticated instruments, which plot dissolution rate curves directly with the help of a dedicated microprocessor. In one of the new versions of this instrument, the Dissolution Rate Monitor of Perkin-Elmer, a

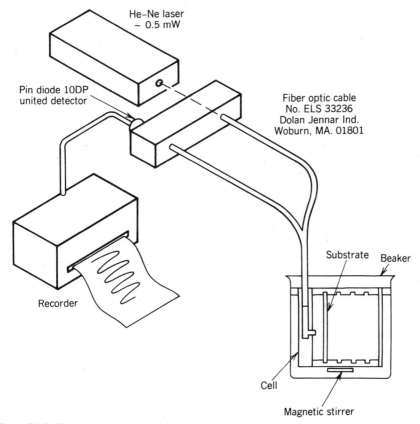

Figure 5-16 Development end-point detector operating on the principle of laser interferometry. The incident beam and the reflected beams are guided by fiber optic conduits, otherwise the comments in the caption to Fig. 5-14 apply. (Reproduced with permission from C. G. Willson [20].)

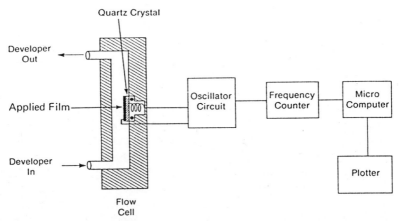

Figure 5-17 Schematic of a quartz crystal microbalance dissolution monitor. The mass of the quartz crystal controls the frequency of the oscillating circuit. As the film coated on the crystal dissolves, the mass of the crystal changes. The frequency meter can be calibrated directly in terms of the mass of the dissolving film. (Reproduced with permission from W. D. Hinsberg et al. [3].)

whole set of locations on a resist coated wafer can be monitored simultaneously and the family of curves shown in Fig. 5-18 can be obtained in a single run. Figure 5-16 shows a simple but highly effective development end-point detector that operates on the same principle.

The Quartz Crystal Microbalance. Laser interferometry is not the only method for measuring film thickness during development. Classical ellipsometry has been adapted successfully to this purpose by Soong et al. [40, 41] and used for a detailed analysis of the dissolution process. Even more recently, an essentially gravimetric method for monitoring resist films has been described [3]. It is based on a quartz crystal microbalance. The resist film is coated onto the crystal and the characteristic frequency of the crystal is measured while the system is immersed in developer. The characteristic frequency depends on the mass of the oscillating unit, of which the resist film is a part. As the film dissolves, the mass changes and so does the frequency. It has been shown that the frequency shift is linearly related to the mass of the cast film and can therefore be used to monitor the dissolution process. Figure 5-17 shows a schematic of the quartz crystal balance described by Hinsberg et al. [3]. The technique has a number of advantages. It can be used to measure very high dissolution rates and it can handle thick films, as well as opaque films and films with an uneven surface, none of which are accessible to interference methods.

Film Thickness versus Development Time Curves. By performing dissolution rate measurements on a series of coatings that have received different radiation

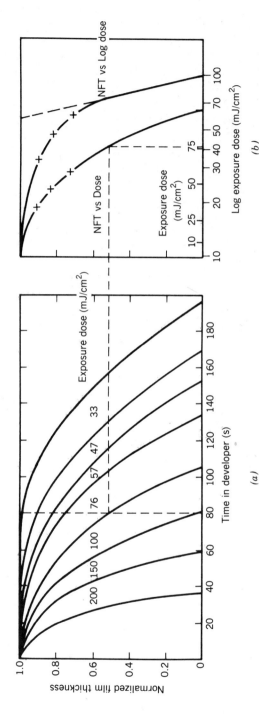

Figure 5-18 (*a*) Normalized film thickness (NFT) versus time in developer curves for identical resist films that have received different exposure doses (indicated on the curves). (*b*) Film thickness versus exposure curve derived from (*a*), and the conventional logarithmic characteristic curve.

201

doses, but are otherwise identical, a family of curves is generated that contains most of the relevant information on the resist–developer pair. Such an ensemble of thickness versus time curves is shown in Fig. 5-18. The diagram makes it possible to choose the correct exposure dose as well as the corresponding optimum time of development, which maximizes image differentiation (i.e., the difference in dissolution rate between exposed and unexposed image areas). The criterion here is that the exposed areas should be completely cleared of resist ($r = 0$), while as little film thickness as possible is lost in the nonimage areas. In the case shown in Fig. 5-18 this is achieved at an exposure dose of 100 mJ/cm^2 and at a development time of 80 s.

The relation between film thickness and exposure dose for a given development time can be read off the curves in Fig. 5-18a. The procedure is indicated in Fig. 5-18b. If the normalized film thickness is plotted against the decadic logarithm of exposure as it is in the right-hand trace of Fig. 5-18b, the conventional characteristic curve of the material is obtained.

RESIST CHARACTERISTICS

The photographic performance of the resist is summarized in the *characteristic curve* (see also Chapter 6), but it must be remembered that for a positive resist where image discrimination is based on a kinetic effect, the characteristic curve applies for a particular set of processing conditions, that is, for a specific

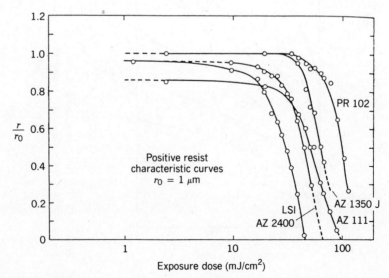

Figure 5-19 Characteristic curves for a group of commercial positive resists. Normalized film thickness is plotted against log exposure dose. (Reproduced with permission from L. F. Thompson and R. E. Kerwin [42].)

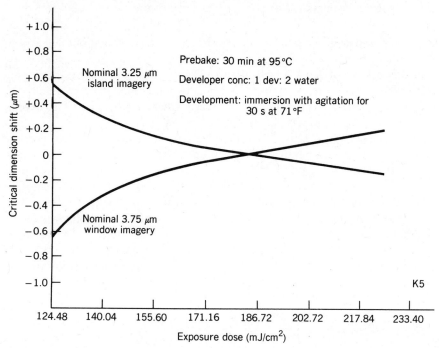

Figure 5-20 Effect of exposure dose on the deviation from the critical dimensions for lines (islands) and spaces (windows) in Kodak Micro Positive Resist 820. Prebake for 30 min at 95°C; immersion development with agitation for 30 s at 71°F. (Reproduced with the permission of KTI Chemicals Inc.)

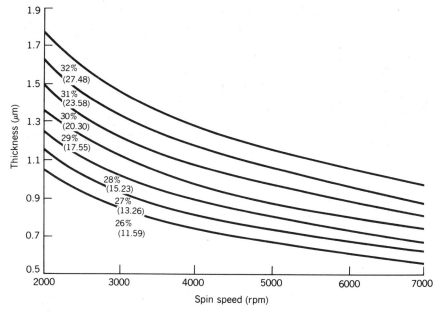

Figure 5-21 Effect of spin speed on final film thickness for a range of solid contents (in wt%) and solution viscosities (in cS). Kodak Micro Positive Resist 820. (Reproduced with the permission of KTI Chemicals Inc.)

203

developer and a given development time. Once these conditions have been established, the characteristic curve is obtained in the usual way, that is, by exposing a strip of resist film behind a photographic step wedge, developing the resist in the agreed manner, and finally measuring the remaining film thickness by interferometry or contact gauge. Characteristic curves for several commercial positive resists [42] are shown in Fig. 5-19.

Figures 5-20 to 5-22 give some practical characteristics of typical positive resists. An important parameter is the *work point*. Just as in negative resists, the work point corresponds to an exposure at which the positive resist reproduces exactly the critical dimensions of the mask. The effect of exposure dose on the critical dimensions for a typical positive resist is illustrated in Fig. 5-20. The working exposure for this system is found at the point where the curves for lines and for spaces (islands and windows) intersect. The *exposure latitude* is then defined as the variation in exposure, which keeps the difference between mask dimensions (ℓ') and resist image dimensions (ℓ) within a specified limit (expressed in percentage of the critical dimension).

$$\text{dimensional latitude} = \left| \frac{\ell - \ell'}{\ell'} \right| < 2.5\% \tag{5}$$

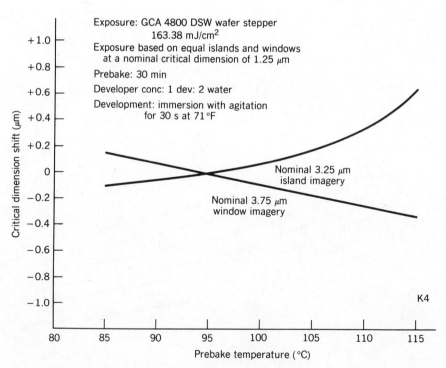

Figure 5-22 Effect of prebake temperature on deviations from mask critical dimension for Kodak Micro Positive Resist 820. Prebake time 30 min; immersion development with agitation, for 30 s at 71°F. (Reproduced with the permission of KTI Chemicals Inc.)

Information on the strictly photographic parameters of the resist is not sufficient to describe the system. During the patterning process the resist has to perform adequately in a number of sequential steps: spin coating, drying, preexposure bake, exposure, postexposure bake, development, hardbake of the resist image or radiative treatment, and finally pattern transfer by wet etching, plasma etching, or by ion implantation. For a complete specification, the characteristics of the resist in each of these steps have to be known.

The behavior of the resist during *spin coating* depends essentially on the viscosity of the coating solution as described in Chapter 2. The information is often presented in the form shown in Fig. 5-21, which indicates the relation between spin speed, solids content of the coating solution, and film thickness for the Kodak Micro Positive Resist 820 [43].

The effect of various thermal treatments and of the temperature of the developer are usually expressed in terms of their effect on the change in critical dimensions. Figure 5-22 is an example.

Resists are marketed in the form of ready-to-use coating solutions of specified solid content and viscosity. They are carefully filtered to comply with the requirements of the clean room, down to particle sizes of 0.5 μm and below. As an example, Table 5-2 gives the manufacturers specifications for some members of the popular AZ 1300 series of positive resists.

TABLE 5-2 Specifications for Members of the AZ 1300 Series of Resists [44]

Property	AZ 1350B/ 1350B-SF	AZ 1370/ 1370-SF	AZ 1350J/ 1350J-SF	AZ 1375
Solids content	16.5% ± 0.7	26.5% ± 1.0	30.5% ± 1.0	37.0% ± 1.5
Kinematic viscosity at 25°C	4.60 ± 0.30 cS	17.0 ± 1.0 cS	30.5 ± 1.5 cS	90 ± 6 cS
Absorptivitiy specification (L/g cm at 398 nm)	0.780 ± 0.040	1.15 ± 0.08	1.35 ± 0.10	1.68 ± 0.10
Specific gravity at 25°C	1.000 ± 0.005	1.025 ± 0.010	1.040 ± 0.005	1.075 ± 0.010
Water content	0.5% max	0.5% max	0.5% max	0.5% max
Principal solvent	2-Ethoxyethyl acetate	2-Ethoxyethyl acetate	2-Ethoxyethyl acetate	2-Ethoxyethyl acetate
Appearance	Clear amber red	Clear amber red	Clear amber red	Clear amber red
Coating Characteristics	Striation Free and Standard Formulations Available			Standard formulation only
Particle count (particles/in.^{2})	< 5	< 5	< 5	< 5
Filterability constant	0.010 max	0.010 max	0.010 max	None
Filtration	0.2 μm abs	0.2 μ abs	0.2 μ abs	0.5 μm nominal

PHENOMENOLOGICAL DESCRIPTION OF IMAGE FORMATION

The generalized treatment on similarity principles, which was so successful with negative resists, is not applicable to positive resists because of the large number of intervening parameters. This makes a phenomenological description of the positive resist process both more difficult and also more desirable from a practical point of view.

The Dill Equations. Dill and his co-workers at IBM [45, 46] have pioneered the mathematical description of the positive resist process. They characterized the optical and the photochemical properties of resist films and used this information to predict the final shape of the resist image [47, 48]. Their analysis is based on the assumption that the rate of decomposition of diazoquinone at any location in the resist film is proportional to the radiation intensity prevailing at that location at the given time. Since only the radiation absorbed by the diazoquinone advances the photoreaction, the changes in the absorbance of the inhibitor and the changes in resist composition are linked. These changes may be described by a set of two simultaneous differential equations, namely,

$$\frac{\partial M(z,t)}{\partial t} = -I(z,t)M(z,t)C \tag{6}$$

and

$$\frac{\partial I(z,t)}{\partial z} = -I(z,t)[AM(z,t) + B] \tag{7}$$

Here $M(z, t)$ is the fraction of inhibitor remaining at depth z in the film at time t, and $I(z, t)$ is the radiation intensity at that location and time. The square bracket in Eq. (7) contains the absorbance of the resist that is made up of two terms: The time-dependent absorbance of the diazoquinone $AM(z, t)$, and the time-independent absorbance B of the base resin. The rate of destruction of the diazoquinone inhibitor under unit radiation intensity is constant and is measured by the dimensionless parameter C, related to the quantum yield of the photoreaction.

The evolution of the absorptive behavior of the film as well as the evolution of local composition, that is, $M(z, t)$, are obtained as the simultaneous solutions of Eqs. (6) and (7), subject to the initial conditions

$$M(z,0) = 1$$

$$I(z,0) = I_0 \exp[-(A + B)z]$$

and the boundary conditions

$$I(0, t) = I_0$$

$$M(0, t) = \exp(-I_0 C t)$$

Although analytical solutions of Eqs. (6) and (7) exist [49], the integration is usually carried out by numerical methods.[1]

The three parameters A, B, and C can be found from the dependence of the transmission of the resist film on the exposure dose. To achieve this, the internal transmission (T) of the film is expressed in the general form

$$T(t) = \exp\left[-\int_0^d \{AM(z, t) + B\} \, dz\right] \qquad (8)$$

which for $t = 0$ and $t = \infty$ reduces to the following:

$$T(0) = \exp[-(A + B)d] \qquad (9)$$

$$T(\infty) = \exp(-Bd) \qquad (10)$$

and the initial slope of the transmission curve is given by the expression

$$\frac{dT}{dt}(0) = T(0)[1 - T(0)]\frac{AI_0 C}{A + B} \qquad (11)$$

Thus the parameters A, B, C may be derived from the initial and the final value of T and from its derivative at time, $t = 0$ (see Fig. 5-23 for an example of such a transmission versus time curve taken from the work of Dill et al. [46]).

Once the optical parameters of the resist are known, the dependence of the dissolution rate on the inhibitor concentration, the so-called development rate curve, can be derived from experimental data. After a film has received a certain exposure, the distribution of inhibitor in the coating, $M(z)$, is fixed and can be calculated from the integrals of Eqs. (6) and (7). At the same time the rate of dissolution at depth z can be derived from an experimental thickness versus time curve. From these two relations the dependence of

---

[1] The analytical methods developed in particular by Babu and his associates (see, for example, Reference 49) provide insight into the physics of the imaging process. However, near a fabrication line expediency is the primary requirement and for this reason computer simulation is in general use.

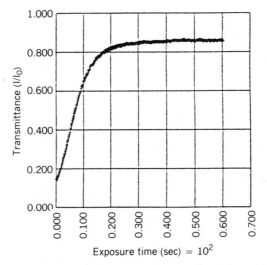

Figure 5-23 Transmission versus exposure time curve for a film of the positive resist AZ 1350 J coated on a glass of matched index of refraction. (Reproduced with permission from F. H. Dill et al. [46]. © IEEE, 1975.)

dissolution rate R on inhibitor concentration (M) can be derived. Figure 5-24 shows a typical example.

If the function $R(M)$ is to be incorporated into other calculations, it has to be presented in analytic form. To do this, Dill and his co-workers have used a polynomial of the form shown in Eq. (12) [46].

$$R = \exp\left(E_1 + E_2 M + E_3 M^2\right) \tag{12}$$

Other expressions relating dissolution rate and inhibitor concentration have been proposed from time to time; see, for example, Ohfuji et al. [50].

Equations (6) to (8), the optical parameters A and B, the photochemical parameter C, and the dissolution rate parameters E_1, E_2, and E_3 characterize the resist system sufficiently for a complete description of its lithographic behavior. Using this method, the depth profile of the resist image at the end of the patterning process can be predicted rather successfully, as shown in Fig. 5-25. The procedure is described in several papers of the Dill group [46, 48] and *computer programs* for its execution have been written, such as the SAMPLE program developed by Neureuther and co-workers [51]. A related program called PROSIM is used in conjunction with the Perkin-Elmer Dissolution Rate Monitor, while a differently conceived simulator, PROLITH, was created by Mack [52]. These programs are now in general use in the semiconductor industry. More up-to-date information on this subject can be

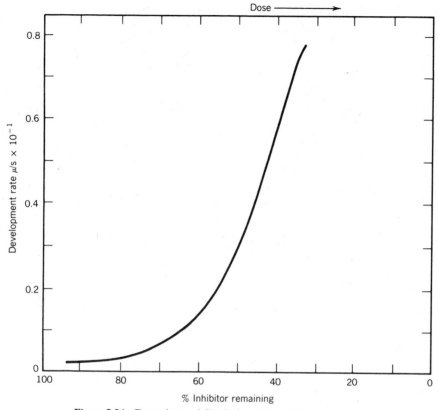

Figure 5-24 Dependence of dissolution rate on inhibitor content.

found in the work of Neureuther and others, for example in References [53–55].

Imaging with positive resists depends on a large number of factors, some of which are hard to control and all of which play a role in determining the final outcome. As a result, the optimization of a microlithographic system is an exercise in multivariant analysis and in that respect too, microlithographic simulator programs have become important [56]. For a given material the optimum exposure and processing conditions correspond to a more or less flat plateau (maximum) on a multidimensional response surface. *Response surface modeling* is a statistical technique by which experimental strategy and data analysis combine to generate a parametric model of the process. If the effect of a single variable can be isolated experimentally or predicted from first principles, the simulator program will help to locate the process window on the multidimensional surface. For example, Fig. 5-26 shows the response surface of a particular resist–developer pair for just two variables, namely, developer temperature and developer concentration. The curves shown are contours of

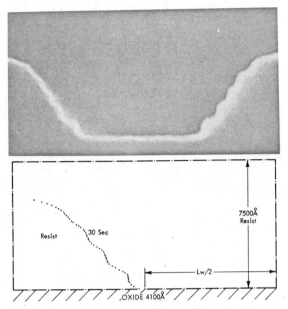

Figure 5-25 Resist profile predicted by computer simulation (below) and observed experimentally (above). (Reproduced with permission from M. A. Narasimham [47]. © IEEE, 1975.)

desirability (1.0 being the highest value), and it can be seen from the contour map that it is more important to control developer strength than it is to control developer temperature.

It would be impractical to explore by trial and error the whole range of parameters for every new system. Here the informed use of a simulator program can take the place of experimentation. In the end, the computer

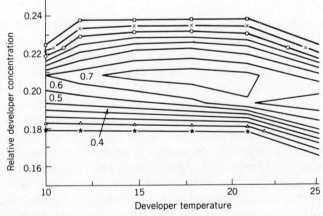

Figure 5-26 Response of resist performance (in terms of *desirability curves*) to changes in developer temperature and developer concentration. Optimum desirability is 1 as defined in Reference 56. (After D. A. Badami and S. E. Knight [56].)

predictions have still to be validated by experiment, but the number of experiments required to do this is substantially reduced.

MOLECULAR MECHANISM OF NOVOLAC-DIAZOQUINONE RESISTS

Although the novolac-diazoquinone resists are at present the most important patterning materials of the semiconductor industry, the molecular mechanism of these systems has only been partly elucidated. The present discussion is concerned first with the mechanism of novolac dissolution, second with the effect of the dissolution inhibitor, and finally with the role of indenecarboxylic acid, which is the product of inhibitor photolysis.

The Mechanism of Novolac Dissolution

The dissolution of high molecular weight polymers was investigated in the 1960s by Ueberreiter and Asmussen [57, 58], who found that dissolution occurs in two stages: In the first stage solvent penetrates into the glassy polymer and forms a gel layer separating the polymer matrix from the pure solvent. In the second stage the polymer coils disentangle from the gel and eventually diffuse into the liquid. Three phases are present in the steady state of the dissolution process:

glassy polymer|swollen gel|solvent or solution

In the systems investigated by Ueberreiter, the diffusion of the solvent across the gel layer is rate determining [59], and in those cases solvent uptake and the inward progress of the glass-gel interface depend on the square root of time (Fickian behavior). In other systems, the events at the glass-gel interface are rate determining and solvent uptake and interface movement are linear functions of time. Alfrey et al. [60] have termed this *case II dissolution* or polymer relaxation controlled mass transfer [61, 62]. Figure 5-27 shows schematic time-concentration profiles for the two dissolution modes.

Except for a brief initial phase, the dissolution of novolac proceeds with a constant rate and follows, therefore, case II mass transfer kinetics. In the low molecular weight novolac the width of the gel layer will be of the order of 100 Å and would be difficult to observe, but a gel layer or penetration zone must exist at the interface. Arcus has demonstrated its existence in a high molecular weight phenolic resin [30]. His observations are summarized in the laser interferogram trace of Fig. 5-28 which shows three separate sets of reflections, corresponding, respectively, to the interface between the polymer matrix and the silicon wafer, the interface between the polymer matrix and the swollen gel, and the interface between the gel layer and the developer solution.

212 POSITIVE RESISTS BASED ON DIAZONAPHTHOQUINONES

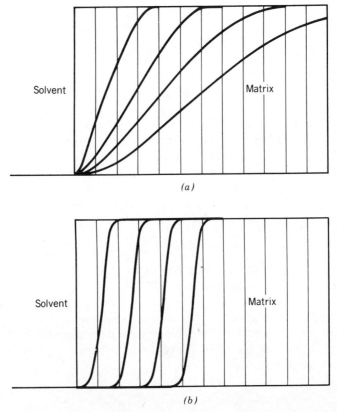

Figure 5-27 Concentration–time profiles (*a*) for Fickian diffusion and (*b*) for Case II mass transfer.

The Empirical Facts of Novolac Dissolution. In the case II dissolution regime the rate determining events occur at the polymer gel interface. The mechanism of the process is a description of these events in molecular terms. The empirical observations for which a satisfactory mechanism will have to account are the following.

A minimum concentration of base is required to make the resin dissolve at a measurable rate, and this minimum concentration varies from one resin to another. Figure 5-29 shows the dependence of dissolution rate on base concentration for several para-substituted phenol novolac resins [63].

The existence of a critical base concentration is linked to the fact that the deprotonation of phenol to phenolate ion must occur at some stage in the dissolution process.

$$\text{Ph-OH} + \text{OH}^- \rightleftharpoons \text{Ph-O}^- + H_2O \qquad (13)$$

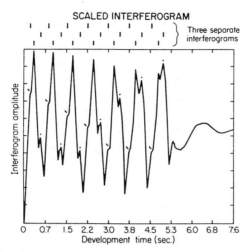

Figure 5-28 Laser interferogram taken during the dissolution of a phenolic base resin of higher molecular weight. The three sets of interferences correspond to reflections from the interfaces between solution and gel layer, gel layer and solid polymer, and polymer and glass. (After R. A. Arcus [30].)

Above that critical concentration the rate of dissolution depends on the base concentration according to a scaling law of the form shown in Eq. (14)

$$R = kc^n \tag{14}$$

Hinsberg and Guttierez [64] found that the dissolution rate is usually increased if a neutral salt of the base cation is added to the developer. In the systems investigated by these authors, the dissolution rate is a linear function of the concentration of base cation, but a strongly supralinear function of the concentration of OH^- ions. For example, the dissolution of a typical unexposed novolac resist in a sodium hydroxide developer can be described by the empirical equation shown below [64].

$$R = 1.3 \times 10^5 [Na^+]^1 [OH^-]^{3.7} \tag{15}$$

In general, the rate of dissolution can be expressed as a function of the concentrations of base cation and of OH^- ions separately in the form shown in Eq. (16)

$$R = k[\text{cation}]^{n^+} [OH^-]^{n^-} \tag{16}$$

The exponents n^+, n^- are here formal reaction orders, describing the effect on the dissolution rate of the concentrations of individual developer components.

The observations of Hinsberg and Guttierez [64] have been confirmed also for other positive resists [65]. It should be noted at this point that although the

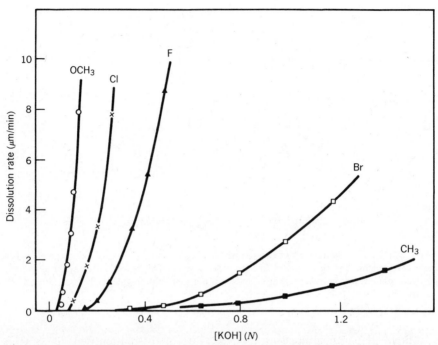

Figure 5-29 Dissolution rate plotted as a function of KOH concentration (mole/L) for a group of ortho–ortho-connected novolacs substituted in the para position of the phenol moiety, as indicated. (After J. P. Huang and T. K. Kwei [65].)

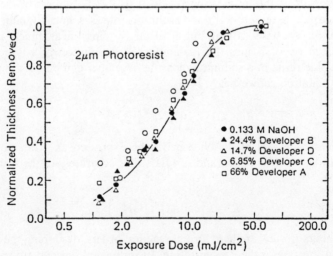

Figure 5-30 Characteristic curves of a novolac–diazoquinone resist developed in four different commercial developers and translated along the exposure dose axis into coincidence. (Reproduced with permission from W. D. Hinsberg and M. L. Guttierez [64].)

TABLE 5-3 Effect of Different Ions on the Dissolution Rate of a Resist [64]

Developer	Dissolution Rate ($\text{Å} \cdot \text{s}^{-1}$)		Ion	Relative Effect in Unexposed Resist
	Unexposed	Exposed		
0.15 M LiOH	21	1380	Li^+	1.00
NaOH	21	1400	Na^+	1.00
KOH	10	860	K^+	0.48
CsOH	1	30	Cs^+	0.05
Me_4NOH	≪ 1	< 1	Me_4N^+	≪ 0.05
0.15 M NaOH	18 (17)[a]	1450	OH^-	1.00[b]
+ 0.1 M Na_2SiO_3	221 (269)	3400	SiO_3^{2-}	0.95
+ Na_3PO_4	135 (346)	2800	PO_4^{3-}	0.78
+ Na_2CO_3	79 (269)	2391	CO_3^{2-}	0.72

[a] Calculated from Eq. (13) for a developer containing OH^- instead of the buffer anions.
[b] Calculated as an efficiency from the ratio [rate with buffer ion/rate with OH] = [efficiency]$^{3.7}$

rate of dissolution changes from one developer to another, the shape of the exposure curve, which is the principal photographic characteristic of the resist, is primarily a property of the polymeric material. This follows from the observation that the characteristic curves of a resist processed in different developers can be brought into superposition with each other by translation along the exposure axis. Fig. 5-30 is an example. This also means that the mechanism of dissolution in all developers of the same type is essentially the same.

Another clue to the nature of the dissolution process may be the observation that in many resist systems the "size" of the cation appears to influence the dissolution rate. This effect was first reported by Hinsberg and Guttierez [64]. Some of their data are assembled in Table 5-3. The effect is shown graphically in Fig. 5-31 which is taken from the work of Huang and Kwei [63].

Mechanistic Models. The empirical observations taken together suggest that the process of novolac dissolution must contain the following steps.

1. The penetration of water and of OH^- into the glassy matrix. This leads to the formation of a narrow penetration zone which is the equivalent of the gel layer observed with high molecular weight polymers.
2. Deprotonation of phenol to phenolate, as shown in Eq. (13).
3. The aquation of phenolate ion.

$$\text{Ph-O}^- + n\,H_2O \longrightarrow \text{Ph-O}^-(H_2O)_n \qquad (17)$$

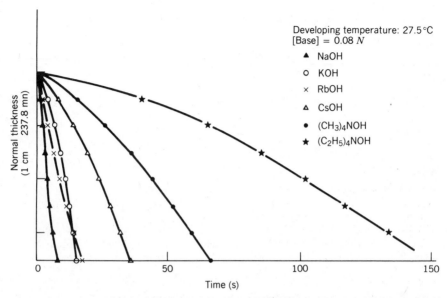

Figure 5-31 Effect of the cation of the developer base. Dissolution curves (resist thickness vs time in developer) of an experimental phenol–novolac resin in 0.08 M solutions of the hydroxides indicated in the figure. (After J. P. Huang and T. K. Kwei [65].)

4. Compensation of the negative charge of the phenolate ion by the base cation.

5. Transfer of the hydrated novolac macromolecule into the developer solution.

In an effort to link these steps into a coherent mechanism Arcus [30] has proposed a model where the interface between resin and developer is viewed as a partially semipermeable membrane which may "differentiate between the ions of aqueous developers due to variations in size, composition and charge... the membrane properties can be modified by chemical treatments, changes in concentration... and most importantly by the photochemistry of the included naphthoquinonediazide." [30]

The *membrane model* covers a great many aspects of the behavior of novolac resins. The membrane would let OH^- ions pass and slow down the diffusion of larger cations. This would make cation transport the rate determining step and explain the increase in dissolution rate by the addition of neutral salts. It also provides a qualitative interpretation of the cation size effect, although in a highly aquated membrane diffusion rates would be expected to correlate with the size of the aquated cations, yet it is the radius of the unhydrated ions derived from crystallographic data which appears to matter (see the data in Table 5-4).

TABLE 5-4 Properties of Hydrated Alkali Ions [66]

	Li^+	Na^+	K^+	Rb^+	Cs^+
Ionic radii (Å) in crystals	0.68	0.98	1.33	1.48	1.67
Radii of hydrated ions	3.40	2.76	2.32	2.28	2.28

The Arcus model implies that the overall rate of the dissolution process is controlled by the rate of diffusion of developer components into the matrix. In contrast to this view Garza et al. [67] have recently developed a model where, at least for some resins, the deprotonation of the phenol groups of the novolac can be the rate determining step. It is the merit of this model that it draws attention to the important role in the dissolution process of the secondary structure of the resin.

The *secondary structure model* is based on earlier work of Templeton et al. [68] who distinguish broadly speaking between structures where intermolecular bonds between novolac molecules predominate and those with predominantly intramolecular hydrogen bonds, (see Fig. 5-32). For resins with the latter structure the Garza model offers an elegant interpretation of the cation size effect [66].

For example, in paracresol-based novolacs (see Fig. 5-32) the secondary structure of the resin brings the OH-groups of the phenols together to such an extent that they are linked by a series of intramolecular hydrogen bonds. When the first phenol group of the ensemble is deprotonated, the negative charge is delocalized via the mobile hydrogens over the whole "nest" of participating phenols.

A small cation arriving at this location will form a tight ion pair with the phenolate and bind the charge to a single center. If, however, a large cation is used, it will form a loose ion pair and part of the negative charge of phenolate will remain delocalized over the other phenols of the ensemble and will make the second deprotonation step within the group much more difficult. This will slow down the deprotonation of the resin and make the chemical reaction the rate determining step in the dissolution process.

The secondary structure model is very successful in interpreting the large differences in dissolution rate between resins based on different cresol isomers, and it has a natural explanation for the cation size effect in paracresol-based novolacs. It has difficulties, however, with the known dependence of the dissolution process on the thermal history of the resin and its free volume status. That is not easily understood in a system controlled by a chemical reaction. Also, if the deprotonation reaction is rate limiting, the high exponent of [OH] in the rate equation is a true kinetic order and implies that the reaction "...involves the *simultaneous* extraction of 3 to 10 protons (from the novolac macromolecule)" [67]. This does not seem a very likely event.

Some of the difficulties of the Garza model can be removed if it is assumed that the rate determining event in novolac dissolution is the last step, namely the detachment of the hydrated novolac molecule from the matrix. A critical degree of deprotonation must be reached before this can occur [69], and in that case the deprotonation equilibrium between phenol and developer base controls the concentration of the deprotonated species P_{n-1} which is just one deprotonation step away from dissolution.

$$K = \frac{[P_{n-1}]}{[\text{NaOH}]^{n-1}} \qquad (18)$$

The dissolution rate is then determined by the kinetics of the last step.

$$R = k_d[P_{n-1}][\text{NaOH}] = k_d K[\text{NaOH}]^n \qquad (19)$$

The Inhibition Effect. The inhibition effect and the effect of the photogenerated acid are even less well understood than resist dissolution. It can be said in general that dissolution inhibition is not specific to diazoquinones, but that any "hydrophobic" component incorporated into the matrix will to some degree slow down dissolution. The only condition is that the inhibitor should be compatible with the resin, that means form a single phase with it.

Although the presence of the inhibitor will affect the chemical potential of all other components of the phase, it is unlikely that bulk thermodynamics play an important role in determining the rate of the dissolution process.

The indenecarboxylic acids photogenerated in the exposed areas of the resist have a lower pK_a than phenol and will therefore be ionized together with the phenol groups or at even lower concentrations of OH^- ions. It was thought earlier that the acid groups simply replace the function of the phenols

TABLE 5-5 Dissolution Rates of Isomeric Novolac Resins [28]

Resin	M_w	Rate ($\text{Å} \cdot \text{s}^{-1}$)
ortho-Cresol resin	2.100	2.7
meta-Cresol resin	15.000	0.7
para-Cresol resin	1.600	3.0
Poly(4-hydroxy styrene)	21.000	1800

and restore, after exposure, the dissolution rate of the uninhibited resin [3]. Recent work by Blum et al. [4], however, indicates that the acid may act as an effective dissolution promoter in the exposed areas of the resist.

Effects of Resin and Inhibitor Structure on Dissolution Rate. The rate of resist dissolution is strongly affected not only by a host of processing conditions, but also by structural properties of the novolac resin and of the inhibitor. The effect of the structure of the dominant cresol isomer in the resist on the dissolution rate of the resin is illustrated in Table 5-5.

Hanabata and co-workers [32, 33] have explored the effect of the cresol isomer and of the position of the methylene link on dissolution rate and some of their results are shown in Fig. 5-11. Analyses such as these have made it possible to optimize resin composition for specific lithographic applications, but they have not offered an explanation as to the mechanism by which resin structure may intervene in the process. The fact that even quite small structural differences can bring about large changes in dissolution rate and in the lithographic behavior of the materials must be connected to a specific dissolution mechanism.

Templeton et al. [68] have recently shown that the steric arrangement of the methylene links in the resin can have a profound effect on dissolution rate and on lithographic performance. These authors have calculated the equilibrium secondary structures of cresol–formaldehyde oligomers using molecular mechanics, backed in some instances with x-ray crystallography [70]. The secondary structure of these molecules determines the relative positions of the hydroxyl groups in the novolac matrix, and hence the possibility of intramolecular hydrogen bonding. Figure 5-32 gives two examples. In the model of the ortho–ortho'-coupled *p*-cresol tetramer all four hydroxyl groups are within reach of each other and intramolecular hydrogen bonding is strongly favored. By contrast, in the secondary structure of the ortho–para'-coupled *m*-cresol–novolac the hydroxyl groups are widely separated and located on the periph-

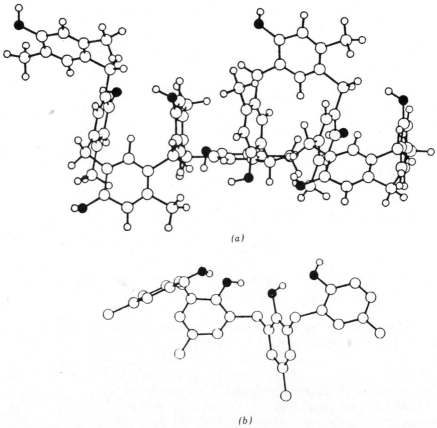

Figure 5-32 (*a*) Secondary structure of an ortho–para coupled *m*-cresol novolac as calculated by molecular mechanics. (*b*) Secondary structure of a *p*-cresol novolac tetramer as determined by x-ray crystallography. (Reproduced with permission from M. K. Templeton et al. [68] and E. Paulus and V. Bohmer [69].)

ery of the molecule. This rules out intramolecular bonding and favors intermolecular network formation. The same applies to poly(hydroxystyrene). The balance between intra- and intermolecular hydrogen bonding affects also the interaction of the resin with the inhibitor, in other words the solubility of the diazoquinone in the novolac.

The effect of the structure of the inhibitor on the dissolution rate of the resist and on its lithographic performance has recently been investigated by Trefonas and Daniels [71], who studied the effect of the number of diazoquinone units attached to a polyfunctional inhibitor. For example, with benzophenone as the central component (balast), six derivatives can be prepared by gradually substituting all six hydroxyl groups of hexahydroxybenzophenone by diazoquinone moieties. The ratio (q) of active groups to ballast can thus assume values between 1 and 6. The commonly used inhibitor is the trisubstituted benzophenone shown here.

DQO—⬡—C(=O)—⬡
DQO   OQD

where DQ is the moiety

—O—SO$_2$—[naphthoquinone diazide]

The DQ groups in the resist absorb light independently and decompose independently, so that in a partially exposed resist film there will be a mixture

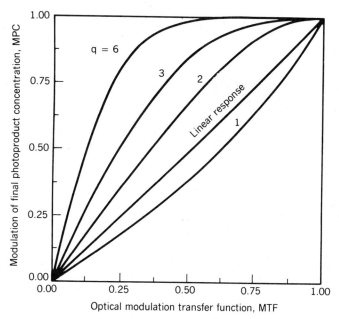

Figure 5-33 Modulation of final photoproduct concentration plotted against the optical MTF at a fixed exposure dose of 89 mJ/cm^2 for a series of proprietary resist compositions (see text). (Reproduced with permission from P. Trefonas and B. K. Daniels [70].)

of photoproducts. It is assumed that the rate of dissolution is the sum of the dissolution rates (r_i) associated with each photoproduct, weighted by the content (m_i) of the product (i) in the partially exposed resist.

$$R = \sum_i m_i r_i \qquad (20)$$

The surprising result of this analysis is the finding that in certain particularly selective novolac resists with a tri-functional inhibitor ($q = 3$), the contribution to the dissolution rate by the fully (i.e., triply) converted inhibitor

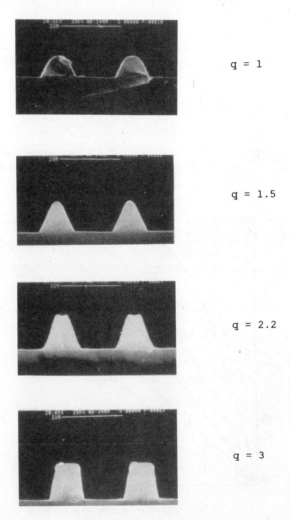

Figure 5-34 Effect of inhibitor functionality on the dimensions and the wall slope of 1.25 μm features in a series of proprietary resist compositions of increasing inhibitor functionality. (Reproduced with permission from P. Trefonas and B. K. Daniels [70].)

is much larger than the contribution of the other components. For example, in the case cited in Reference 71, the rate contribution (r_0) of the fully converted inhibitor ($q = 0$) and those of the others, (r_1, r_2, r_3) are given in the following list.

$$r_0 = 1690 \ (\text{Å} \cdot \text{s}^{-1})$$
$$r_1 = 24$$
$$r_2 = < 1$$
$$r_3 = 0$$

This means that the dissolution of the resist is controlled almost exclusively by the fully exposed (triply converted) component, the concentration of which becomes important in the later stages of irradiation. Dissolution of the resist will therefore be slow in the beginning, but will grow quite suddenly when the third component starts to dominate the composition of the photoproduct. The resist will therefore have a sharp exposure threshold and high photographic contrast. That is indeed what Trefonas and Daniels have observed [71]. Figure 5-33 shows the ratio of the modulation transfer function of the photoproduct image to the optical transfer function (see Chapter 6) for resists of functionality ratios from 1 to 6.

The very real effect of inhibitor functionality on resist performance is demonstrated in Fig. 5-34, which shows scanning electron micrographs of 1.25 µm wide features, reproduced in resists of increasing functionality. The increase in image quality, dimensional accuracy, and wall slope in this series is remarkable.

While the practical performance of positive photoresists is being steadily developed on an empirical basis, the functional mechanism of the system is still not fully understood. The studies reported here throw at least some light on the fundamentals of the process.

REFERENCES

1. D. Meyerhofer, *IEEE Trans. Electron. Devices*, **ED-27**, 921 (1980).
2. A. C. Ouano, *Polym. Eng. Sci.*, **18**, 306 (1978); also in *Polymers for Electronics* (T. Davidson, ed.), pp. 79–90. Am. Chem. Soc., Washington, D.C., 1984.
3. W. D. Hinsberg, C. G. Willson, and K. K. Kanazawa, *Proc. SPIE*, **539**, 6 (1985).
4. L. Blum, M. E. Perkins, and A. W. McCullough, *Proc. SPIE*, **771**, 148 (1987).
5. G. Koegel and H. Neuenhaus, Ger. Patent 376,385 (1917).
6. G. Koegel and H. Neuenhaus, *Z. Wiss. Photogr.*, **24**, 171 (1926).
7. H. Zollinger, *Diazo and Azo Compounds*, Interscience, New York, 1961.
8. J. Kosar, *Light Sensitive Systems*, Wiley, New York, 1965.
9. Kalle AG, Ger. Patent 879,205 (1949).
10. Kalle AG, Ger. Patents 865,109 (1956) to 894,959 (1961).

11. Azoplate Corp., U.S. Patents 2,766,118 (1956) to 3,046,122 (1962).
12. O. Suess, *Annalen*, **556**, 65 (1944).
13. O. Suess, *Annalen*, **557**, 237 (1947).
14. J. Pacansky and D. Johnson, *J. Electrochem. Soc.*, **124**, 862 (1977).
15. A. D. Erlikh, N. P. Protsenko, L. N. Kurovskaja, and G. N. Rodionova, *Zh. Vses. Khim. Ova.*, **20**, 593 (1975).
16. J. Pacansky and J. R. Lyerla, *IBM J. Res. Dev.*, **23**, 42 (1979).
17. T. Itoh, Y. Yamashita, R. Kawazu, K. Kawamura, S. Ohno, T. Asano, K. Kobayashi, and G. Nagamatsu, *Polym. Eng. Sci.*, **26**, 1105 (1986).
18. S. A. MacDonald, R. D. Miller, C. G. Willson, G. M. Feinberg, R. T. Gleason, R. M. Halverson, M. W. McIntire, and M. T. Motsiff, *Kodak Microelectron. Semin. Interface '82, San Diego, CA, 1982.*
19. H. Moritz and G. Paal, U.S. Patent 4,104.070 (1978).
20. C. G. Willson, in *Introduction to Microlithography* (L. F. Thompson, C. G. Willson, and M. J. Bowden, Eds.), *ACS Symp. Ser.*, **219**, 117 (1984).
21. M. L. Long, *Proc. SPIE*, **469**, 189 (1984).
22. R. M. R. Gijsen, H. J. J. Kroon, F. A. Vollenbroek, and R. Vervoordeldonk, *Proc. SPIE*, **631**, 108 (1986).
23. W. E. Feely, J. C. Imhof, and C. M. Stein, *Polym. Eng. Sci.*, **26**, 1101 (1986); W. E. Feely, *Proc. SPIE*, **631**, 48 (1986).
24. D. W. Johnston, *Proc. SPIE*, **469**, 72 (1984).
25. A. Stein, *The Chemistry and Technology of Positive Resists*, A Waycoat Tutorial, Philip A. Hunt Chemical Co., Palisades Park, NJ 07650, 1980.
26. C. G. Willson, R. Miller, D. McKean, N. Clecak, T. Tompkins, D. Hofer, J. Michl, and J. Downing, *Polym. Sci. Eng.*, **23**, 1004 (1983).
27. R. D. Miller, D. R. McKean, T. L. Tompkins, N. Clecak, C. G. Willson, J. Michl, and J. Downing, *ACS Symp. Ser.*, **242**, 1982 (1984).
28. T. R. Pampalone, *Solid State Technol.*, **27**(6), 115 (1984).
29. J. M. Shaw and M. Hatzakis, *IEEE Trans. Electron Devices*, **ED-25**, 425 (1978).
30. R. A. Arcus, *Proc. SPIE*, **631**, 124 (1986).
31. M. Hanabata, Y. Uetani, and A. Furuta, *Proc. SPIE*, **920**(43) (1988).
32. A. Furuta, M. Hanabata, and Y. Uemura, *J. Vac. Sci. Technol.*, B, **4**(1), 430 (1986).
33. M. Hanabata, A. Furuta, and Y. Uemura, *Proc. SPIE*, **631**, 76 (1986).
34. W. H. L. Ma, *Proc. SPIE*, **333**, 19 (1982).
35. P. A. Ruggiero, *Solid State Technol.*, **27**(3), 165 (1984).
36. A. Zampini, P. Trefonas, B. K. Daniels, and M. J. Eller, *Proc. SPIE*, **920**(28) (1988).
37. K. L. Konnerth and F. H. Dill, *IEEE Trans. Electron Devices*, **ED-22**, 453 (1975).
38. K. L. Konnerth and F. H. Dill, *Solid-State Electron.*, **15**, 371 (1972).
39. F. Rodriguez, P. D. Krasicky, and R. J. Groele, *Solid State Technol.*, **28**(5), 125 (1985).
40. D. S. Soong, *Proc. SPIE*, **539**, 2 (1985).

41. W. W. Flack, J. S. Papanu, D. W. Hess, D. S. Soong, and A. T. Bell, *J. Electrochem. Soc.*, **131**, 2200 (1984).
42. L. F. Thompson and R. E. Kerwin, *Annu. Rev. Mater. Sci.*, **6**, 267 (1975).
43. "Kodak Micro Positive Resist Specifications," Kodak Publ. G-103, Eastman Kodak Co., Rochester, NY 14650, 1982.
44. "Information Bulletin," American Hoechst Corp., 3070 Highway 22 West, Somerville, NJ 08876, 1985.
45. F. H. Dill, *IEEE Trans. Electron Devices*, **ED-22**, 440 (1975).
46. F. H. Dill, W. P. Hornberger, P. S. Hauge, and J. M. Shaw, *IEEE Trans. Electron Devices*, **ED-22**, 445 (1975).
47. M. A. Narasimham, *IEEE Trans. Electron Devices*, **ED-22**, 478 (1975).
48. F. H. Dill, A. R. Neureuther, J. A. Tuttle, and E. J. Walker, *IEEE Trans. Electron Devices*, **ED-22**, 456 (1975).
49. S. V. Babu and V. Srinivasan, *Proc. SPIE*, **539**, 36 (1985).
50. T. Ohfuji, K. Yamanaka, and M. Sakamoto, *Proc. SPIE*, **920**(26) (1988).
51. W. G. Oldham, S. Nandgaonkar, A. R. Neureuther, and M. M. O'Toole, *IEEE Trans. Electron Devices*, **ED-26**, 717 (1979).
52. C. A. Mack, *Proc. SPIE*, **538**, 207 (1985).
53. A. R. Neureuther, *IEEE Trans. Electron Devices*, **ED-30**, 308 (1983).
54. D. J. Kim, W. G. Oldham, and A. R. Neureuther, *IEEE Trans. Electron Devices*, **ED-31**, 1730 (1984).
55. A. R. Neureuther and W. G. Oldham, *Solid State Technol.*, **28**(5), 139 (1985).
56. D. A. Badami and S. E. Knight, *Electron. Mater. Conf.*, Fort Collins, CO Pap. N-2 (1982).
57. K. Ueberreiter and F. Asmussen, *J. Polym. Sci.*, **57**, 187 (1962).
58. F. Amussen and K. Ueberreiter, *J. Polym. Sci.*, **57**, 199 (1962).
59. K. Ueberreiter, in *Diffusion in Polymers* (J. Crank and S. Park, Eds.). Academic, New York, 1968.
60. T. Alfrey, Jr., E. F. Gurnee, and W. O. Lloyd, *J. Polym. Sci, Part C*, **12**, 249 (1966).
61. G. C. Sarti, *Polymer*, **20**, 827 (1979).
62. N. L. Thomas and A. H. Windle, *Polymer*, **23**, 529 (1982).
63. J. P. Huang and T. K. Kwei, unpublished data, 1987.
64. W. D. Hinsberg and M. L. Guttierez, *Proc. SPIE*, **469**, 57 (1984).
65. M. J. Hanrahan and K. S. Hollis, *Proc. SPIE*, **771**, 128 (1987).
66. R. S. Berry, S. A. Rice, and J. Ross, *Physical Chemistry*, p. 422, Wiley, New York, 1980.
67. C. M. Garza, C. R. Szmanda, and R. L. Fischer, Jr., *Proc. SPIE*, **920**(41) (1988).
68. M. K. Templeton, C. R. Szmanda, and A. Zampini, *Proc. SPIE*, **771**, 136 (1987).
69. T. K. Kwei, private communication, 1988.
70. E. Paulus and V. Bohmer, *Makromol. Chem.*, **185**, 1921 (1984).
71. P. Trefonas, III and B. K. Daniels, *Proc. SPIE*, **771**, 194 (1987).

6 THE RUDIMENTS OF IMAGING SCIENCE

SENSITOMETRY OF IMAGING MATERIALS

Many of the basic ideas and concepts of imaging science were developed in silver halide photography [1]. For example, from the very beginnings the characterization of photographic sensitivity was a practical necessity. Hurter and Driffield [2] dealt with the problem of sensitometry in a systematic way. They devised a sector wheel that allowed the application of a range of exposures to different parts of a photographic plate in a single take. The plate was then developed and fixed and the degree of blackness (the optical density[1]) of the exposed parts of the plate was measured. Hurter and Driffield plotted developed optical density against the decadic logarithm of exposure and termed the resulting graph the *sensitometric or characteristic curve* of the material.

Figure 6-1 shows the sensitometric curve of a photographic negative. It has the shape of the response curve of a light detector: in the toe region (A–C) the illuminance (stimulus) is near the sensitivity threshold of the device and there is little response, in the shoulder region (B–D) the detector response gradually saturates and can no longer faithfully follow an increase in stimulus. Between threshold and shoulder there is an almost linear part of the curve where the

---

[1]Optical density has now been replaced by the term "absorbance," although optical density is still used sometimes in photographic texts. Absorbance (D) is defined as the logarithm of the ratio of incident to transmitted light intensity (or radiation intensity).

$$D = \log(I_0/I)$$

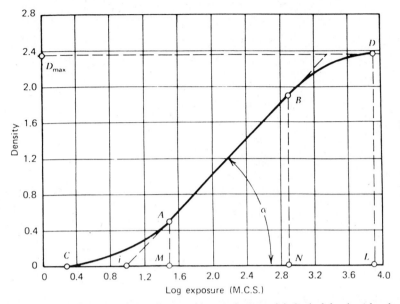

Figure 6-1 Sensitometric curve of a photographic negative material. Optical density (absorbance) is here plotted against log exposure (measured in meter-candle-seconds). Points A–C are the toe region of the curve, points B–D are the shoulder, and the slope, $tg\alpha$ of the linear part A–B of the curve, is called the contrast of the photographic material. (Reproduced with permission from B. H. Carroll et al. [1]. © Wiley, 1980.)

response is proportional to exposure. The gradient of this linear part, that is the tangents of the angle α, is called the *photographic contrast*. The origin (i) of the linear portion of the curve is a measure of photographic sensitivity and was so used in the early days of photographic science. Today, practical *sensitivity ratings* are defined differently. For example, the American Standards Association (ASA) rating of a photographic film is the exposure dose, in meter-candle-seconds, which produces an optical density of 0.1 under certain well-defined conditions of development [3]. (The ASA rating of the material in Fig. 6-1 is about 6, its contrast is $1.75/1.9 = 0.92$.)

Speed and contrast are the most important characteristics of an imaging material, but their significance and interpretation vary from system to system. In conventional silver halide photography, speed determines whether the material can be used in a camera or as a copying or printing material. In extending the concept of a sensitometric curve to polymeric imaging materials [4], the optical density has to be replaced by some other measure of "quantity of image," and that measure depends on the type of application considered. In microlithography the quantity of image is expressed in terms of the *normalized film thickness*, r/r_0, that is, the fraction of the original coating thickness remaining after exposure and development. The illuminance exposure of silver halide photography, which is uniquely linked to visible light, must be replaced

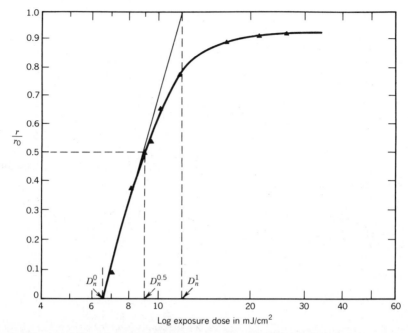

Figure 6-2 Sensitometric curve of a negative photoresist (Kodak KTFR). The term r/r_0 is the normalized film thickness after exposure and development, D is the exposure dose, the subscript n indicates the negative tone of the system, and the superscripts indicate the fraction of film thickness retained after development.

in polymer photography by some other measure of exposure dose, either *incident energy density*, for UV radiation and x-rays, or incident charge density for electrons and ions.

Figure 6-2 shows the sensitometric curve of a typical negative resist, where the normalized film thickness is plotted against radiation dose, millijoules per square centimeter. The curve is characterized by a sensitivity threshold (D_n^0) which is closely related to the gel point exposure mentioned in Chapter 2, and by an extrapolated saturation dose D_n^1, which corresponds to the intersection of the tangent to the curve with the $r/r_0 = 1$ line. The contrast, γ, of the material is defined as the slope of the initial tangent to the sensitometric curve.

$$\gamma = \frac{1}{\log D_n^1 - \log D_n^0} = \log \frac{D_n^0}{D_n^1} \tag{1}$$

The definition of contrast is the same for positive resists, but the sign is reversed to give in every case a positive value of contrast.

Proposals have been made [5, 6] to unify the presentation of the characteristic curves of resists, but none has been widely adopted. Most agree that the term "dose" should be used instead of the traditional photographic term

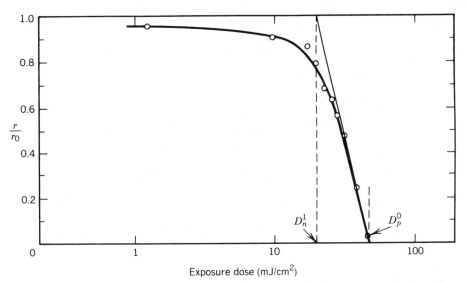

Figure 6-3 Sensitometric curve of a typical positive resist (AZ 2400). The exposure dose is D, the subscript p indicates the positive tone of the material, and the superscript indicates the fraction of film thickness retained after development.

"exposure," to indicate that we are not dealing exclusively with light. It is suggested that the "tone" (negative or positive) of the material be indicated by the subscript n or p, and that the normalized coating thickness retained after development be indicated by a superscript. These conventions are illustrated in Fig. 6-2 and in Fig. 6-3, which show the characteristic curves of typical negative and positive resist materials [7].

The choice of a *work point* on the characteristic curve that defines a satisfactory image exposure dose depends on the particular application envisaged. In graphic arts it will not be sufficient to expose a negative resist to the gel point D_n, but on the other hand it may not be necessary to retain all the original film thickness in the final image. A work point, say $D_n^{0.5}$, may be a practical compromise. In microlithography the working exposure is that for which the critical dimensions of the mask pattern are correctly reproduced. (The critical dimensions are the dimensions of the smallest opaque and clear features of a pattern.) The correct exposure is determined by monitoring the width of the lines and spaces in the resist image as a function of exposure and finding the exposure for which they coincide with the critical dimensions of the mask [4]. The procedure is indicated in Fig. 6-4.

In the case of positive resists, the work exposure is simply D_p, which is the exposure necessary to clear the resist completely from the substrate. The equality between the critical dimensions of the resist image and that of the mask is achieved by selecting development conditions that bring the two to coincidence for the exposure dose D_p^0.

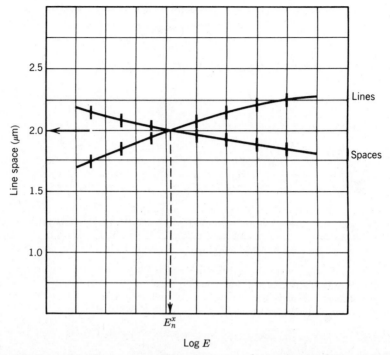

Figure 6-4 Determining the correct working exposure by matching the critical dimensions of the resist image (lines and spaces) with those of the mask. Nominal dimensions of mask features are here 2 μm.

The dependence of feature size on exposure determines *exposure latitude*, which is an important practical characteristic of the resist. It is defined as the variation in exposure that keeps the difference between mask dimensions and resist image dimensions within a specified limit (expressed as a fraction of the critical dimensions).

$$\text{dimensional latitude} = \left| \frac{\ell' - \ell}{\ell'} \right| < 0.05$$

Here ℓ is the linewidth of the resist image and ℓ' is the linewidth of the corresponding mask feature.

In polymer photography, the principal significance of photographic speed is that it sets the time that the coating or the device have to spend in the exposure tool. This codetermines the throughput of items (wafers, wire boards, etc.) through the fabrication line, or the preparation time of a printing master. The significance of contrast is that it determines image quality. In graphic arts applications image quality is measured by edge acuity, in microlithography it is gauged by the resolution that can be achieved with the material and by the depth profile of the resist image.

The Characteristics of Composite Systems

Without exception, practical imaging systems are composed of several subsystems and the final image produced by the composite is the result of a number of distinct sequential steps. This *imaging chain* is illustrated in the following diagram on the example of the conventional photographic process. In the making of a photograph, the three-dimensional scene is illuminated by light and, in the first step, the camera optics form a two-dimensional aerial image of the scene in the plane of the photographic film.

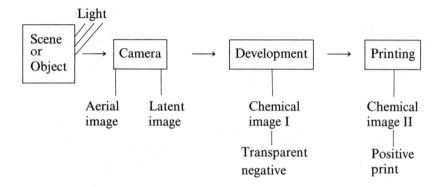

A latent image is formed in the photographic emulsion by exposure to the aerial image. Development with a reducing agent turns the latent image into a chemical image that is visible as the black silver image of the negative. Finally, the negative is transformed into a positive silver image by a sequence of operations summarized in the term "printing."

Each individual imaging step may be described by its characteristic curve, which links the output of the step to its input. For example, the optical density of the photographic negative is linked to the illuminance of the aerial image by the sensitometric curve of the negative material. The overall characteristic that describes the performance of the composite system is the product of the characteristics of the subsystems.

$$F = F_1 \cdot F_2 \cdot F_3 \cdot \cdots \tag{2}$$

If the characteristic curves of the subsystems, F_1, F_2, and so on, are known, the overall characteristic F can be derived by a graphic procedure introduced by Lloyd Jones [8]. In this method the characteristic curve of each subsystem is joined to that of the others in a way that makes the output of one imaging step the input of the next. The method is demonstrated in Fig. 6-5.

Three subsystems are considered: (1) the camera, (2) the negative film, and (3) the positive printing material. The characteristic curve of the camera is the "flare curve" in quadrant 1, which adds to the scene luminance (which is the signal of interest), the stray light scattered in the camera optics. The near

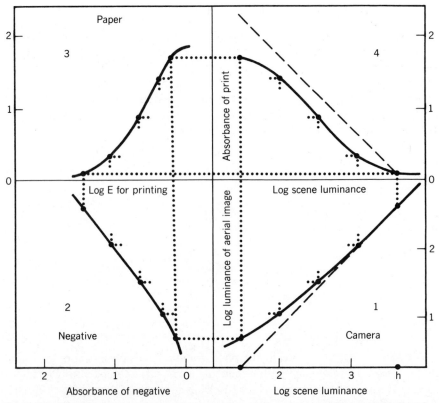

Figure 6-5 The Lloyd Jones diagram for a negative–positive photographic system (see text for detailed explanation). (After T. H. James, *The Theory of the Photographic Process*, 4th ed., Macmillan, 1977. Reproduced with permission.)

linear shape of the flare curve indicates that the camera transforms the luminance of the scene into the illuminance of the aerial image almost without distortion of tone. The illuminance distribution of the aerial image on the ordinate of quadrant 1 is then the input of the next exposure and development step. The sensitometric curve of the negative material in the adjoining quadrant 2 transforms illuminance into the optical density of the developed negative. The optical density of the developed negative in turn controls the image exposure of the printing operation, and the sensitometric curve of the photographic paper on which the print will be formed is placed in quadrant 3. Finally, the overall characteristic curve of the composite system is formed in quadrant 4. It determines the distribution of optical density that arises in the final print in response to the luminance distribution of the original scene.

Imaging systems involving photopolymers can be treated in a similar manner. For example, in the microlithographic patterning process of a negative resist with a projection printer, the image is transferred from an original

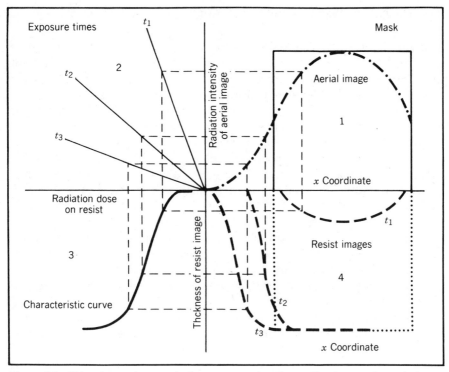

Figure 6-6 Using the quadrant method to construct the edge profile of a negative resist image produced with a projection printer. The nominal mask image and the aerial image of the mask projected onto the resist plane are in quadrant 1; the characteristic curve of the resist is in quadrant 3, the resist profile resulting from exposure and development is shown in quadrant 4 for three different exposure times.

mask via the printers optics onto the resist surface. Consider the edge of the mask pattern shown in Fig. 6-6. The corresponding aerial image projected by the optics onto the surface of a negative resist is shown as a function of position (x) in the broken curve in quadrant 1 of the figure. The aerial image controls the intensity of irradiation at the resist surface. Radiation intensity is converted into exposure dose by the exposure lines in quadrant 2, different lines corresponding to different exposure times.[2] Exposure of the resist and development results in a spatial distribution of retained film thickness. The two are linked by the characteristic curve of the resist shown in quadrant 3. By correlating retained thickness with position (x) in quadrant 4, profiles of the resist images can be constructed for the exposure doses t_1, t_2, and t_3.

---

[2] We ignore here the depth distribution of light intensity.

IMAGE DEGRADATION AND THE SPREAD FUNCTIONS

The final aim in imaging is the faithful reproduction of an original pattern, and the question arises to what extent truly faithful reproduction is achievable [9]. Consider the simplest image element, a point. Experience shows that it cannot be imaged without some distortion or degradation. In fact, in all real imaging systems the image of a point is more "spread out" than the original object and the performance of an imaging system may be described by a *point spread function*, as shown in Fig. 6-7 [3].

The physical causes of point spread vary from system to system. For example, even the performance of the most perfect optics is limited by light diffraction and a point will therefore be imaged in the form of a diffraction pattern, the so-called Airy disc. In photographic emulsions point spread is the result of light scattering by the silver halide grains, in diffusion transfer it is caused by the lateral diffusion of image dyes and other reactive components. In resist imaging, interference of monochromatic wave trains within the resist layer and resist swelling in the developer are the chief causes of image degradation; in electron beam lithography it is the emission and scattering of secondary electrons.

Since a line can be regarded as a row of points, a *line spread function*, $L(x)$, can be defined on the basis of the point spread functions $P(x, y)$ (see Fig. 6-8a [10]). The line spread function depends only on one spatial variable (x) and is closely related to the modulation transfer function, to be introduced

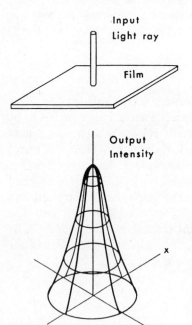

Figure 6-7 Representation of the point spread function of a given imaging process. (Reproduced with permission from T. H. James, *The Theory of the Photographic Process*, 4th ed., Macmillan, 1977.)

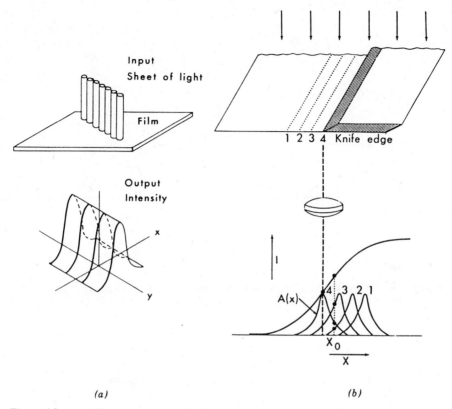

Figure 6-8 (*a*) Schematic representation of a line spread function; (*b*) the knife edge as a superposition of line spread functions. (Reproduced with permission from T. H. James, *The Theory of the Photographic Process*, 4th ed., Macmillan, 1977.)

shortly. Experimentally, the line spread function of an imaging system may be derived by imaging a knife edge and monitoring the image intensity in the vicinity of the edge coordinate. Since an edge may be pictured as a superposition of line spread functions (see Fig. 6-8*b*), the intensity distribution, $I(x)$, in the x coordinate of the image may be expressed in the form of the integral

$$I(x) = \int_{\text{near edge}} L(x)\, dx \qquad (3)$$

and the line spread function $L(x)$ may be determined as the derivative of the intensity distribution at the knife edge.

$$L(x) = \frac{dI(x)}{dx} \qquad (4)$$

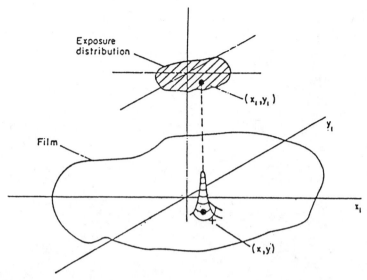

Figure 6-9 Imaging a distribution of luminance (exposure) in terms of point spread functions. (Reproduced with permission from J. C. Dainty and R. Shaw [9]. © Academic Press, 1977.)

No imaging system is free of the limitations defined by the spread functions. Any object, viewed as an ensemble of points, will always be imaged as an ensemble of point spread functions, as indicated in Fig. 6-9. Image degradation is not an attribute of poor imaging systems, but a fundamental reality of imaging. The harmonic analysis of images offers a systematic method of dealing with this situation [9].

Harmonic Analysis and the MTF of Imaging Systems

Fourier has shown (in 1798) that any periodic function can be approximated to any degree of accuracy by the superposition of a weighted set of sine and cosine functions with frequencies that are integer multiples of a fundamental frequency ν.

$$F(\nu) = a_0 + a_1\cos\nu + a_2\cos 2\nu + a_3\cos 3\nu + \cdots$$
$$+ b_1\sin\nu + b_2\sin 2\nu + b_3\sin 3\nu + \cdots \quad (5)$$

or

$$F(x) = \sum_{n_0}^{\infty}(a_n\cos n_x + b_n\sin nx) \quad (6)$$

In Eqs. (5) and (6) the coefficients a_n and b_n are the weights that indicate the

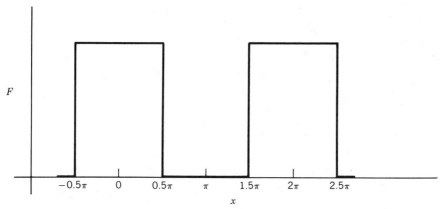

Figure 6-10 A crenelate bar pattern (square wave) plotted in terms of trigonometric angle coordinates.

contributions to $F(x)$ of the individual cosine or sine waves. For example, the simple bar pattern in Fig. 6-10 [3], which is highly relevant to microlithography, can be approximated as closely as required by the set of cosine functions indicated in Eq. (7).

$$F(x) = \frac{1}{2} + \frac{2}{\pi}\cos x - \frac{2}{3\pi}\cos 3x + \frac{2}{5\pi}\cos 5x - \cdots \qquad (7)$$

Figure 6-11 shows how this approximation succeeds with the use of cosines of increasing frequency [11]. It will be noted that the edges of the pattern represent high frequency features, in as much as frequencies at least 10 times the fundamental frequency of the *square wave* are needed for a reasonable approximation of the edges of the pattern. Obviously, the approximation improves with the use of even higher frequencies.

It may be concluded from this that the imaging capabilities of a system depend on the frequency range that the system is able to handle. With respect to accuracy of reproduction the system will be characterized by its frequency response. To describe this, the concept of *modulation*, borrowed from acoustics and from electrical engineering, is used. The modulation of a sine wave signal is a measure of its amplitude. It is defined as the ratio

$$M = \frac{I'' - I'}{I'' + I'} \qquad (8)$$

where I'' and I' are the maximum and the minimum signal intensities. Using the concept of modulation, the performance of a signal transducer (and

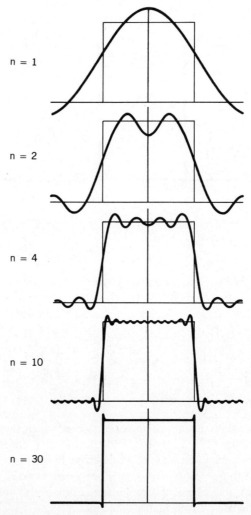

Figure 6-11 Approximating the bar pattern with cosine functions of increasing frequency. (Reproduced with permission from M. Schwartz and L. Shaw [11]. © McGraw-Hill, 1975.)

imaging systems are such transducers) is measured by comparing the modulation of the output signal to the modulation of the input. The ratio of the two is termed the *modulation transfer* (*MT*) *factor* of the device (Fig. 6-12).

$$\text{MT} = \frac{M_{out}}{M_{in}} \tag{9}$$

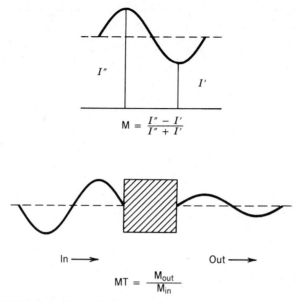

Figure 6-12 Modulation and modulation transfer of a "black box" transducer.

The frequency of a pure sine wave will not change on passage through the transducer, but its modulation will change. When sine wave signals with a wide range of frequencies are offered to the transducer the device will be able to follow the modulation of the input signal up to a certain frequency, but not much beyond it. The frequency response of the transducer may be represented by a plot of its MT factor as a function of frequency. This is termed the *modulation transfer function* (MTF) of the device. It defines the range of frequencies that the device can handle. Figure 6-13 shows the MTF of a microlithographic projection printer fitted with a high quality Zeiss lens [12].

The MTF of an imaging system and its line spread function are linked as a Fourier transform pair, the MTF being the modulus of the Fourier transform of the line spread function.

$$\text{MTF}(\nu) = \left| \int_{-\infty}^{+\infty} L(x) \exp(-2\pi i \nu x)\, dx \right| \qquad (10)$$

Figure 6-14 illustrates in a primitive graphic form the nature of this relationship.

The standard method of determining the MTF of an imaging system is to offer it a test object with a range of spatial frequencies, such as the Sayce chart in Fig. 6-15. This figure also shows the result of printing the variable frequency

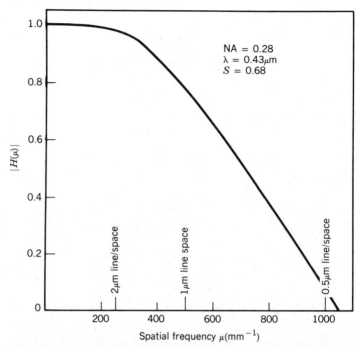

Figure 6-13 The modulation transfer function of the Zeiss optics of a 4800 DSW 10:1 wafer stepper (projection printer). (Reproduced with permission from R. K. Watts and J. H. Bruning [12].)

pattern onto a photographic emulsion. Where the MTF value of the material is low, the system can no longer resolve the test pattern; the cutoff of the MTF defines the limit of the resolving power of the imaging system.

Just as the characteristic curve of a composite imaging system is the product of the sensitometric curves of its components, so the MTF of the overall system is the product of the modulation transfer functions of the individual subsystems. For example, for the photographic system considered in Fig. 6-5, the overall MTF can be expressed in the following form.

$$\text{MTF(system)} = \text{MTF(camera)} \times \text{MTF(negative)} \times \text{MTF(paper)} \quad (11)$$

Figure 6-16 is a demonstration of this *cascading principle*. It shows the overall MTF as well as the modulation transfer function of the three components of the operation, the camera optics, the negative film material, and the photographic paper. It can be seen that the resolving power of the imaging system is chiefly limited by the subsystem with the lowest MTF, in this case the photographic paper.

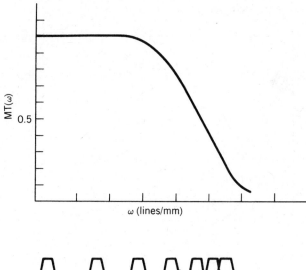

Figure 6-14 Schematic representation of the MTF of an imaging system. The MTF decreases as the line spread function of the device causes the images of adjoining features to overlap.

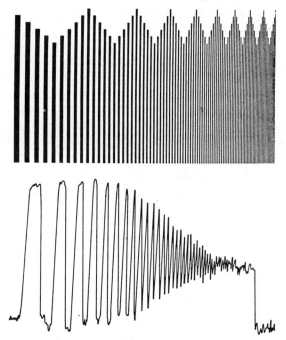

Figure 6-15 Determining the MTF of an imaging material using a variable frequency test object (Sayce chart). The curve below is a microdensitometer trace of the image of the chart copied onto a photographic negative of moderate resolving power. (Reproduced with permission from J. C. Dainty and R. Shaw [9]. © Academic Press, 1977.)

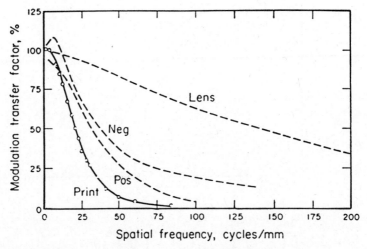

Figure 6-16 "Cascading" the modulation transfer functions of the camera optics, the negative film and the photographic paper to find the overall MTF of the compound imaging system. (Reproduced with permission from B. H. Carroll et al. [1]. © Wiley, 1980.)

RESOLUTION IN OPTICAL PATTERN TRANSFER

In silver halide photography, the graininess of the imaging material is the most serious limitation to image resolution. In optical microlithography, however, the resolving power of the resist is usually far greater than that of the printer and pattern transfer in the optics is the limiting factor.

Two types of pattern transfer are of interest here. In the *shadow printing* mode the original mask (usually in the form of a chromium pattern on glass or on quartz) is brought into contact with the resist layer (Fig. 6-17). The minimum line width, b, that can be imaged in hard contact of the two elements is given by the expression [13]

$$2b = 3\sqrt{0.5\lambda d} \qquad (12)$$

where λ is the wavelength of the radiation used in the transfer, d is the thickness of the resist layer, and $2b$ is the repeat period of the line–space pattern. In principle, hard contact printing produces the most faithful pattern transfer. For example, with $\lambda = 0.4$ μm and $d = 1$ μm, lines of 0.7 μm width can still be imaged. *Contact printing* is optically effective, but alignment of mask and wafer in hard contact damages both and creates debris that mars the performance of the system.

To reduce the mechanical damage of hard contact, a small gap between mask and resist surface can be introduced. In this so-called *proximity printing*

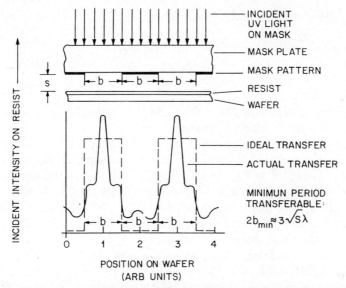

Figure 6-17 Schematic of the light distribution profiles on a photoresist surface in an arrangement for proximity printing. The mask is a pattern of equal lines and spaces. (After M. J. Bowden, *J. Electrochem. Soc.*, **128**, 196c (1981). Reproduced with the permission of the Electrochemical Society, Inc.)

mode (see Fig. 6-17) the resolution is given by the expression [13]

$$2b = 3\sqrt{\lambda(s + 0.5d)} \simeq 3\sqrt{s\lambda} \qquad (13)$$

where s is now the gap between mask and resist. A minimum gap of $s = 10$ μm has to be used in practical proximity printing. Inserting the previous values into the equation one finds that the minimum feature size, b, that can be imaged in this way has now grown to 3 μm.

It is not easy in practice to maintain a sufficiently small gap between mask and resist over the whole area of a wafer. For this reason *projection printing* with all its attending optical difficulties is now the preferred pattern transfer method.

The maximum resolving power of a projection system is given by the Raleigh criterion.

$$\delta = k\frac{\lambda}{NA} \qquad (14)$$

Here δ is the minimum distance of two points or lines that can still be distinguished as two separate entities, λ is the wavelength of the radiation used for imaging, k is a constant (it is 0.5 in Raleigh theory), and NA is the numerical aperture of the lens, that is, the ratio between the radius ($r = 0.5$

diameter D) of the lens and its focal length.

$$\mathrm{NA} = \frac{\tfrac{1}{2}D}{f} \qquad (15)$$

For a standard lens with NA = 0.25 and a wavelength of $\lambda = 0.4$ μm the value of the Raleigh criterion is about 1 μm.

A lens with a resolving power of 1 μm can distinguish two lines placed at that distance, but it cannot reproduce their tonal shape. In fact, a bar pattern of 1.5 μm linewidth will be imaged by this lens as a pure sine wave. The situation is illustrated in Fig. 6-18. This result is also understandable in terms of the MTF of the printer lens shown in Fig. 6-13 [13]. With a numerical aperture of about 0.3 this lens has a frequency cutoff at 900 lines/mm. To truly image the tonal shape of a bar pattern of 1.5 μm linewidth with some degree of accuracy, a lens would be needed that could handle frequencies of the order of 5000 lines/mm. That is clearly not possible with the lens of Fig. 6-13. Indeed, even the first harmonic, 1300 lines/mm, is out of its range and the aerial image of the bar pattern on the surface of the resist will be the

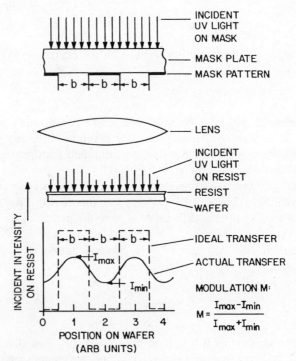

Figure 6-18 Schematic representation of optical image transfer in a 1:1 projection printer operating at its resolution limit. [After D. Widman and K. U. Stein, *Eur. Solid State Circuits Conf.*, 2nd, pp. 29–47 (1976).]

simple sine wave with a fundamental frequency of 667 lines/mm corresponding to a feature size of 1.5 μm.

Imaging at the Resolution Limit of the Optics

The resolution limit of conventional optics using visible light (typically the line of the Hg arc at 436 nm), is in the region of 1 μm. As was noted previously, resolution in these terms means the ability to distinguish two neighboring elements of a pattern; information on tonal values of the pattern that was contained in the higher frequencies of the harmonic representation is irretrievably lost. If one wanted to image an unknown intensity distribution with the present arrangement, this would be impossible. In microlithography, however, the tonal shape of the mask pattern is *a priori* known to be a step function, and the task of the imaging system will be to restore this step function from the sine wave signal of the aerial image. This is possible, in principle, by using the contrast properties of the resist. In this context, the role of the resist is to compensate for the image degradation caused by the optics.

This qualitative argument demonstrates the importance of resist contrast and justifies current interest in high contrast resists. Let us now put it in more quantitative form. Since the optics of the projector and the photoresist cooperate in producing the final image, the necessary condition for complete image restoration, is

$$\text{MTF} = 1$$

That means that the product of the modulation transfer functions of the optics and of the resist must be unity. Since the MTF of the resist is the ratio of the modulation of film thickness to the modulation of exposure dose, and since the modulation of film thickness is desired to be maximum (i.e., unity), it follows that for optimum reproduction the MTF of the optics simply equals the modulation of dose.

$$\text{MTF(optics)} = \frac{1}{\text{MTF(resist)}} = M(\text{dose}) \qquad (16)$$

The modulation of dose required to achieve maximum modulation of the image can be expressed in terms of the exposure parameters of the resist film. For a positive resist these are simply D_p^1 and D_p^0. (For a definition of D_p^1 and D_p^0 see Fig. 6-4.) This leads to an expression for the dose modulation.

$$M(\text{dose}) = \frac{D_p^0 - D_p^1}{D_p^0 + D_p^1} \qquad (17)$$

By introducing the contrast $\gamma = \log(D_p^0/D_p^1)$, the modulation can be written

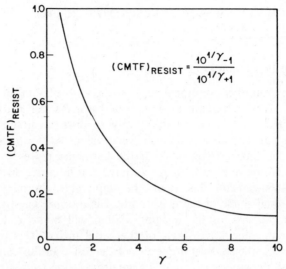

Figure 6-19 The critical MTF of a resist as a function of resist contrast. (Reproduced with permission from K. L. Tai et al. [14].)

in the form

$$M(\text{dose}) = \frac{10^{1/\gamma} - 1}{10^{1/\gamma} + 1} = \text{CMTF} \tag{18}$$

This means that to achieve full restoration of a line pattern of given spatial frequency (feature size), with a resist of contrast γ, the optics of the printer need to have an MTF equal to M(dose). This is also termed the *critical MTF (CMTF) of the resist* [14].

The relation between the CMTF and resist contrast is plotted in Fig. 6-19. It shows that with a high contrast resist a modest MTF of the optics is sufficient for the satisfactory imaging of a step function [14].

Contrast Enhancement Techniques. High contrast recording materials are required to deal with the low contrast aerial image of the optics when these operate at the limits of their resolving power. In recent years a number of resists have been reported with contrast values well in excess of 2, and these will be described in Chapters 7 and 8.

Another effective route to high contrast is by the use of a timed inhibition effect, whereby the function of the system is suppressed in the early stages of exposure. An example is the *oxygen inhibition of a bis-azide resist* based on p-diazidostilbene–poly(vinyl pyrrolidone), which was designed in the Hitachi laboratories [15]. The inhibition of the photogenerated nitrene by aerial oxygen is used here to produce low-intensity reciprocity failure. At low

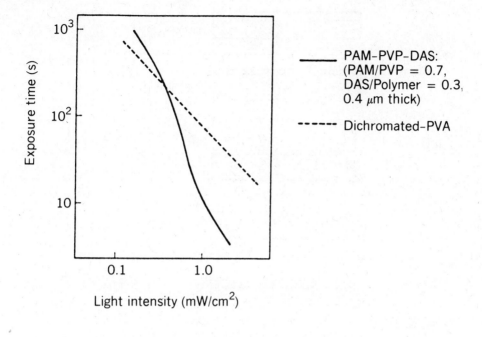

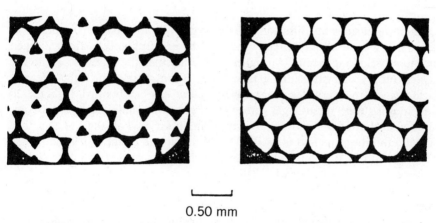

0.50 mm

Figure 6-20 Contrast enhancement by oxygen inhibition of crosslinking in a bis-azide resist. Compare the reciprocity failure curve of the inhibited resist with that of dichromated poly(vinyl alcohol). The lower part shows the black matrix of a color TV screen fabricated with the dichromated system (left) and with the oxygen-inhibited resist (right). (Reproduced with permission from T. Kohashi et al. [15].)

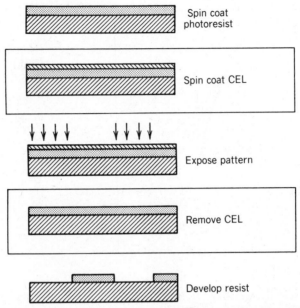

Figure 6-21 Processing steps in the contrast enhancement technique of Griffing and West. [Reproduced with permission from P. R. West et al., *Proc. SPIE*, **394**, 33 (1983).]

radiation intensities the diffusion rate of oxygen into the system is sufficient to insure that almost all the nitrenes produced by azide photolysis are scavenged and no crosslinking occurs. At higher light intensities the nitrenes swamp the oxygen supply and crosslinking takes over. By increasing the light intensity during exposure a very steep characteristic curve can be obtained. In Fig. 6-20 the exposure time versus light intensity curve of the inhibited resist is compared with that of the traditional dichromate–poly(vinyl alcohol) system. The oxygen inhibition method is used in the manufacture of the black matrix for color television CRTs (cathode ray tubes); the improvement in image quality by increased contrast is clearly visible in the lower part of Fig. 6-20.

A more general approach to the problem of contrast enhancement has recently been devised by Griffing and West [16, 17]. They revived an idea of Wiebe [18], originally intended for edge enhancement in silver halide photography and made it into a successful imaging tool. The method is based on the use of a *bleachable dye*, which is coated as a separate contrast enhancement layer (CEL) over the resist (see Fig. 6-21). When the combined system is exposed to radiation, the area of the coating receiving high radiation intensity is bleached earlier and uncovers the imaging layer to the radiation, while the parts that receive lower intensities are still shielded by the dye.

West et al. [19] designed a whole class of dyes on the basis of the aryl nitrones that have high extinction coefficients and at the same time fade rapidly on exposure to near-UV radiation. The generic structure is shown below.

$$R-N=CH-\text{---}R'$$
$$\uparrow$$
$$O$$

After exposure the dye layer is washed off and the resist is developed in the usual way. The improvement in the fidelity of pattern transfer that can be achieved in this way is shown in Fig. 6-22.

West et al. have now extended the range of their fugitive dyes into the mid UV [20]. Workers at Hitachi have used a simple diazonium salt in combination with poly(vinyl pyrrolidone) as a contrast enhancement system [21].

The polysilanes introduced by Miller and Stover [22] are resists with a built-in contrast enhancement function. They are described in Chapter 10.

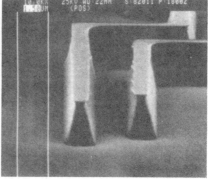

Figure 6-22 Scanning Electron Micrograph of a 1.5 μm line/space pattern printed in 2.2 μm thick resist, left: without contrast enhancement; right: with contrast enhancement. (Reproduced with permission from B. F. Griffing and P. R. West [16].)

Brown and Mack have recently studied the problem of contrast enhancement by computer simulation [23].

Imaging Beyond the Resolution Limit of Conventional Optics

It was noted previously that the minimum linewidth capable of being imaged by an optical pattern transfer process is determined by the Raleigh criterion, Eq. (11). It is proportional to the wavelength of the radiation used and inversely proportional to the numerical aperture (NA) of the optics. In Raleigh theory the constant k is 0.5, in practice, the proportionality constant for the overall imaging process depends on the resist, on the mechanics of the device, and even on the skill of the operator. In the laboratory or in a pilot line k would be about 0.65, in production it is nearer 0.8. For example, with a printer using the G-line of the mercury arc (436 nm) and with optics of a standard numerical aperture of 0.28 it is possible in the laboratory to image lines and spaces of about 1 μm width [22]. To reproduce pattern with substantially smaller feature sizes the conventional transfer process has to be modified in one of two ways: Either the NA of the optics has to be increased, or radiation of shorter wavelength has to be used as information carrier.

It is possible to increase the numerical aperture of the optics, and lenses of higher NA values have recently been produced [24]. The technical problems are considerable and the high value of NA is paid for not only by a high manufacturing price, but also by a decreasing depth of focus (d_f). The depth of focus, which is important in high resolution lithography depends inversely on the square of the numerical aperture.

$$d_f = \pm \frac{\lambda}{2(\mathrm{NA})^2} \qquad (19)$$

Lenses of NA = 0.35 are not rare anymore and lenses with numerical apertures as high as 0.6 exist and are being tried in several laboratories [25].

The other route to higher image resolution, namely, shortening of the radiation wavelength, has led to deep-UV lithography, and to the lithographies using electrons, x-rays, and ion beams. These are described in some detail in Chapters 7 through 9.

REFERENCES

1. B. H. Carroll, G. Higgins, and T. H. James, *Introduction to Photographic Theory*, Chap. 4, Wiley, New York, 1980.
2. F. Hurter and W. C. Driffield, *J. Soc. Chem. Ind.*, **9**, 455 (1960).
3. *The Theory of the Photographic Process* (T. H. James, Ed.), 4th ed., Macmillan, New York, 1977.

4. *Introduction to Microlithography, Theory, Materials and Processing* (L. F. Thompson, C. G. Willson, and M. J. Bowden, Eds.), *ACS Symp. Ser.*, **219**, Chap. 3 (1983).
5. G. N. Taylor, *Solid State Technol.*, **27**(6), 105 (1984).
6. D. B. Novotny, *Photogr. Sci. Eng.*, **21**, 351 (1977).
7. L. F. Thompson and R. E. Kerwin, *Annu. Rev. Mater. Sci.*, **6**, 267 (1976).
8. L. A. Jones, *J. Franklin Inst.*, **190**, 39 (1920).
9. J. C. Dainty and R. Shaw, *Image Science, Principles, Analysis and Evaluation of Photographic Type Imaging Processes*, Academic, New York, 1974.
10. A. Kriss, in Reference 2, Chap. 21.
11. M. Schwartz and L. Shaw, *Signal Processing*, Chap. 2, McGraw-Hill, New York, 1975.
12. R. K. Watts and J. H. Bruning, *Solid State Technol.*, **24**(5), 99 (1981).
13. J. H. Bruning, *J. Vac. Sci. Technol.*, **17**, 1147 (1980).
14. K. L. Tai, R. G. Vadimski, C. T. Kemmerer, J. S. Wagner, V. E. Lamberti, and A. G. Timko, *J. Vac. Sci. Technol.*, **17**, 1169 (1980).
15. T. Kohashi, M. Agaki, S. Nonogaki, H. Hashimoto, N. Hayashi, and T. Tomita, *Photogr. Sci. Eng.*, **23**, 168 (1979).
16. B. F. Griffing and P. R. West, *Solid State Technol.*, **28**(5), 152 (1985).
17. B. F. Griffing and P. R. West, *IEEE Electron Device Lett.*, **EDL-4**, 14 (1983).
18. A. F. Wiebe, U.S. Patent 3,511,653 (1970).
19. P. R. West, G. C. Davis, and B. F. Griffing, *J. Imaging Sci.*, **30**, 65 (1986).
20. P. R. West, G. C. Davis, and K. A. Regh, *Proc. SPIE*, **920**(10) (1988).
21. T. Iwayanagi, M. Hashimoto and S. Nonogaki, *Polym. Eng. Sci.*, **23**, 935 (1983).
22. V. Miller and H. L. Stover, *Solid State Technol.*, **28**(1), 127 (1985).
23. T. Brown and C. A. Mack, *Proc. SPIE*, **920**(48) (1988).
24. T. Omata, *Solid State Technol.*, **27**(9), 173 (1984).
25. D. A. Markl, *Solid State Technol.*, **27**(9), 159 (1984).

7 DEEP-UV LITHOGRAPHY

THE PHYSICS OF DEEP-UV LITHOGRAPHY

Resolution Limits

We have seen in Chapter 6 that the resolution achievable in pattern transfer depends on the wavelength of the information carrier used. One way of improving resolution is, therefore, to use shorter wavelength radiation. The transition from visible light to mid UV and deep UV does not seem a dramatic change, but it does achieve a gain in resolution that is real and worthwhile [1–3]. The listing in Table 7-1 shows the theoretical (Raleigh) resolution for some characteristic spectral lines and for a standard lens with a numerical aperture of 0.28. The real line/space resolution achievable at these wavelengths on a pilot line and in mass production are also listed. The spectral bands indicated are representative of the four subregions of the UV spectrum: near UV (between 400 and 365 nm), mid UV (between 365 nm and 300 nm), deep UV (from 300 nm down to 200 nm), and far UV (below 200 nm).

Line/space dimensions of 1.25 μm are now commonly achieved in good production units with printers using the mercury G-line at 437 nm. Deep UV has reached the 0.5 μm level, which is an important technological goal in the production of very high speed integrated circuits (VHISICs) and other advanced devices [4].

Radiation Sources

From the point of view of practical realization, *near-UV* lithography is the easiest to achieve. The I line at 365 nm is the strongest line of the mercury arc

TABLE 7-1 Resolution in Different Regions of the Mercury Spectrum

Spectral Region (nm)		$\delta_{Raleigh}$ (μm)	δ_{pilot} (μm)	$\delta_{production}$ (μ)
Visible	437	0.74	1.01	1.26
	400	0.72	0.93	1.15
Near UV				
	365	0.65	0.85	1.02
Mid UV				
	300	0.54	0.70	0.86
Deep UV	250	0.45	0.59	0.72
	200	0.36	0.47	0.57
Far UV				

and is therefore a readily available source. Many resists used with the *G*-line work even better with the *I*-line. Also, at this wavelength it is still possible to use refractive glass optics.

In the *mid-UV* region the conventional mercury lamp is not efficient. In that part of the spectrum there is only one reasonably intense line, at 313 nm; the weak line at 303 nm and the very weak line at 334 nm are not usable for lithography. High pressure mercury arcs have a more intense output in the mid UV as can be seen in Fig. 7-1, which shows the emission of the 1600 W radiation source of the Perkin-Elmer Micralign 500. The situation is worse in the *deep UV*. Here the traditional sources are the deuterium lamp and the low

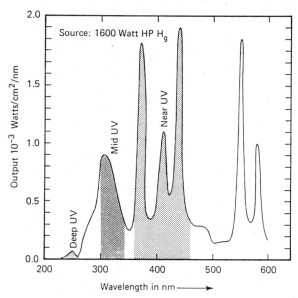

Figure 7-1 Spectrum of the optical output at wafer plane of the Perkin-Elmer Micralign 500 Projection Printer. Courtesy of Perkin-Elmer.

pressure mercury (germicidal) lamp. The luminosity of the deuterium lamp is not adequate as a photochemical irradiation source. The germicidal lamp is stronger and is used in laboratory tests of deep-UV resists, but it emits only about 0.2 mW per cm of length of tube. The deep-UV power available in a 1 kW high pressure mercury arc is of the order of 20 mW and that too is barely useful for mass production. However, with the advent of the rare gas halide excimer lasers deep-UV lithography has become a realistic production possibility.

Excimer lasers [5] are based on the emission properties of rare gas halogen mixtures. The rare gases and halogens do not appreciably interact in the ground state, but on excitation they form excimers that have comparatively long lifetimes. These excimers are the result of the interaction of a ground state halogen atom (produced on excitation from a halogen molecule) with an excited rare gas atom where one of the peripheral electrons is promoted into a higher (Rydberg) orbital. The electron vacancy that arises is filled by the odd electron of the halogen and an excited complex is formed, which on deactivation emits fluorescence.

$$Cl_2 \rightsquigarrow Cl + Cl \tag{1}$$

$$Xe \rightsquigarrow Xe^* \tag{1a}$$

$$Xe^* + Cl \rightarrow (XeCl)^* \rightarrow Xe + Cl + h\nu \tag{1b}$$

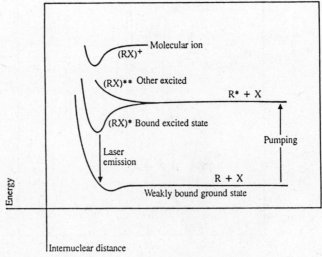

Figure 7-2 Potential energy diagram of a rare gas/monohalide molecular system showing excimer formation and laser transition. (Reproduced with permission from K. Jain et al. [6].)

TABLE 7-2 Properties of Excimer Lasers

Type	λ (nm)	Power Average (W)	Energy/Pulse (mJ)
XeF	351, 353	15	400
I_2	342	3	
XeCl	308	20	500
ClF	284		
KrF	245	45	1000
KrCl	222	30	
ArF	193	20	500
F_2	157	1	

While the excimer state is bonded strongly enough to have a radiative lifetime of 10^{-8} to 10^{-6} s, the lower state reached by the radiative transition is not bonded at all and dissociates immediately (within a single vibration, about 10^{-12} s) into free atoms that are again available for excitation. Because of the large difference in lifetimes, population inversion between the two states (Fig. 7-2) is readily achieved.

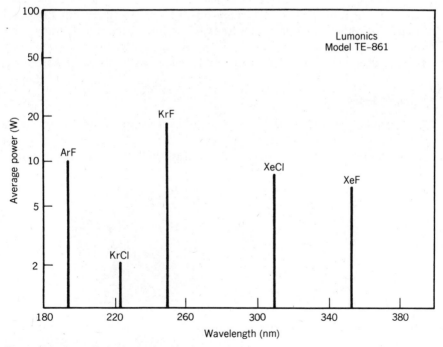

Figure 7-3 Laser lines obtained, with the appropriate gas filling, using the Lumonics Model TE-861 excimer laser unit. Courtesy Lumonics, Inc.

There are several methods by which energy can be pumped into the system, but the most usual excitation mode is by a direct electric discharge [6]. A gas mixture is pumped through an electrode assembly at several atmospheres pressure. By discharging a condenser across the electrodes a laser pulse of high intensity and short duration (10–100 ns) is created. Repetition rates of up to 1000 cps can be achieved, but the laser is usually run at low frequencies to allow for the gas in the electrode space to be replaced between pulses. Table 7-2 gives a listing of excimer lasers, including three pure halogen lasers that operate on the same principle [7, 8].

Fig. 7-3 shows the laser lines that are obtained (with the appropriate gas filling) with a Lumonics Model TE-861 excimer laser instrument.

Possibly the most serious problem of the excimer laser is the degradation of the gas mixture brought about by some unwanted photochemistry. This used to necessitate frequent gas changes, but with the installation of recirculators and purification units it is now possible to operate, for example, the XeCl laser on a single gas filling for nearly 10^7 pulses without appreciable loss of power.

The coherence properties that are the basis of many other laser applications are highly undesirable in laser lithography. As it happens, the mode structure of the transverse discharge used in excimer lasers is so complex that the spatial coherence of the stimulated radiation is largely lost and excimer lasers do not produce speckle [6]. This is important, because speckle would appear as intense image noise on the resist surface. The related problem of the interference of the radiation beam incident on the resist surface with the beam reflected from the substrate, which leads to the formation of a standing wave pattern in optical lithography and in x-ray lithography is not completely absent in deep-UV laser lithography, but is much less troublesome here.

Printing Modes

Commercial excimer lasers have rectangular beam shapes of considerable cross section. For example, the Lumonics TE-290 type has beam dimensions of 25 mm × 30 mm with excellent uniformity of the radiation intensity across the

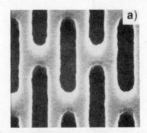
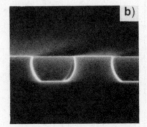

Figure 7-4 Scanning electron micrograph of pattern defined by contact printing in the positive photoresist AZ 2400. (*a*) 0.5 μm lines and spaces, obtained with a XeCl laser at dose 500 mJ/cm²; (*b*) 1.5 μm lines and spaces obtained with a KrF laser at 248 nm, dose 125 mJ/cm². (Reproduced with permission from K. Jain et al. [6].)

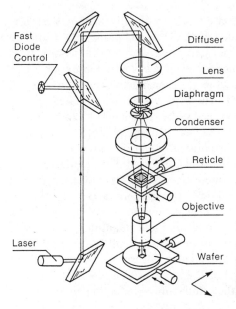

Figure 7-5 Early KrF excimer laser projection printing setup. (After G. M. Dubroeucq et al. [11b].)

whole beam area. This makes it possible to use the excimer laser directly as a radiation source in a contact printing mode. Results reported by Jain et al. [6] are shown in Fig. 7-4 and demonstrate the excellent image quality that can be achieved in this way, down to feature sizes of 0.5 μm. More recent work, reviewed by Rothchild and Ehrlich, has achieved line–space resolution approaching 0.2 μm and has defined isolated lines of 0.13 μm.

The great advantage of the excimer laser is the shortening of the exposure times by a factor of 100 or more. Images that require several minutes exposure to the deep-UV radiation of a high pressure mercury arc are adequately exposed by the laser within a couple of seconds. Excimer lasers, in particular the KrF laser at 248 nm and the ArF laser at 193 nm, have been used in a proximity printing mode [9] to produce 0.2 μm lines with high aspect ratios.

It is also possible to use the excimer laser as the illuminant in projection systems (e.g., the Canon Deep-UV Printer PLA-520 FA [10]). Glass optics are obviously out of the question, but quartz lenses have sufficient transparency down to 200 nm. The drawback is that there is no suitable material to combine with quartz in making compound optics.[1] Reflection optics can deal with narrow-band radiation and, to a degree, with broad-band radiation. Several projection printers and steppers based on reflection optics have been found suitable for conversion to deep-UV operation [11a]. Figure 7-5 shows an early KrF laser projection printing setup [11b].

---

[1]Special fluorophosphate glasses with low oxygen content have recently been described that in a thickness of 5 mm transmit 90% of the 248 nm radiation of the KrF laser. These materials could lead to chromatically corrected lenses for deep-UV radiation [12].

Finely focused laser beams are currently used to directly write pattern into resists in a maskless patterning mode [13]. This technique has attracted considerable attention and a laser driven pattern generator [14] has now been successfully operated in IBM Germany for over two years. Lasers have also been reported to induce chemical reactions at or near semiconductor surfaces in the presence of various gases [15]. These processes are initiated by the photodissociation of gas molecules and lead either to chemical vapor deposition or to etching, both on a spatial scale of 0.5 μm or less [16].

One concern on changing from a conventional radiation source to a laser in carrying out photochemical processes is the likelihood of a strong dependence of the quantum yield on radiation intensity. Calculations of Albers and Novotny [17] have indicated that at power densities of about 5 kW/cm^2 saturation of many photoreactions should set in (due to complete conversion of the reactants within their diffusion zone) and the quantum yield should approach zero. The exposure experiments of Jain and others [6] appear to contradict this. Comparing the lithographic effect of laser exposures at power densities of 10 MW/cm^2 with conventional exposures at a few milliwatts per square centimeter (a factor of 10^9) they find only a reduction of at most a factor of 3 at the highest radiation fluxes. It seems that from a practical point of view reciprocity failure is not a problem here. Nevertheless, the discrepancy between these findings and the reasonable arguments of Albers and Novotny indicate that the photochemistry induced by laser irradiation is different from the photochemistry under conventional low power irradiation [18].

NEGATIVE DEEP-UV RESISTS

Bis-azide Resists. The earliest resist specifically formulated for use in the deep-UV is a bis-azide-cyclized rubber composition described by Iwayanagi et al. [19]. The traditional formula for azide resists is retained, but the usual bis-azide (optimized for absorption at 365 nm) is replaced by a bis-azide that absorbs at shorter wavelength. From a whole group of candidates the most successful is 4,4'-diazidodiphenyl sulfide

$$N_3-\bigcirc-S-\bigcirc-N_3$$

4,4'-Diazidodiphenyl sulfide

which has an absorption maximum at 273 nm. It was used at a concentration of 1% by weight with cyclized poly(*cis*-isoprene) and was developed in hexane–xylene mixtures. This resist produced useful lithographic results on exposures of 20 s to the deep-UV of a 2 kW Xe–Hg lamp. The resist swells

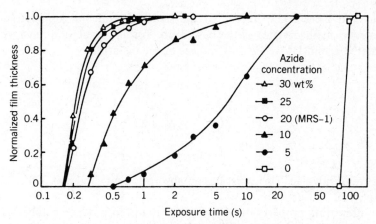

Figure 7-6 The effect of azide content on the characteristic curve of bis-azide/phenolic resin resists. Intensity of exposure source 70 mW/cm^2; immersion development in dilute alkali. (Reproduced with permission from T. Iwayanagi et al. [20]. ©IEEE, 1981.)

during development, however, and as a result its resolution does not even approach 1 μm.

A system that combines some of the virtues of the positive novolac resists, such as high resolution and good resistance to plasma etching, with the speed advantage of a negative resist is a bis-azide–phenolic resin composition developed by Nonogaki and co-workers at Hitachi [20]. The components are poly(vinyl phenol) as the base resin and 3,3′-diazidodiphenyl sulfone as the crosslinking agent.

3,3′-Diazidodiphenyl sulfone

This bis-azide is chosen because of its solubility in the phenolic resin matrix. Figure 7-6 shows the effect of azide content on the characteristic curve of the resist. It can be seen that the base resin on its own is a negative resist of low sensitivity. The addition of the azide increases the photographic speed of the material up to an optimum concentration of 20% by weight. Two aspects of this system are new and important: It is a negative resist developable in aqueous alkali, and it does not swell in the developer, but dissolves by an etch-type mechanism. The resist, termed MRS-1, is now being sold as RD 2000N. It has good sensitivity (lithographic speed is about 20 mJ/cm^2) and excellent resolution. Line–space patterns down to 0.2 μm have been claimed for this resist [21].

Han and Corelli [22] have used the same bis-azide in conjunction with poly(methyl methacrylate) as the base resin. Toriumi et al. [23] have recently formulated a mid-UV resist on the basis of the bis-azide

$$N_3-\text{C}_6\text{H}_3(\text{OCH}_3)-\text{C}_6\text{H}_3(\text{OCH}_3)-N_3$$

and a cresol novolac. This resist has an absorption maximum at 320 nm and a sensitivity of 150 mJ/cm^2.

An interesting property of MRS-1 is its ability to produce highly undercut profiles that are desirable in lithographic lift-off processes (see Chapter 10). The degree of overhang produced in these pattern can be controlled accurately by development time, as illustrated in Fig. 7-7 [24, 25]. The exact functional mechanism of MRS-1 is not completely clear. It appears that the photoreaction produces primarily an increase in the overall molecular weight of the polymer, without reaching the stage of an interlinked three-dimensional gel. The resin has an absorption band at 280 nm and on its own is also insolubilized by deep-UV irradiation, presumably by the formation of crosslinks via photogenerated radicals. The nitrenes, which are produced by azide photolysis, will facilitate this process by abstracting hydrogen either from the —OH groups of the phenols or directly from the backbone.

An indication that the base resin is the principal site of the crosslinking radicals is provided by recent observations in the OKI and Fuji laboratories [26] where it was found that azonaphthoquinone sulfonate groups bound to a cresol–novolac resin

$$\left[\text{Ar(OR)(CH}_3\text{)-CH}_2\right]_m\left[\text{Ar(OH)(R)-CH}_2\right]$$

$$R = -SO_2-\text{(naphthoquinone diazide)}$$

promote solubility in alkali and make the resin function as a negative resist in

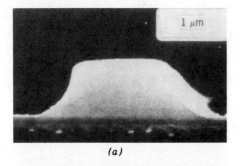

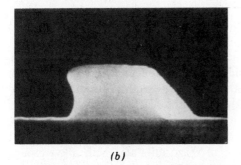

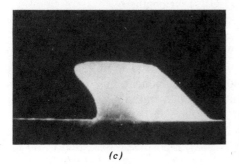

Figure 7-7 Scanning electron micrograph of a 2 μm line imaged in the MRS-1 resist and developed for (a) 55; (b) 75; and (c) 95 s. (Reproduced with permission from T. Matsuzawa et al. [24]. © IEEE, 1982.)

organic developers, although it is a conventional positive resist on irradiation at 365 nm.

Pampalone [27] has observed crosslinking in conventional novolac based positive diazoquinone resists on prolonged deep UV irradiation, and pulses of intense deep-UV have been used for the hardening of resist images before reactive ion etching or ion implantation [28]. In the absence of water the ketene that is a product of diazoquinone photolysis may react with the OH groups of the phenol or abstract hydrogen from methyl groups of the cresol novolac.

$$\diagdown_{\diagup}C=C=O + ROH \longrightarrow -\underset{|}{\overset{COOH}{\underset{|}{C}}}-R + RO^{\cdot} \qquad (2)$$

$$\diagdown_{\diagup}C=C=O + R-CH_3 \longrightarrow \cdot\underset{|}{\overset{CHO}{\underset{|}{C}}}- + R-\dot{C}H_2 \qquad (3)$$

POSITIVE DEEP-UV RESISTS

Modified Diazoquinone Resists. Some of the classical diazoquinone resists, like AZ 24110, can be used in the mid UV to good effect. The basic condition that the resists must fulfill is to absorb at the wavelength of irradiation and to be bleached (i.e., become transparent) as the photoreaction progresses. Willson and co-workers [29, 30] have shown that the optical properties of diazoquinone resists can be modified to fit more closely with the required wavelength of irradiation. In particular, substitution of diazosulfonates with alkyl groups shifts an absorption band, which in the unsubstituted diazoquinones is located at 300 to 330 nm. This is a useful mid-UV wavelength.

In designing a deep-UV resist to be used with 250 nm radiation, the problem is not lack of absorption, but rather the opposite, excessive absorption of the base resin in this region. What is required here is a base resin with an absorbance window at 250 nm so that the radiation may reach the photosensitive component. Furthermore, this component must be bleached on exposure so that the radiation may gradually penetrate the entire thickness of the resist film.

All these problems are neatly resolved in diazo-Meldrum's acid [31a]. This system is based on the principle of the classical positive resists, that is, on the removal of a dissolution inhibitor by a photoreaction. The photolabile inhibitor itself, diazo-Meldrum's acid, is reminiscent of diazoquinone. On irradiation it decomposes according to Eq. (4).

Diazo-Meldrum's Acid

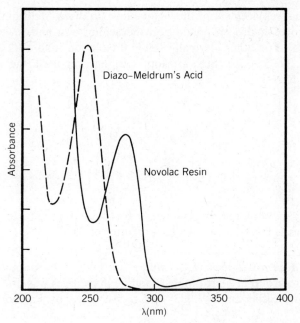

Figure 7-8 Absorption spectrum of 5-diazo-Meldrum's acid and of a Varcum 6000 novolac base resin. (Reproduced with permission from B. D. Grant et al. [31a]. © IEEE, 1981.)

Diazo-Meldrum's acid has an absorption peak almost exactly at 250 nm, and it is obviously bleachable. Together with a suitable novolac resin it makes an optically ideal resist for deep-UV exposure (see Fig. 7-8). Its sensitivity (about 50 mJ/cm^2) compares favorably with the sensitivity of conventional positive resists (about 100 mJ/cm^2).

Diazo-Meldrum's acid has one serious problem: it sublimates out of the coating during the inevitable prebake. Considerable effort has been invested to find analogous chromophores that would be retained more permanently in the system. Diazopyrazolidine dione (DPD), diazotetramic acid, and diazopiperidine dione were found to have that property

Diazopyrazolidine dione   **Diazotetramic acid**   **Diazopiperidine dione**

It appears that replacement of the lactone link (in diazo-Meldrum's acid) by a lactam improves hydrogen bonding in the acidic resin and lowers its vapor pressure [31b]. For other thermally stable dissolution inhibitors of that category see Reference 31c. Schwartzkopf [32] has reported the synthesis of 2-diazodimedones

which are highly photoreactive and which are said to be completely retained in the coated resist layer up to bake temperatures of 95°C.

Resists Based on *o*-Nitrobenzyl Chemistry. Intramolecular photooxidation by an ortho-nitro group in *o*-nitrotoluene and its derivatives was first reported by Ciamician and Silber in 1901 [33] and rediscovered by Patchornik in 1973 [34, 35].

$$\text{o-C}_6\text{H}_4(\text{CH}_2\text{OH})(\text{NO}_2) \xrightarrow{h\nu} \text{o-C}_6\text{H}_4(\text{CHO})(\text{NO}) + \text{H}_2\text{O} \tag{5}$$

In some instances the reaction can bring about large changes in solubility between reactants and products and it has been used in the design of several interesting deep-UV resists.

Barzynski and Saenger [36] have utilized the Patchornik reaction to produce a photofunctional change and hence a change in solubility by unhooking a protective group from a side chain and exposing a free carboxylic acid.

$$[-\text{CH}_2-\text{CH}(\text{COOCH}_2\text{-}o\text{-C}_6\text{H}_4\text{NO}_2)-]_n \xrightarrow{h\nu} [-\text{CH}_2-\text{CH}(\text{COOH})-]_n + \text{o-C}_6\text{H}_4(\text{CHO})(\text{NO}) \tag{6}$$

While the unexposed polymer is solvent soluble, but not soluble in water, the exposed resist dissolves in dilute alkali by reason of the free carboxyl groups. In this system, image discrimination is not based on kinetics, but on a thermodynamic difference, namely, the change from a truly insoluble to a soluble substance.

Another imaging system based on *o*-nitrobenzyl chemistry was reported by Reichmanis and co-workers [37]. In trying to design a dissolution inhibitor with the highest possible contrast, they reasoned that resist contrast, the change in film thickness corresponding to a small change in radiation dose, will depend on the molar volume of the inhibitor that can be made hydrophilic by a single photon. They decided to prepare an inhibitor that would make it possible to unhook a very large protecting group via the *o*-nitrobenzyl reaction. After an extensive search they chose substituted cholic acids as the most suitable moieties. The structures of the inhibitors are shown here.

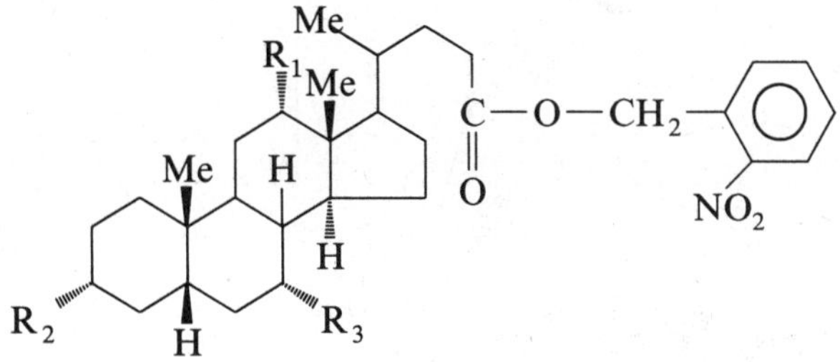

1(a) $R_1, R_2, R_3 = OH$

(b) $R_1, R_2 = OH, R_3 = H$

(c) $R_1, R_3 = H, R_2 = OH$

(d) $R_1, R_2, R_3 = H$

2(a) $R_1, R_2, R_3 = OSi(CH_3)_3$

(b) $R_1, R_2, R_3 = OCCF_3$ (with C=O)

(c) $R_1, R_2, R_3 = OCC(CH_3)_3$ (with C=O)

(d) $R_1, R_2, R_3 = OCCH_3$ (with C=O)

In mixtures with novolac (20 wt% of inhibitor) or with copolymers of methyl methacrylate and methacrylic acid (7:1) it was possible to formulate resists with exceptionally high photographic contrast (see Fig. 7-9). These resists have good absorption in the vicinity of 250 nm and have a sensitivity comparable to that of diazoquinone resists (about 100 mJ/cm^2).

The *o*-nitrobenzyl reaction has also been used to bring about scission of the polymer backbone. Petropoulos [38] has shown that irradiation of the *o*-nitrobenzyl-substituted polyether in Eq. (7) leads to backbone fragmentation and results in a lowering of the molecular weight of the polymer. This eventually causes a change in solubility sufficient to be useful in a deep-UV resist.

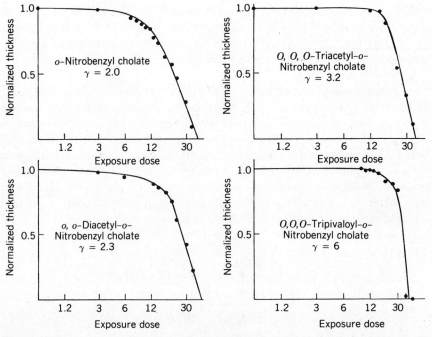

Figure 7-9 Characteristic curves of resists obtained from a copolymer of PMMA with poly(methacrylic acid), PMAA 7:3 and derivatives of *o*-nitrobenzyl cholate. (Reproduced with permission from E. Reichmanis et al. [37].)

$$-\text{CH}-\text{CH}_2-\text{CH}-\text{O}-\text{CH}-\text{O}-\cdots$$

o-Nitrobenzyl-substituted polyether (7)

$$\downarrow h\nu$$

$$-\text{CH}-\text{CH}_2-\text{C}=\text{O} \;+\; \text{HO}-\text{CH}-\text{O}-\cdots$$

MacDonald and Willson [39] have achieved a similar result using the *o*-nitrobenzyl reaction

$$\cdots-\overset{\text{O}}{\overset{\|}{\text{C}}}-\overset{\text{R}}{\underset{}{\text{N}}}-\text{Ar(NO}_2)\xrightarrow[350\text{ nm}]{h\nu} \cdots-\overset{\text{O}}{\overset{\|}{\text{CH}}} + \text{photoproducts} \quad (8)$$

which they applied to a group of poly(N-alkyl-o-nitroamides), such as the m-poly(nitroanilide) in Eq. (9).

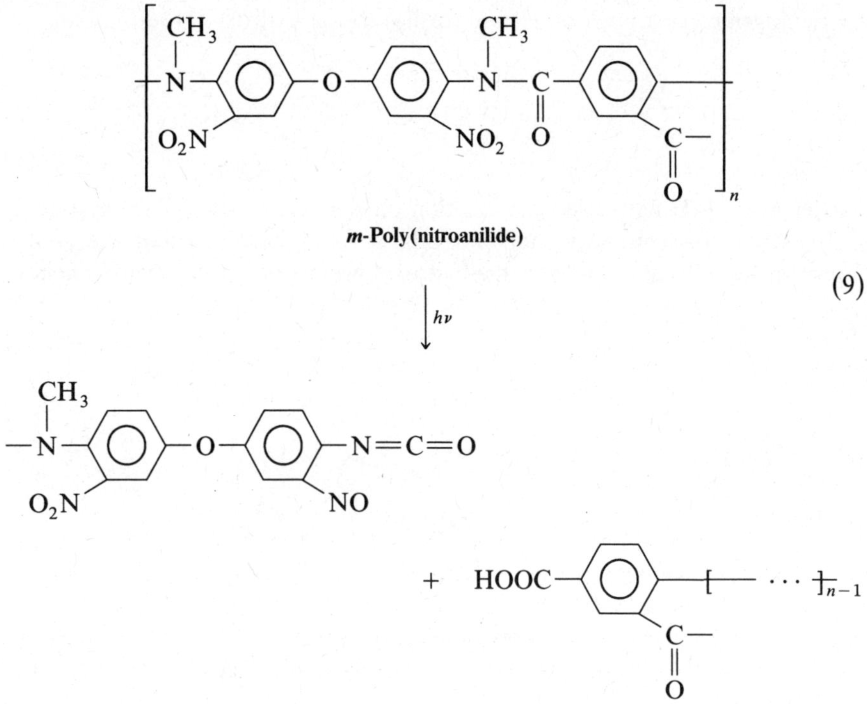

$$\left[-N(CH_3)-\underset{O_2N}{\bigcirc}-O-\underset{NO_2}{\bigcirc}-N(CH_3)-\underset{O}{\overset{\|}{C}}-\bigcirc-\underset{O}{\overset{\|}{C}}- \right]_n$$

m-Poly(nitroanilide)

$\downarrow h\nu$

(9)

$$-N(CH_3)-\underset{O_2N}{\bigcirc}-O-\underset{NO}{\bigcirc}-N=C=O$$

$$+ \; HOOC-\bigcirc-\{\cdots\}_{n-1}$$
$$\overset{\|}{C}-$$
$$\phantom{+ \; HOOC-\bigcirc\overset{\|}{C}}O$$

In this system the solubility change is caused by two effects: by the lowering of the molecular weight as a result of backbone scission, and by the appearance of a free acid group on half of the fragments. This allows development in dilute aqueous alkali. The sensitivity of the material is similar to that of the conventional diazoquinone resists (about 100 mJ/cm^2).

Houlihan et al. [40] have used the o-nitrobenzyl photochemistry to uncouple p-toluenesulfonic acid from its ester

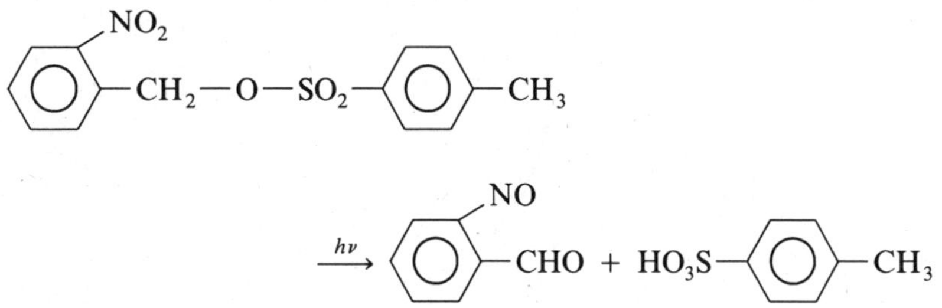

$$\underset{NO_2}{\bigcirc}-CH_2-O-SO_2-\bigcirc-CH_3$$

$$\xrightarrow{h\nu} \underset{NO}{\bigcirc}-CHO \; + \; HO_3S-\bigcirc-CH_3 \quad (10)$$

and use this strong acid to catalyze the deprotection of poly(tert-butoxycarbonyl styrene), PBOCST (see p. 279) used here as the base resin of the resist [40].

The Photo-Friess Rearrangement. Another mechanism that can be used to produce polarity changes in polymers is the photo-Friess rearrangement. This is a classical photoreaction whereby the acid component of a phenolic ester or amide migrates to a position vicinal (ortho) to the hydroxyl group.

$$\text{Ph-O-CO-R} \xrightarrow{h\nu} \text{Ph(OH)(COR)} \tag{11}$$

Tessier et al. [41] have used the reaction on derivatives of *p*-hydroxystyrene, such as poly(*p*-acetoxystyrene) and others. The reaction takes the course shown in Eq. (12) and uncovers the hydroxyl groups of the poly(vinyl phenol).

$$\cdots-CH_2-CH(C_6H_4-O-COCH_3)-\cdots \xrightarrow{h\nu} \cdots-CH_2-CH(C_6H_3(OH)(COCH_3))-\cdots \tag{12}$$

As a result the polymer becomes soluble in alkali, or insoluble in common organic solvents, and can be used as a positive or as a negative resist, depending on the developer. Poly(*p*-acetoxystyrene) is rather insensitive, but its homolog, poly(*p*-formyloxystyrene) has reasonable speed (70 mJ/cm^2) and good contrast.

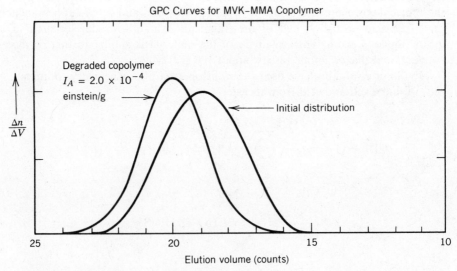

Figure 7-10 Molecular weight distribution (GPC elution curve) of methyl vinyl ketone–methyl methacrylate copolymer before and after irradiation in benzene solution; radiation dose 2.0×10^{-4} einstein/g. (Reproduced with permission from Y. Amerik and J. E. Guillet [42].)

Resists Based on Backbone Degradation by C—C Bond Scission

Deep-UV photochemistry is the transition from photochemistry to radiation chemistry. The incident photons are still absorbed by specific electronic systems since the molecules have well-defined absorption bands in the deep-UV region, and that is a characteristic of photochemistry. On the other hand, the deep-UV photons carry energy in excess of the binding energy of common C—C bonds. For example, photons of 250 nm radiation have an energy of 114 kcal/mole, C—C bond energies are in the range between 80 and 90 kcal/mole. As a result, deep-UV irradiation can lead directly to the scission of C—C bonds and to the formation of carbon centered radicals, which is a characteristic of radiation chemistry. In particular deep-UV radiation, in contrast to near UV, can lead to the degradation of the polymer backbone and bring about a change in solubility by lowering the molecular weight. Figure 7-10 shows the change in the molecular weight distribution of a methyl methacrylate–methyl vinyl ketone copolymer brought about by exposure to 313 nm radiation [42].

Poly(methyl methacrylate) and Its Derivatives. Poly(methyl methacrylate) (PMMA) is the classic resist for ionizing radiations, but it can also be used with deep-UV. Backbone scission occurs here as a secondary process following cleavage of a C—C bond adjacent to carbonyl. This so-called Norrish-type I process can be described for PMMA by the following reaction scheme.

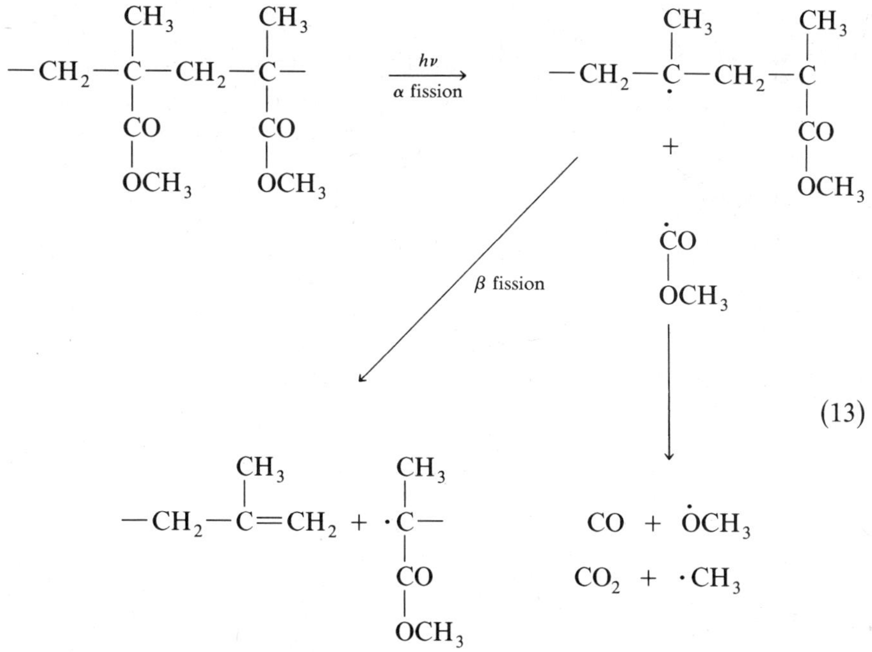

$$\begin{array}{c}\text{reaction scheme (13)}\end{array}$$

(13)

Norrish-type I degradation is not efficient in PMMA, which makes this polymer a rather insensitive deep-UV resist; in fact, at 240 nm, 3400 mJ/cm^2 are required to make PMMA developable. Nevertheless, the physical properties of PMMA and its high resolution capability are so outstanding that great efforts have been made to find derivatives or copolymers that would retain the good qualities of PMMA and yet be more sensitive to radiation.

Photochemical studies by Guillet and co-workers [42, 43] have indicated that the cause for the inefficiency of the process is the α-fission step in the sequence. Attempts to improve the photographic speed of the system were therefore directed towards weakening the α C—C bond. An early attempt in this direction is represented by poly(fluorobutyl methacrylate). Its sensitivity was reported as 480 mJ/cm^2 at 240 nm by Mimura et al. [44].

$$\left[\text{CH}_2-\underset{\underset{\underset{OC_4F_9}{|}}{\underset{CO}{|}}}{\overset{\overset{CH_3}{|}}{C}}- \right]_n$$

Copolymers of glycidyl methacrylate with methyl methacrylate degrade on deep UV irradiation at a faster rate than PMMA alone [45].

$$\left[\text{CH}_2-\underset{\underset{COOCH_3}{|}}{\overset{\overset{CH_3}{|}}{C}}- \right]_m \left[\text{CH}_2-\underset{\underset{COO-CH_2-CH\overset{O}{\diagdown\diagup}CH_2}{|}}{\overset{\overset{CH_3}{|}}{C}}- \right]_n$$

One of these copolymers has an exposure requirement of 250 mJ/cm^2 when irradiated with a deuterium lamp (250 nm). It will be noted that on deep-UV irradiation poly(glycidyl methacrylate) functions as a positive resist, while on exposure to an electron beam or x-rays the epoxy ring opens, the material crosslinks, and functions as an efficient negative resist (see Chapter 8).

The low sensitivity of PMMA to deep UV is caused in part by the low absorption coefficient of the material in this region and by the poor match of its absorption spectrum with the emission of the Xe–Hg lamp. This deficiency was removed by incorporating acyloximino groups into the base polymer. These chromophores, first introduced by Delzenne et al. [46], have a strong absorption around 225 nm and are highly photoreactive [47].

The reaction that occurs in the copolymeric system is indicated in reaction (14).

$$\begin{array}{c}
\text{—CH}_2\text{—C(CH}_3\text{)—CH}_2\text{—C(CH}_3\text{)—} \\
\text{| | } \\
\text{CO \quad CO} \\
\text{| | } \\
\text{OCH}_3 \quad \text{O} \\
\text{| } \\
\text{N=C(CH}_3\text{)—C(=O)CH}_3
\end{array} \xrightarrow{h\nu} \begin{array}{c}
\text{—CH}_2\text{—C(CH}_3\text{)—CH}_2\text{—C(CH}_3\text{)—} \\
\text{| | } \\
\text{CO \quad CO} \\
\text{| | } \\
\text{OCH}_3 \quad \text{O·}
\end{array} + CH_3CN + CH_3\dot{C}O$$

$$\text{—CH}_2\text{—}\overset{\text{CH}_3}{\underset{\text{CO—OCH}_3}{\text{C·}}} + \text{CH}_2\text{=}\overset{\text{CH}_3}{\text{C—}} \longleftarrow \text{—CH}_2\text{—}\overset{\text{CH}_3}{\underset{\text{CO—OCH}_3}{\text{C}}}\text{—CH}_2\text{—}\overset{\text{CH}_3}{\underset{\cdot}{\text{C}}}\text{—} + CO_2$$

(14)

The improvement in photographic speed is substantial and depends on the contents of 3-oximino-2-butanone methacrylate (OMMA) in the copolymer. For example, 16 mol% of OMMA result in an exposure requirement of about 80 mJ/cm² at 240 nm. The sensitivity of the system can be further enhanced by preparing a terpolymer containing methacrylonitrile as the third component. The most sensitive material prepared in this way contains the three components in the ratio (70 : 15 : 15).

$$\left[\text{—CH}_2\text{—}\underset{\text{CO—OCH}_3}{\overset{\text{CH}_3}{\text{C}}}\text{—} \right]_{0.70} \left[\text{—CH}_2\text{—}\underset{\underset{\underset{\text{CH}_3}{\text{N=C—C(=O)—CH}_3}}{\text{O}}}{\overset{\text{CH}_3}{\text{C}}}\text{—} \right]_{0.15} \left[\text{—CH}_2\text{—}\underset{\text{CN}}{\overset{\text{CH}_3}{\text{C}}}\text{—} \right]_{0.15}$$

Its monochromatic sensitivity at 240 nm is given as 40 mJ/cm^2. Further improvement in the speed of these materials is obtained by sensitization. The best results were achieved with *p-tert*-butylbenzoic acid (see Fig. 7-11). The sensitization process appears to involve energy transfer from the excited singlet state of the sensitizer [48].

A material of high sensitivity was also obtained by the copolymerization of methyl methacrylate with indenone [49]. This copolymer has a strong absorption in the 230 to 300 nm region (see Fig. 7-12) and exposure requirements as low as 20 mJ/cm^2 at 240 nm.

The reaction that occurs in the system is described by Eq. (15).

$$\text{(15)}$$

The high radiation sensitivity of the system is attributed in part to the fact that a comparatively stable (highly conjugated) radical is formed in the primary step.

Polymeric Ketones. Guillet et al. have made a thorough study of the photolysis and radiolysis of polymeric ketones [50]. Figure 7-13 shows some results for a copolymer of methyl vinyl ketone with methyl methacrylate irradiated in

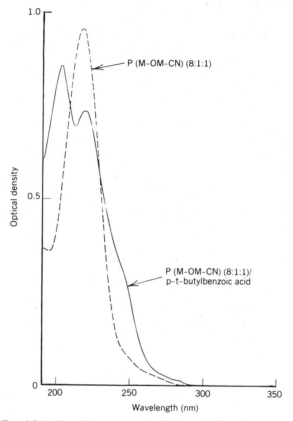

Figure 7-11 Ultraviolet absorption spectrum of the terpolymer PMMA–PMOM–PMCN 8:0.6:1.4 (broken line) and of the same terpolymer sensitized with 10% of *p-tert*-butylbenzoic acid (full line). (Reproduced with permission from E. Reichmanis et al. [49].)

benzene solution at 313 nm. It can be seen from this figure that the reciprocal of the molecular weight is accurately proportional to exposure dose [42, 43].

From experiments of this kind the quantum yield of the degradation process can be derived via the expression

$$\phi_{sc} = \frac{w}{M_n^0} \frac{d\left(\dfrac{M_n^0}{M_n} - 1\right)}{dE} \qquad (16)$$

where M_n^0 and M_n are the number average molecular weight before and after irradiation, w is the sample weight, and E is the absorbed radiation dose in einsteins per unit weight of polymer. Table 7-3 gives some data for aliphatic ketones and acetophenones with different lengths of the hydrocarbon chain.

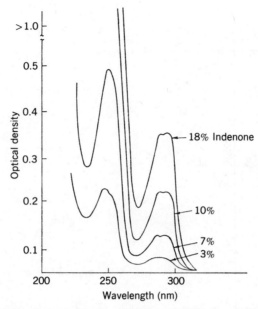

Figure 7-12 Absorption spectra of 1 μm thick films of different copolymers of MMA and indenone. (Reproduced with permission from E. A. Chandross et al. [45].)

The only practical resist based on this photochemistry is poly(methyl isopropenyl ketone) PMIPK [51]. The photolytic process that occurs in this material can be described schematically by the reaction (17).

$$\begin{array}{c}
\text{CH}_3 \quad\quad \text{CH}_3 \\
| \quad\quad\quad | \\
-\text{CH}_2-\text{C}-\text{CH}_2-\text{C}- \\
| \quad\quad\quad | \\
\text{CO} \quad\quad \text{CO} \\
| \quad\quad\quad | \\
\text{CH}_3 \quad\quad \text{CH}_3
\end{array} \xrightarrow{h\nu} \begin{array}{c}
\text{CH}_3 \quad\quad \text{CH}_3 \\
| \quad\quad\quad | \\
-\text{CH}_2-\overset{\cdot}{\text{C}}-\text{CH}_2-\text{C}- \\
\quad\quad\quad\quad\quad | \\
\quad\quad\quad\quad\quad \text{CO} \\
\quad\quad\quad\quad\quad | \\
\quad\quad\quad\quad\quad \text{CH}_3
\end{array} + \text{CO} + \cdot\text{CH}_3$$

$$\downarrow$$

$$\begin{array}{c}
\text{CH}_3 \quad\quad\quad \text{CH}_3 \\
| \quad\quad\quad\quad\quad | \\
-\text{CH}_2-\text{C}=\text{CH}_2 + \cdot\text{C}- \\
\quad\quad\quad\quad\quad\quad | \\
\quad\quad\quad\quad\quad\quad \text{CO} \\
\quad\quad\quad\quad\quad\quad | \\
\quad\quad\quad\quad\quad\quad \text{CH}_3
\end{array}$$

(17)

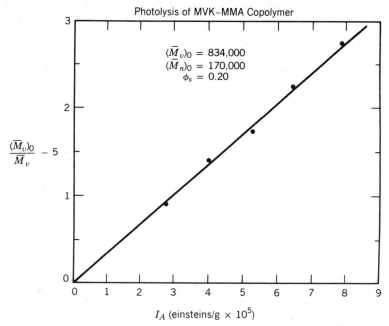

Figure 7-13 Plot of the ratio of viscosity average molecular weight $[(\overline{M}_v/\overline{M}_v) - 1]$ against irradiation dose for a copolymer of methyl vinyl ketone and methyl methacrylate, exposed to 313 nm radiation in benzene solution. (Reproduced with permission from Y. Amerik and J. E. Guillet [42].)

Although PMIPK has only a weak absorption at its absorption peak, (ε = 20 L/mole · cm at 280 nm), the sensitivity of the material is at least five times higher than that of PMMA and has been reported as 700 mJ/cm² at 280 nm. Tsuda [52] who made a detailed study of this system has suggested that the increased speed may be the consequence of a Norrish-type II process where internal hydrogen abstraction is the primary reaction step.

TABLE 7-3 Quantum Yield of Scission as a Function of Chain Length for a Group of Ketones [43]

R (number of C atoms)	R—CO—R	Ph—CO—R
4	0.11	0.31
5		0.31
6	0.092	0.25
7	0.080	0.30
8		0.29
11	0.072	0.31
21	0.059	0.31

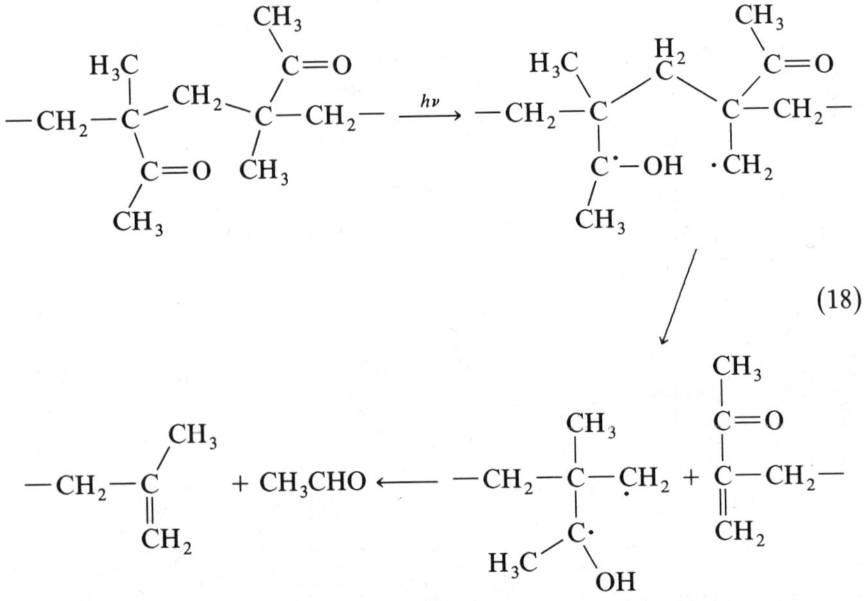

$$\text{—CH}_2\text{—}\underset{\underset{\text{CH}_3}{|}}{\overset{\overset{\text{H}_3\text{C}}{|}}{\text{C}}}\text{—}\underset{\underset{\text{CH}_3}{|}}{\overset{\overset{\text{CH}_2}{|}}{\text{C}}}\text{—CH}_2\text{—} \xrightarrow{h\nu} \text{—CH}_2\text{—}\underset{}{\overset{\overset{\text{H}_3\text{C}}{|}}{\text{C}}}\underset{\underset{\text{CH}_3}{|}}{\overset{\overset{\text{H}_2}{\text{C}}}{}}\underset{\underset{\cdot\text{CH}_2}{|}}{\overset{\overset{\text{CH}_3}{|}}{\text{C}}}\text{—CH}_2\text{—}$$

(where middle carbon bears C=O/CH$_3$ groups and right side has C=O, CH$_3$; product has C$^\cdot$–OH)

(18)

$$\text{—CH}_2\text{—}\underset{\underset{\text{CH}_2}{||}}{\overset{\overset{\text{CH}_3}{/}}{\text{C}}} + \text{CH}_3\text{CHO} \longleftarrow \text{—CH}_2\text{—}\underset{\underset{\underset{\text{OH}}{|}}{\overset{\text{C}^\cdot}{\diagdown}}}{\overset{\overset{\text{CH}_3}{|}}{\text{C}}}\text{—CH}_2\cdot + \underset{\underset{\text{CH}_2}{||}}{\overset{\overset{\text{C=O}}{|}\overset{\text{CH}_3}{|}}{\text{C}}}\text{—CH}_2\text{—}$$

Another system that functions by backbone degradation and that has been used successfully in deep-UV lithography is the well-known electron resist (see Chapter 8) poly(butene sulfone), PBS. On irradiation with deep UV at 185 nm it has a reported sensitivity of 5 mJ/cm^2 [53].

Table 7-4 gives an overview of the exposure requirements of PMMA derivatives and related systems.

Polymers Containing the Triphenylsulfonium Ion in Their Backbone. Triphenylsulfonium salts and other *onium salts* act as photoinitiators in cationic polymerization (see Chapter 4) and as photogenerators of acid (p. 278). On irradiation they undergo facile bond cleavage.

$$\text{Ph}_3\text{S}^+\text{X}^- \xrightarrow{h\nu} \text{Ph}\cdot + \text{Ph}_2\text{S}^+\text{X}^- \tag{19}$$

and yet, in the absence of light they are stable up to 400°C. Crivello et al. [54] have recently shown that the triphenylsulfonium moiety can be built into various polyimides, for example, into the following structure.

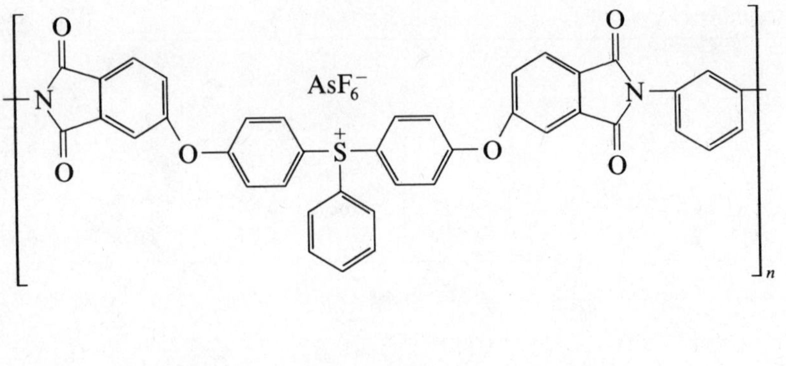

TABLE 7-4 Exposure Requirements of Deep-UV Resists Based on PMMA Derivatives and Related Systems

Polymers	Exposure Dose (mJ/cm^2)	λ (nm)
Poly(methyl methacrylate) Derivatives		
PMMA	3400	240
Poly(fluorobutyl acrylate)	480	240
Poly(MMA-co-glycidyl MA)	250	240
Poly(MMA-co-3-oximino-2-butanone MA)	80	240
Poly(MMA-co-OM-co-MACN)	40	240
Polyketones		
Poly(methyl isopropenyl ketone) PMIPK	680	250
Polysulfones		
Poly(5-hexene-2-one sulfone)	500	220–400
Poly(butenesulfone)	5	189

On irradiation, cleavage of one or the other of the C—S bonds occurs. Opening of the bonds in the main chain of the polyimide leads to backbone scission according to Scheme 1.

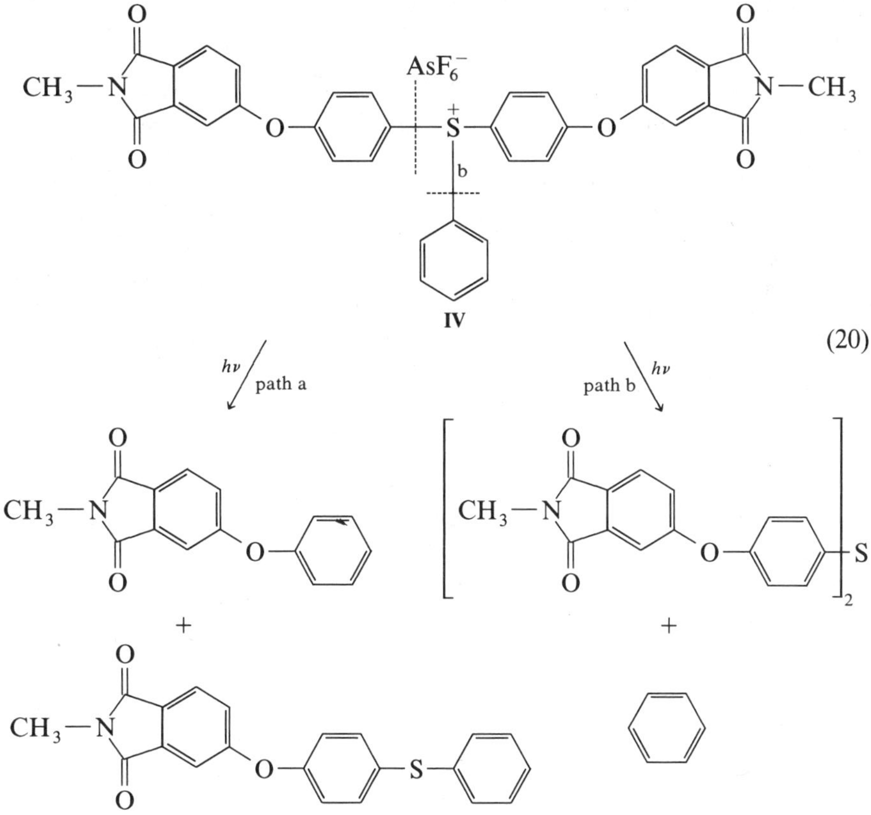

(20)

Scheme 1

Although reaction path (b) is favored over path (a) in a 3:1 ratio, viscosity measurements in solution show a large decrease in molecular weight on irradiation. It remains to be seen whether this idea can be developed into a useful resist, but it is a novel approach that deserves to be reported.

Resist Systems with Chemical Amplification

The sensitivity of an imaging system is determined by the amount of "detectable" change that is brought about by a single quantum of radiation. The effective sensitivity of an imaging system depends therefore not only on the molecular change on which the expression of the image is based, but also on the sensitivity of the detection system. Nonetheless, the unifying concept that allows a comparison of chemical recording systems is the quantum yield. In the majority of photochemical imaging processes the incident quantum will at best affect a single molecule. For example, in a photochromic system, the absorbed quantum will bring about the isomerization of one molecule of spiropyran to merocyanine (Chapter 3, p. 94). In this and in similar systems the quantum yield of the imaging process is less than unity and the photographic sensitivity is low. In imaging systems where the overall quantum yield exceeds unity, a secondary process must operate and amplify the primary photochemical event. For example, in silver halide photography the number of silver atoms produced per absorbed photon after development is of the order of 10^8. Image formation is here a two-step process, where in the first step a small number of quanta produces a latent image center, and this is subsequently amplified in the development step where all the 10^8 or so silver ions in a silver halide crystal are turned into metallic silver and contribute to the visible photographic image. The process by which the development centers of the latent image are formed does not have to be particularly efficient; the sensitivity of the system is caused by the fact that the development centers act as catalysts and control the thermodynamically driven amplification process, the dark reaction between silver halide and the developer.

Photogeneration of a Catalyst. The idea of generating a catalyst by a photoreaction and thereby controlling the rate of thermodynamically driven thermal process has now been applied to organic systems. For example, in cationic ring opening polymerization (see Chapter 4), acid acts as a catalyst; it initiates the polymerization chain without taking part in the propagation process. As a result, attention has focused on the efficient photogeneration of acid.

Pappas [55] has recently published a review on the photogeneration of acids from *onium salts*. The following schemes (Eqs. 21–21b) describe the photolytic reaction for diazonium, iodonium, and sulfonium salts.

$$ArN_2MX_n \xrightarrow{h\nu} ArX + N_2 + MX_{n-1} \qquad (21)$$

$$Ar_wIMX_n \xrightarrow{h\nu} ArI + HMX_n + \text{others} \qquad (21a)$$

$$Ar_3SMX_n \xrightarrow{h\nu} Ar_2S + HMX_n + \text{others} \qquad (21b)$$

$$MX_n = BF_4, PF_6, AsF_6, SbF_6$$

Crivello [56] has shown that onium salts can be used as photocatalysts (or photoinitiators) for a range of acid catalyzed processes [56]. Following this lead, Dubois et al. [57] at Thompson CSF in France have formulated a negative deep UV and electron resist based on the photopolymerization of epoxides with onium salts. By a similar process, Ito and Willson [58] have produced excellent negative images in a commercial epoxy resin, EPI-Rez-SU-8 (Celanese), using n-hexyloxydiazonium hexafluorophosphate as photoinitiator.

$$CH_3-(CH_2)_5-O-\langle O \rangle-N=N^+ \ PF_6^-$$

Deprotection of Poly(tert-butoxycarbonyloxystyrene). The photogeneration of acid has led to two further interesting resist systems. The first [41, 59, 60] is based on the catalytic deprotection of poly(p-*tert*-butoxycarbonyl oxystyrene), PBOCST, which is poly(vinyl phenol) protected with *tert*-butoxycarbonyl groups, a versatile protecting agent used in peptide chemistry. The polymer PBOCST can be deprotected thermally by heating to about 200°C. Figure 7-14 shows the result of the thermogravimetric analysis of PBOCST.

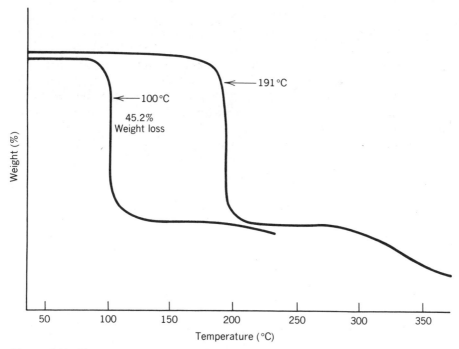

Figure 7-14 Thermogravimetric analysis of poly(*tert*-butoxycarbonyl oxystyrene), PBOCST, showing partial loss of material sharply at 191°C in the absence of acid, and a similar decomposition curve at about 100°C in the presence of catalytic quantities of acid. (With permission after J. M. J. Frechet et al. [61].)

The loss of 45% of the original weight at 200°C corresponds to the disappearance of the volatile components formed in the reaction, namely, CO_2 and isobutylene. In the presence of catalytic quantities of acid, the same reaction occurs at a much lower temperature (100°C) during a short (1 min) postexposure bake.

$$-CH_2-CH(C_6H_4-O-CO-O-C(CH_3)_3)- \xrightarrow{H^+, \Delta} -CH_2-CH(C_6H_4-OH)- + CO_2 + CH_2=C(CH_3)_2 \quad (22)$$

The poly(vinyl phenol) left behind is freely soluble in aqueous alkali; as a result, a positive tone image is obtained on aqueous development, a negative tone image on development in nonpolar solvents (which dissolve PBOCST but not the phenolic resin). In combination with an onium salt acid generator PBCOST is a very sensitive deep UV resist. It can be sensitized to near UV (365 nm) and it also responds to electron exposure (see Chapter 8).

In a different approach acid sensitive carbonates were found to function as dissolution inhibitors of novolac. In the presence of onium salts, irradiation of novolac–carbonate films produced hydrolysis of the carbonates, thus replacing the inhibitor with an alkali soluble phenol. After a short postexposure bake at 90°C standard development in alkali produced positive relief images at a dose of 3.5 to 5 mJ/cm^2 [62]. Similarly, naphthalene-2-carboxylic acid–*tert*-butyl ester can be used as dissolution inhibitor in novolac. It is catalytically deblocked on irradiation in the presence of onium salts when it produces the free carboxylic acid [63].

$$\text{Naphthalene-COO-}t\text{-Bu} \xrightarrow{H^+, \Delta} \text{Naphthalene-COOH} + HO-C(CH_3)_3 \quad (23)$$

The same basic chemistry has been applied to esters of poly(vinyl benzoic acid) [60, 63a]. The alcohol moieties of the esters were chosen to be bulky so as to impart some steric instability to the ester. The deprotection reaction of some of these is indicated in Scheme 2.

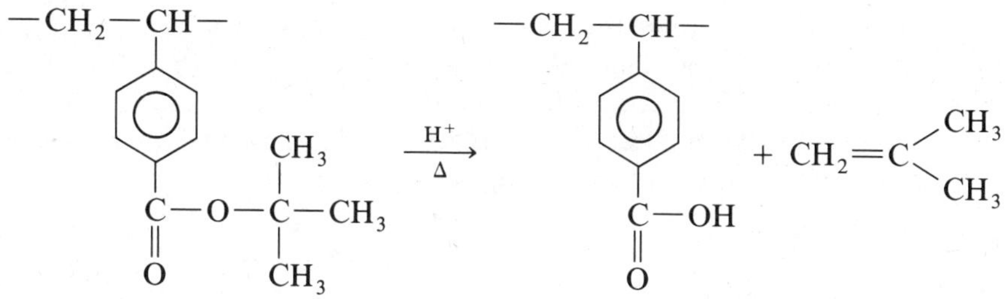

(24)

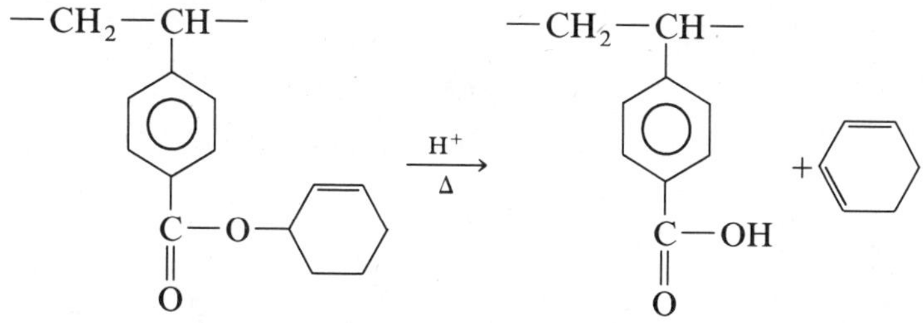

(24a)

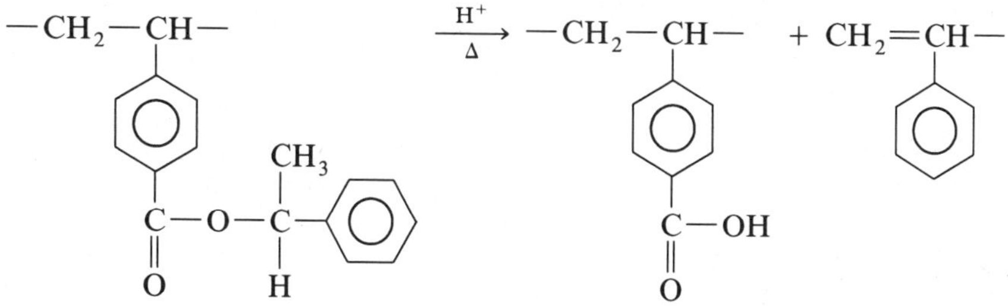

(24b)

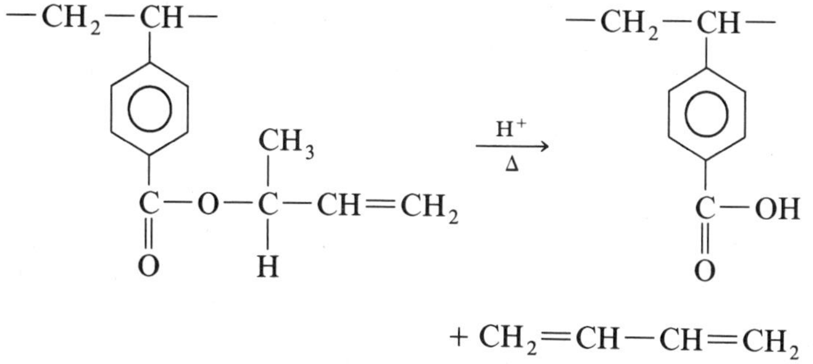

(24c)

Scheme 2

These materials have the advantage of a high T_g and therefore fulfill the condition of high thermal stability that has now become important.

The hydrolysis of esters or carbonates of tertiary alcohols can also be used to degrade a polymer backbone. Frechet et al. [64, 65] have prepared thermally depolymerizable polycarbonates, such as the structure shown in reaction (25) [64], which exhibit great thermal lability and undergo multiple backbone scission when heated to a critical temperature (usually 210 to 250°C). At that temperature the polymers decompose suddenly and cleanly into volatile products and no solid residue is left behind.

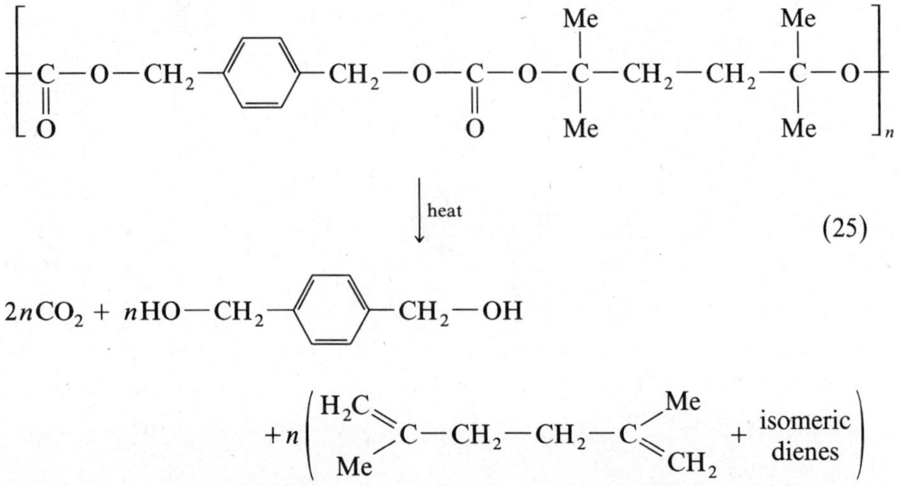

$$\left[\begin{array}{c}\text{C}-\text{O}-\text{CH}_2-\!\!\bigcirc\!\!-\text{CH}_2-\text{O}-\overset{\text{O}}{\underset{\|}{\text{C}}}-\text{O}-\overset{\text{Me}}{\underset{\text{Me}}{\text{C}}}-\text{CH}_2-\text{CH}_2-\overset{\text{Me}}{\underset{\text{Me}}{\text{C}}}-\text{O}\end{array}\right]_n$$

$\downarrow$ heat (25)

$2n\text{CO}_2 + n\text{HO}-\text{CH}_2-\!\!\bigcirc\!\!-\text{CH}_2-\text{OH}$

$+ n \left(\begin{array}{c} \text{H}_2\text{C}\!\!\diagdown \\ \phantom{\text{H}_2}\text{C}-\text{CH}_2-\text{CH}_2-\text{C} \\ \text{Me}\!\!\diagup \end{array} \begin{array}{c} \diagup\!\!\text{Me} \\ \diagdown\!\!\text{CH}_2 \end{array} + \begin{array}{c} \text{isomeric} \\ \text{dienes} \end{array} \right)$

The important observation in this reaction is that in the presence of merely catalytic quantities of acid much lower decomposition temperatures are observed [66]. On the same lines, Narang and Attarwala [67] have described a positive resist based on the following polycarbonate.

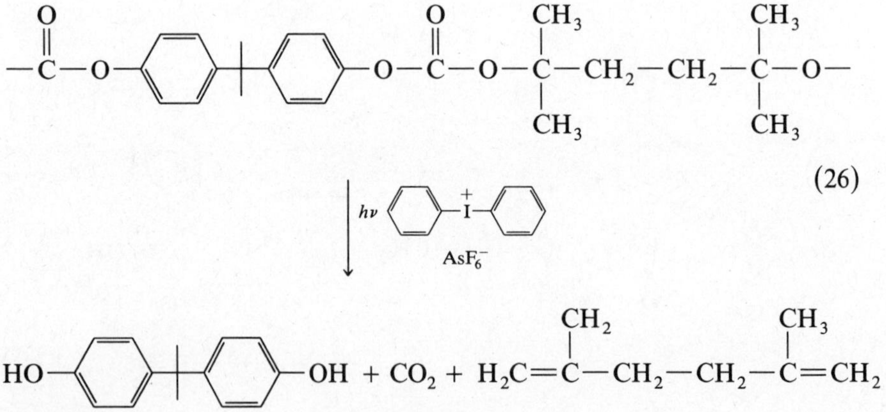

$$-\overset{\text{O}}{\underset{\|}{\text{C}}}-\text{O}-\!\!\bigcirc\!\!\!-\!\!\!\bigcirc\!\!-\text{O}-\overset{\text{O}}{\underset{\|}{\text{C}}}-\text{O}-\overset{\text{CH}_3}{\underset{\text{CH}_3}{\text{C}}}-\text{CH}_2-\text{CH}_2-\overset{\text{CH}_3}{\underset{\text{CH}_3}{\text{C}}}-\text{O}-$$

(26)

$\downarrow h\nu \quad \bigcirc\!\!-\!\!\overset{+}{\text{I}}\!-\!\!\bigcirc$
$\quad\quad\quad\quad \text{AsF}_6^-$

$\text{HO}-\!\!\bigcirc\!\!\!-\!\!\!\bigcirc\!\!-\text{OH} + \text{CO}_2 + \text{H}_2\text{C}=\overset{\text{CH}_2}{\underset{}{\text{C}}}-\text{CH}_2-\text{CH}_2-\overset{\text{CH}_3}{\underset{}{\text{C}}}=\text{CH}_2$

On irradiation in the presence of diphenyliodonium hexafluoroarsenate the resin decomposes during a short (1 min) postbake at 120°C. Under these conditions one of the products, bisphenol A, does not evaporate, but development in n-butanol produces positive tone images of good quality.

Frechet and co-workers have substantially expanded the range of materials available for acid catalyzed thermolytic depolymerization and have developed the phenomenon into a valuable microlithographic procedure [68].

Photosensitization of Onium Salts. The photogeneration of acid as a catalyst in polymer degradation and in cationic polymerization opens up a new field of polymer imaging with an effective radiation sensitivity orders of magnitude higher than that of conventional resists. Unfortunately, the onium salts that are the most effective components in acid generation do not absorb in a convenient spectral range. The more general exploitation of the process hinges therefore on the possibility of sensitizing these systems into the near UV or into the visible region of the spectrum (see Crivello [56] and Pappas [55]).

Early attempts at sensitization using the classical sensitizers (acetophenone, benzophenone, etc.) were not successful. Although energy transfer from acetophenone and benzophenone to the onium salts does occur (Gatechair and Pappas [69]) the resulting triplet excited states of the salts are fairly unreactive and of little practical use. In contrast, a number of polynuclear aromatic hydrocarbons are effective. Since these could not possibly act by energy transfer it was thought that sensitization occurs by electron transfer, and that has been proven to be the case [55].

For a diaryliodonium salt and anthracene, for example, the sensitization process can be described by reactions (27)–(31).

$$\text{[anthracene]}^* + Ar_2I^+ X^- \longrightarrow \text{[anthracene]}^{+\cdot} + X^- + Ar_2I\cdot \quad (27)$$

$$Ar_2I\cdot \longrightarrow ArI + Ar\cdot \quad (28)$$

$$\text{[anthracene]}^{+\cdot} + H_2O \longrightarrow \text{[H, OH}_2^+\text{-anthracene}]\cdot \quad (29)$$

$$\text{[H, OH}_2^+\text{-anthracene}]\cdot \xrightarrow{-e} \text{[OH-anthracene]} + 2H^+ \quad (30)$$

$$H^+ + X^- \rightleftharpoons H^+X^- \quad (31)$$

Several points are noted:

1. The catalyst (the initiating species) comes here from the photosensitizer, while it originates from the onium moiety in the absence of a sensitizer.
2. The role of water can be, and often is, taken by another nucleophile.
3. The role of the complex anion of the onium salt is to control the concentration of free protons in the system via Eq. (31).
4. The radical Ar· produced in Eq. (28) is able to initiate radical polymerization.

All these points have been confirmed by experiment [55]. The dramatic effect of the anion of the onium salt on the efficiency of cationic polymerization is illustrated by Fig. 4-16. The photodegradation reaction can be carried out in the total absence of water, and onium salts are indeed effective (although expensive) photoinitiators of radical polymerization.

The electron transfer character of the sensitization process is further supported by the correlation of sensitization efficiency with the free energy of charge transfer. Rehm and Weller [70] have shown that the free energy of charge transfer (CT), ΔG_{CT}, between a photoexcited electron donor and an electron acceptor in the ground state is given by Eq. (14) of Chapter 3. For the case under discussion this may be written in the form

$$\Delta G_{CT} = E^{ox}(\text{sensitizer}) - E^{red}(\text{onium}) - \Delta E^*(\text{sensitizer}) + 10 \text{ kcal} \quad (32)$$

Here E^{ox}(sens.) is the oxidation potential of the sensitizer, E^{red}(onium) is the reduction potential of the onium salt, ΔE^* is the optical excitation energy of the sensitizer, and the 10 kcal stand for a (constant) coulombic term. Crivello [56] has shown that only those sensitizer–onium pairs are successful where the free energy of charge transfer is negative (see Table 7-5).

TABLE 7-5 Free Energy of Electron Transfer and Photosensitization of Triphenylsulfonium hexafluoroarsenate [56]

Photosensitizer	E^{ox} (sensitization) (kcal/mole)	E^* (kcal/mole)	ΔG_{CT} (kcal/mole)	Photopolymerization
Anthracene	31	76	−17	+
Perylene	32	66	−15	+
Phenothiazine	6	57	−23	+
Acetophenone	62	74	16	−
Thioxanthone	39	66	1	−
Benzophenone	55	69	14	−
1,2-Benzanthracene	23	75	−24	+
Coronene	28	67	−11	+
Pyrene	31	77	−18	+
Tetracene	22	61	−11	+

Radiation-Induced Depolymerization. To the more traditional amplification mechanisms just described a new one was recently added: The radiation induced depolymerization of a labile polymer chain. Here the driving force is the thermodynamic instability of the polymer.

In principle, the formation of polymer from monomer is a reversible process.

$$\text{RM}_n + \text{M} \underset{k_{dp}}{\overset{k_p}{\rightleftharpoons}} \text{RM}_{n+1} \tag{33}$$

The direction of the process will be determined by the free energy change, ΔG, associated with it. This can be written as the sum of the free energies of activation of the forward and the backward reaction.

$$\Delta G = \Delta G_p^\ddagger - \Delta G_{dp}^\ddagger \tag{34}$$

where

$$\Delta G = \Delta H - T \Delta S \tag{35}$$

so that

$$\Delta G = \left(\Delta H_p^\ddagger - \Delta H_{dp}^\ddagger \right) + T \left(\Delta S_{dp}^\ddagger - \Delta S_p^\ddagger \right) \tag{36}$$

In Eq. (36) ΔH_p is more negative than ΔH_{dp} since it is associated with the formation of a single bond from a double bond. At the same time the entropy change on depolymerization is much more positive than the entropy of activation of polymerization.

$$\Delta S_{dp}^\ddagger \gg \Delta S_p^\ddagger \tag{37}$$

At low temperatures, the negative enthalpic term in Eq. (36) will dominate, but as the temperature increases a point will be reached where the two terms become equal and above which the positive entropy term will take over

$$T \left(\Delta S_{dp}^\ddagger - \Delta S_p^\ddagger \right) \geq \left(\Delta H_p^\ddagger - \Delta H_{dp}^\ddagger \right) \tag{38}$$

The temperature at which this occurs is termed the ceiling temperature T_c. At the ceiling temperature the rates of the forward and the backward reaction balance and the net rate of polymerization is zero (see Fig. 7-15).

It is not possible to form a polymer from its repeat units above T_c. If the polymer is prepared well below its ceiling temperature and then brought to a temperature above T_c in the presence of traces of the polymerization catalyst,

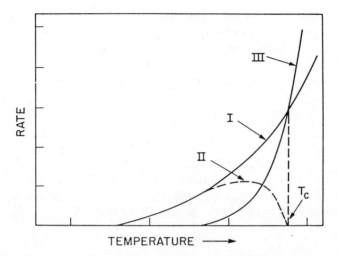

Figure 7-15 Effect of temperature on the propagation rate (I), the depropagation rate (III), and the net polymerization rate (II). Here T_c is the ceiling temperature of the polymer. (Reproduced with permission from M. J. Bowden [71].)

for example a free radical,[2] it will spontaneously depolymerize. It is this phenomenon of spontaneous depolymerization that has been used by Ito and Willson [72] in the design of a highly sensitive resist system based on poly(phthalaldehyde).

The ceiling temperature of practical polymers is well above room temperature, and their rate of depolymerization in normal circumstances is negligible. However, the polyaldehydes and some polysulfones, for example, have low ceiling temperatures and must be prepared under cryogenic conditions. If these materials are endcapped (with alkyl or acyl groups, for example) while still at low temperature, they are often found to be stable at room temperature and beyond (see footnote). Poly(phthalaldehyde) first described by Vogl [73] is an example. The ceiling temperature of the polymer is at $-40°C$, but if the polymer is endcapped in these conditions it is stable up to some $180°C$ and can be spin coated and otherwise handled in the usual way. Willson and co-workers prepared numerous samples (molecular weights 100,000–500,000) by anionic polymerization in dry ice conditions ($-78°C$), using n-butyl lithium and similar reagents as catalysts and endcapping the polymer finally with acetyl groups. These materials are stable up to $198°C$ (see Fig. 7-16).

---

[2] The necessity of the presence of the polymerization catalyst in the depolymerization process is a consequence of the principle of microscopic reversibility. If, for example, all free radicals are removed from the system by endcapping the polymer, the thermodynamic equilibrium that demands depolymerization cannot be reached and the system appears to be stable.

To trigger spontaneous depolymerization of poly(phthalaldehyde) it was necessary to find a photoreaction that would remove the cap or cause a break in the chain. In their first experiments Ito and Willson [58] used the o-nitrobenzyl reaction for this purpose. They copolymerized phthalaldehyde with o-nitrobenzaldehyde. On irradiation, the backbone of the copolymer is scissioned by the radiation induced intramolecular oxidation of the C—O—C link.

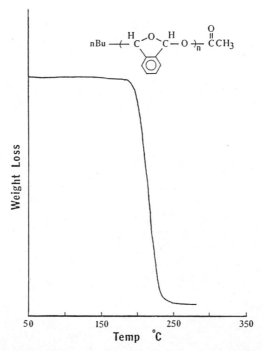

Figure 7-16 Thermogravimetric analysis of poly(phthalaldehyde) synthesized by butyllithium initiation and endcapped with acyl groups while still at low temperature. Although the polymer is well above its ceiling temperature, it is stable up to 180°C. (Reproduced with permission from C. G. Willson et al. [72].)

Progress of the reaction can be monitored by IR spectroscopy. Results are shown in Fig. 7-17 where the disappearance of the NO_2 absorption from the spectrum is clearly visible.

Films of this material were exposed to deep UV, to electrons, and to x-rays: In the exposed areas a loss of film thickness of up to 50% was observed. The reason for this partial success is the fact that after scission, one of the chains ends in a hemiacetal and does depolymerize as expected, while the other chain is effectively endcapped by *o*-nitrobenzoate and is stable.

Willson and Ito looked for a mechanism that would leave both chain ends unprotected. They found it in the hydrolytic scission process brought about by photogenerated acid. Up to 10% by weight of various onium salts were incorporated into poly(phthalaldehyde) films; when the films were exposed to deep UV or to electrons, the irradiated areas simply evaporated. Here was the first resist system that self-developed reliably at room temperature without any further operations or special conditions [72].

The exposure requirements for complete self-development depend on the degree of loading with the onium salt and on the nature of the salt itself, as can be seen from the listing in Table 7-6. These data refer to deep-UV

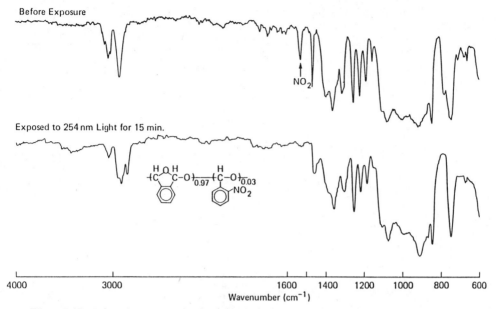

Figure 7-17 Infrared spectrum of poly(phthalaldehyde-co-2-nitrobenzaldehyde) before and after exposure to 254 nm radiation. Note disappearance of NO_2 absorption at 1510 cm^{-1}. (Reproduced with permission from C. G. Willson et al. [72].)

exposure. A 1 μm thick film can also be imaged with 20 kV electrons at a dose of about 1 μC/cm².

The mechanism of the depolymerization process appears to be well documented. The acid attacks the lone electron pair on oxygen and brings about the effective reversal of polymerization.

$$S \xrightarrow{h\nu} H^+ \qquad (41)$$

$$B \!-\!\!\left[\begin{matrix} H \\ | \\ C-O \\ | \\ R \end{matrix}\right]_n\!\!-\!Ac \xrightarrow{H^+} -\ddot{O}-\overset{H}{\underset{R}{C}}-\overset{H^+}{\underset{}{O}}-\overset{H}{\underset{R}{C}}- \longrightarrow nRCHO \qquad (41a)$$

TABLE 7-6 Exposure Requirements for Poly(phthalaldehyde) Films

Loading (wt%)	Onium Salt	Exposure Dose (mJ/cm²)
2	$Ph_3S^+AsF_6^-$	24
	$Ph_2I^+AsF_6^-$	72
	$p\text{-}C_6H_{13}OPhN_2^+BF_4^-$	> 200
10	$Ph_3^+AsF_6^-$	1.5–5
	$Ph_2^+AsF_6^-$	5
	$p\text{-}C_6H_{13}OPhN_2^+BF_4^-$	6.5

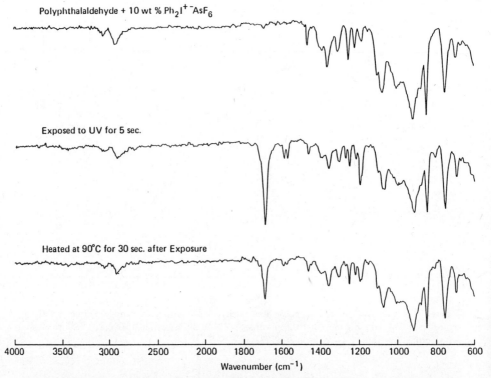

Figure 7-18 Infrared spectrum of films of poly(phthalaldehyde) containing a diphenyliodonium salt initiator, before exposure, after exposure to 254 nm radiation, and after exposure and 1 min heating to 90°C. Note the appearance of the C=O absorption at 1700 cm^{-1} and its partial disappearance in the heating step. (Reproduced with permission from C. G. Willson et al. [72].)

S is here the substrate and H$^+$ signifies the acid. The reversion to the monomeric aldehyde and its gradual evaporation are illustrated by the IR spectra of the film before exposure and after exposure (note the appearance of the carbonyl band of the free phthalaldehyde at 1700 cm^{-1} and its partial loss after a short heating period) (Fig. 7-18).

The acid catalyzed depolymerization of phthalaldehyde has been used in conjunction with novolac. Onium salts on their own can stand in for diazoquinones in novolac resists, allowing the formulation of a very useful deep-UV resist of fairly high sensitivity (25 mJ/cm^2). The addition of poly(phthalaldehyde) to the mixture, which also acts as a dissolution inhibitor, dramatically increases the sensitivity of the system to 2 mJ/cm^2 for full development. A typical formula consists of novolac Varcum 6000-7, 8.8% by weight of poly(phthalaldehyde) and 1.8% Ph$_3$S$^+$ SbF$_6^-$ [74]. Steinmann followed a similar strategy in designing the top imaging layer over a soluble polyimide or novolac planarizing layer, formulating it from a silylated poly(phthalaldehyde)

$$\left[\begin{array}{c} \text{HC}\overset{\text{O}}{\diagdown}\text{CH} \\ \diagup\diagdown \\ \text{Me}_3\text{Si}\text{SiMe}_3 \end{array} \right]_n$$

together with a suitable onium salt [75].

REFERENCES

1. B. J. Lin, *J. Vac. Sci. Technol.*, **12**, 1317 (1975).
2. J. H. Bruning, *J. Vac. Sci. Technol.*, **16**, 1925 (1979).
3. T. Iwanayagi, T. Ueno, S. Nonogaki, H. Ito, and C. G. Willson, *IBM Res. Rep.*, No. RJ 5826 (1987).
4. C. A. Deckert and D. L. Ross, *J. Electrochem. Soc.*, **127**, 45 (1980).
5. J. Ewing, *Phys. Today*, **31**(5), 32 (1978).
6. K. Jain, C. G. Willson, and B. J. Lin, *IBM J. Res. Dev.*, **26**, 151 (1982).
7. T. McGrath, *Solid State Technol.*, **26**(12), 165 (1983).
8. D. A. Markle, *Solid State Technol.*, **27**(9), 159 (1984).
9. E. Cullman, *Lambda Phys. Highlights*, Apr., p. 9 (1987).
10. S. Iwamatsu and K. Asanami, *Solid State Technol.*, **23**(5), 81 (1980).
11. (a) K. Jain and R. T. Kerth, *Appl. Opt.*, **23**, 648 (1984). (b) G. M. Dubroeucq and D. Zahorsky, *Proc. Int. Conf. Microcircuit Eng.*, Grenoble, Fr. p. 73 (1982).
12. L. M. Cook, M. J. Liepmann, and A. J. Marker, *SPIE Symp. Halide Glasses, Monterey, CA, Jan. 1987*.
13. K. Jain, *Laser Appl.*, **9**, 49 (1983).
14. U. Boettiger and B. Hafner, *VDI Ber.*, No. 621, 59 (1986).
15. D. J. Ehrlich and I. Y. Tsao, *J. Vac. Sci. Technol.*, B, **1**, 969 (1983).
16. D. J. Ehrlich, *Solid State Technol.*, **28**(12), 81 (1985).
17. J. Albers and D. B. Novotny, *J. Electrochem. Soc.*, **127**, 1400 (1980).
18. N. J. Turro, M. Aikawa, J. A. Butcher, Jr., and G. W. Griffith, *J. Am. Chem. Soc.*, **102**, 5127 (1980).
19. T. Iwayanagi, T. Kohashi, and S. Nonogaki, *J. Electrochem. Soc.*, **127**, 2759 (1980).
20. T. Iwayanagi, T. Kohashi, S. Nonogaki, T. Matsuzawa, K. Douta, and H. Yanazawa, *IEEE Trans. Electron Devices*, **ED-28**, 1306 (1981).
21. T. Matsuzawa, A. Kishimoto, T. Iwayanagi, H. Yamazawa, and H. Obayashi, *IEEE Trans. Electron Devices*, **ED-30**, 1780 (1983).
22. C. C. Han and J. C. Corelli, *J. Vac. Sci. Technol.*, B, **6**, 219 (1988).

23. M. Toriumi, T. Ueno, T. Iwayanagi, M. Hashimoto, N. Moriuchi, and S.-I. Shirai, *Proc. SPIE*, **920**(04) (1988).
24. T. Matsuzawa, A. Kishimoto, and H. Tomioka, *IEEE Electron Device Lett.*, **EDL-3**, 58 (1982).
25. T. Matsuzawa and H. Tomioka, *IEEE Electron Device Lett.*, **EDL-2**, 90 (1981).
26. Y. Yamashita, R. Kawazu, K. Kawamura, S. Ohno, T. Asano, K. Kobayashi, and G. Nakamatsu, *J. Vac. Sci. Technol.*, *B*, **3**, 314 (1985).
27. T. R. Pampalone, *Solid State Technol.*, **27**(6), 115 (1984).
28. P. A. Ruggiero, *Solid State Technol.*, **27**(3), 165 (1984).
29. C. G. Willson, R. Miller, D. McKean, N. Klecak, T. Tompkins, D. Hofer, J. Michl, and J. Downing, *Polym. Sci. Eng.*, **23**, 1004 (1983).
30. R. D. Miller, D. R. McKean, T. L. Tompkins, N. Klecak, C. G. Willson, J. Michl, and J. Downing, *ACS Symp. Ser.*, **242**, 25 (1984).
31. (a) B. D. Grant, N. J. Klecak, R. J. Twieg, and C. G. Willson, *IEEE Trans. Electron Devices*, **ED-28**, 1300 (1981). (b) C. G. Willson, R. D. Miller, D. R. McKean, and L. A. Pedersen, *Proc. SPIE*, **771**, 2 (1987). (c) S. R. Turner, K. D. Ahn, and C. G. Willson, *ACS Symp. Ser.*, **346**, 200 (1987).
32. G. Schwartzkopf, *Proc. SPIE*, **920**(07) (1988).
33. G. Ciamician and P. Silber, *Ber.*, **34**, 2040 (1901).
34. B. Amit and A. Patchornik, *Tetrahedron Lett.*, p. 2205 (1973).
35. A. Amit, U. Zehavi, and A. Patchornik, *Isr. J. Chem.*, **12**, 103 (1974).
36. H. Barzynski and D. Saenger, *Angew. Makromol. Chem.*, **93**, 131 (1981).
37. E. Reichmanis, C. W. Wilkins, Jr., and E. A. Chandross, *J. Vac. Sci. Technol.*, **19**, 1338 (1981).
38. C. C. Petropoulos, *J. Polym. Sci., Polym. Chem. Ed.*, **15**, 1637 (1977).
39. S. A. MacDonald and C. G. Willson. *IBM Res. Rep.*, No. RJ 3155 (38802) (1981).
40. F. M. Houlihan, A. Shugard, R. Gooden, and E. Reichmanis, *Proc. SPIE*, **920**(09) (1988).
41. T. G. Tessier, J. M. J. Frechet, C. G. Willson, and H. Ito, *ACS Symp. Ser.*, **266**, 269 (1985).
42. Y. Amerik and J. E. Guillet, *Macromolecules*, **4**, 375 (1971).
43. E. Dan and J. E. Guillet, *Macromolecules*, **6**, 230 (1973).
44. Y. Mimura, T. Ohkubo, T. Takeuchi, and K. Sekihawa, *Jpn. J. Appl. Phys.*, **17**, 541 (1978).
45. E. A. Chandross, E. Reichmanis, C. W. Wilkins, Jr., and R. L. Hartless, *Solid State Technol.*, **24**(9), 81 (1981).
46. G. A. Delzenne, U. Laridon, and H. Peeters, *Eur. Polym. J.*, **6**, 933 (1970).
47. C. W. Wilkins, Jr., E. Reichmanis, and E. A. Chandross, *J. Electrochem. Soc.*, **127**, 2510 (1980).
48. E. Reichmanis and C. W. Wilkins, Jr., *Org. Coat. Plast. Chem.*, **43**, 243 (1980).
49. E. Reichmanis, C. W. Wilkins, Jr., and E. A. Chandross, *J. Electrochem. Soc.*, **127**, 2514 (1980).
50. J. E. Guillet, S. K. L. Li, and H. C. Ng, *ACS Symp. Ser.*, **266**, 165 (1984).
51. A. Levine, *Soc. Plast. Eng., Prepr.*, p. 106 (1973).

52. M. Tsuda, *Photogr. Sci. Eng.*, **23**, 290 (1979).
53. Y. Ohnishi, M. Itoh, K. Mizumo, H. Gokan, and S. Fujiwara, *J. Vac. Sci. Technol.*, **19**, 1141 (1981).
54. J. V. Crivello, J. L. Lee, and D. A. Conlon, *J. Polym. Sci., Polym. Chem. Ed.*, **25**, 3293 (1987).
55. S. P. Pappas, *J. Imaging Technol.*, **11**, 146 (1985).
56. J. V. Crivello, *Adv. Polym. Sci.*, **62**, 1 (1984).
57. J. C. Dubois, A. Eranian, and E. Datmanti, *Proc. Symp. Electron Ions Beam Sci., Int. Conf., 8th*, p. 940 (1978).
58. H. Ito and C. G. Willson, *Polym. Sci. Eng.*, **23**, 1012 (1983).
59. J. M. J. Frechet, F. Bouchard, F. M. Houlihan, B. Kryczka, E. Eichler, N. Clecak, and C. G. Willson, *J. Imaging Sci.*, **30**, 59 (1985).
60. C. G. Willson, H. Ito, J. M. J. Frechet, and F. Houlihan, *Proc. IUPAC Macromol. Symp., 28th, Amherst, MA, July*, p. 448 (1982).
61. J. M. J. Frechet, E. Eichler, H. Ito, and C. G. Willson, *Polymer*, **24**, 995 (1983).
62. D. R. McKean, S. A. MacDonald, C. G. Willson, and N. J. Clecak, *Proc. SPIE*, **920**(08) (1988).
63. M. J. O'Brien and J. V. Crivello, *Proc. SPIE*, **920**(06) (1988).
63a. H. Ito, C. G. Willson, and J. M. J. Frechet, *Proc. SPIE*, **771**, 24 (1987).
64. J. M. J. Frechet, F. M. Houlihan, F. Bouchard, B. Kryczka, and C. G. Willson, *J. Chem. Soc., Chem. Commun.*, p. 1514 (1985).
65. F. M. Houlihan, F. Bouchard, J. M. J. Frechet, and C. G. Willson, *Macromolecules*, **19**, 13 (1986).
66. J. M. J. Frechet, F. Bouchard, F. M. Houlihan, E. Eichler, B. Kryczka, and C. G. Willson, *Makromol. Chem. Rapid Commun.*, **7**, 121 (1986).
67. S. C. Narang and S. T. Attarwala, *ACS Polym. Prepr.*, **26**(2), 323 (1985).
68. J. M. J. Frechet, E. Eichler, M. Stanciulescu, T. Iizawa, F. Bouchard, F. M. Houlihan, and C. G. Willson, *ACS Symp. Ser.*, **346**, 138 (1987).
69. L. R. Gatechair and S. P. Pappas, *Proc. Org. Coat. Appl. Polym. Sci. Div.*, **46**, 707 (Mar. 1982).
70. D. Rehm and A. Weller, *Ber. Bunsenges. Phys. Chem.*, **73**, 834 (1969).
71. M. J. Bowden, in *Macromolecules, An Introduction to Polymer Science* (F. A. Bovey and F. H. Winslow, Eds.), p. 45, Academic, New York, 1979.
72. C. G. Willson, H. Ito, J. M. J. Frechet, T. G. Tessier, and F. M. Houlihan, *J. Electrochem. Soc.*, **133**, 181 (1986).
73. O. Vogl, *J. Polym. Sci.*, **46**, 241 (1960).
74. H. Ito, *Proc. SPIE*, **920**(05) (1988).
75. A. Steinmann, *Proc. SPIE*, **920**(02) (1988).

8 ELECTRON BEAM LITHOGRAPHY

THE PHYSICS OF THE ELECTRON BEAM

Electron beam lithography is one of the newer lithographic patterning methods which use ionizing radiation (electrons, x-rays, or ions) as information carrier. Electron beam lithography is the most developed of these techniques. It has several important advantages over optical lithography; not only are diffraction effects virtually absent, electrons can be focused into very fine points (0.1 μm and less) and deflected by electrostatic and by magnetic fields and are therefore easily scanned and controlled by computer. The combination of an electron beam instrument with a computer makes electronic pattern generation possible and that, in fact, is the method by which the great majority of integrated circuit (IC) patterns are formed at present.

Resolution Limits

While diffraction is insignificant in electron beam lithography, there are other sources of image degradation, foremost the scattering of electrons in the resist film, the random emission of secondary electrons, and the backscatter of electrons from the denser substrate on which the resist is coated [1]. Figure 8-1 shows a computer simulation of the trajectories of electrons arriving at the surface of a resist film that is coated on silicon [2]. This kind of image degradation can again be described empirically [3] by a line spread function such as the one shown, for example, in Fig. 8-2 taken from the work of Thompson [4].

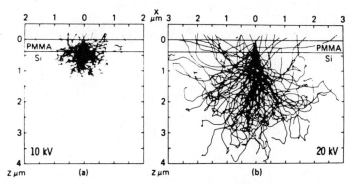

Figure 8-1 Monte Carlo simulation of the trajectories of 10 and 20 kV electrons in a PMMA film coated on silicon. (Reproduced with permission from D. Kyser and N. S. Viswanathan [2].)

Although electron scattering broadens the incoming beam considerably, electron lithography can nevertheless achieve higher resolution than optical patterning methods [5]. In Fig. 8-3 modulation transfer functions of an optical printer are compared with those of an early electron exposure tool [6].

Exposure Tools

Electron beam exposure tools have been developed from the electron microscope [7]. A source of electrons, the electron gun, is arranged on a vertical column, together with electrostatic and magnetic electron optics, blanking

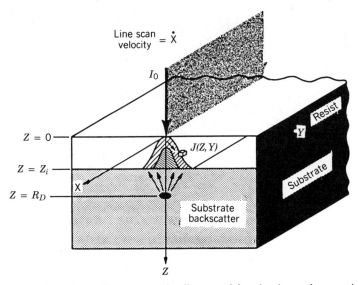

Figure 8-2 Two-dimensional electron scattering line spread function in a polymer resist coated on a dense substrate. (Reproduced with permission form L. F. Thompson [4].)

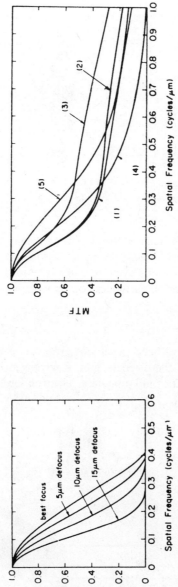

Figure 8-3 On the left is the MTF of a 1:1 F/1.5 optical projection printer, patterning a resist film of 1 μm thickness coated on SiO$_2$; on the right, the MTF of a electron beam instrument patterning the same material, with varying accelerating potentials and spot sizes r.

1. 20 keV $r = 0.5$ μm   film thickness 1 μm
2. 20 keV $r = 0.25$ μm film thickness 1 μm
3. 20 keV $r = 0.5$ μm   film thickness 0.5 μm
4. 10 keV $r = 0.5$ μm   film thickness 1 μm
5. 10 keV $r = 0.5$ μm   film thickness 0.5 μm

(Reproduced with permission from R. K. Watts and J. H. Bruning [5].)

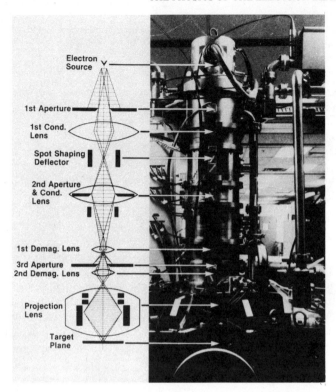

Figure 8-4 Schematic of the imaging concept of the electron beam column of the IBM exposure tool EL-3, and photograph of the actual instrument. (Reproduced with permission from H. C. Pfeiffer [9].)

plates, and beam shapers. The electron gun contains either a thermionic or a field emitting source of electrons. The first is either a heated tungsten wire, or a wire covered with a material such as LaB_6, which has a low work function for electron emission. The whole arrangement, including a mobile stage, which carries the wafer, is enclosed in a carefully managed vacuum. Figure 8-4 shows a schematic of the electron optics and a photograph of the current IBM EL-3 exposure tool [8, 9].

Almost all of these machines expose the resist film in serial fashion by scanning the electron beam over the surface of the resist. Scanning is accomplished by moving the wafer in one direction and scanning the electron beam via the deflector plates in a direction perpendicular to the stage motion. Both movements have to be controlled to fine tolerances of the order of less than one fourth of the width of the smallest line feature in the pattern. This control is achieved by interferometrically matching two optical grids in the corners of the scan field. In this way overlay and butting errors have been reduced down to 0.1 μm.

TABLE 8-1 Performance Parameters of Representative Electron Beam Lithography Tools [7]

Performance Parameters	EL-1 1975	EL-2 1977	EL-3 1980	EL-3 1982
Minimum image	2.5 μm	1 μm	1 μm	0.5 μm
Field size	6.5 mm	8 mm	10 mm	5 mm
Overlay accuracy	0.75 μm	0.4 μm	0.4 μm	0.1 μm
Wafer size	100 mm	100 mm	165 mm	165 mm

Design Rule (μm)	Throughput (wafer/h)			
2.5	10	15		
1.0		5	10–20	
0.5				2–5

Depending on the scanning strategy, electron beam instruments can be divided into two groups: those using a raster scan mode and those using a vector scan. In the *raster scan* mode every point on the film surface is addressed. The stage moves in rows and the electron beam exposes the wafer in columns. To convey an idea of the data stream in an electron beam exposure tool the following points are noted: In a frequently used production instrument there are 512 addresses per column, and the pattern information comes from a shift register at the rate of 40 MHz (it has now grown to 100 MHz and more). This translates into a scan rate of about 10 cm^2/min, using a Gaussian exposure spot of 0.5 μm diameter.

The first practical electron exposure tool (1975) was the EBES machine of Bell Laboratories [10]. It was soon followed by the highly automated EL-1 machine of IBM [11, 12]. Table 8-1 indicates the evolution of some of the performance parameters of the IBM EL system over seven years.

These machines have been highly successful as mask making devices, to the point where the time required for writing a mask has become shorter by more than two orders of magnitude compared with the time required by an optical pattern generator. Also, the writing accuracy both in feature definition and in the avoidance of writing errors is much higher in the electron beam instruments [7]. Toshiba Machine Co. has designed (1987) an electron beam reticle (mask) writing system that can cope with a data rate of 80 MHz and which is able to write an 80 mm × 90 mm reticle with a 0.25 μm smallest address size in 40 min. This corresponds to several million features per reticle [13]. The Hitachi Company has developed a nanometric electron lithographic system that will pattern a standard wafer at 0.1 μm resolution in 1 h. The technical breakthrough here is a titaniated tungsten thermal field emitter that is 10 times brighter than conventional electron sources [13, 14].

The one inherent weakness of a beam scan device is the low throughput compared with a large area exposure device. To combat the low throughput, designers have looked to ways for cutting exposure time. Since in many integrated circuit patterns only some 20 to 30% of the wafer area carry information, considerable time can be saved by sequentially addressing only those areas that need to be exposed. This so-called *vector scan* method entails some mechanical complications, for example, the stage has to be moved rapidly with great precision in a step and repeat fashion. Nonetheless, several successful machines have been built on this principle, such as the Philips

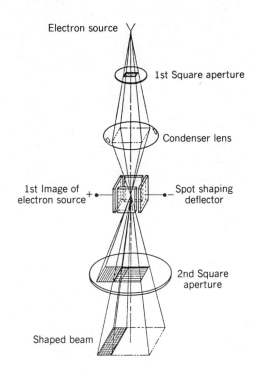

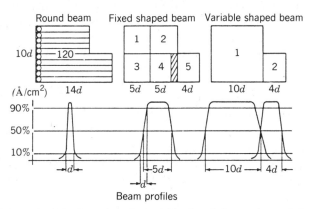

Figure 8-5 Schematic of the formation of a variable shaped electron beam; below, principle of pattern generation with varying beam profiles. (Reproduced with permission from B. P. Piwczyk and A. E. Williams [16] and R. D. Moore [17].)

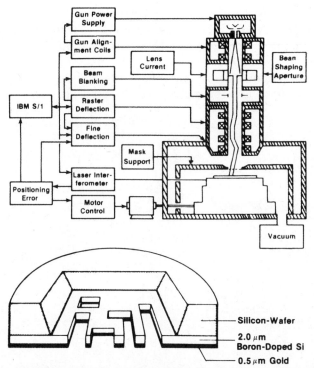

Figure 8-6 Block diagram of electron beam proximity printing system, and below, schematic cross section of electron beam transmission mask. (Reproduced with permission from H. Bohlen et al. [21].)

Beamwriter, the Cambridge Instruments EBMF2, the Texas Instruments EBSP, the IBM Vector Scan I (VSI), and the Waferwriter of the Electron Beam Corporation of San Diego.

These designs are using a fixed Gaussian beam to write the pattern. It occurred to some designers that even more exposure time could be saved by using a square-shaped beam [15], which would expose a whole address in a single flash. Even better, a *variable shaped beam* could do the same for whole local areas. Figure 8-5 shows the schematic of the electron optics of a variable shaped electron beam [15], as well as the principle of pattern generation with fixed and variable beam shapes [16, 17].

The ultimate in exposure time saving would be achieved with an *electron projection* system where a broad (10 mm × 10 mm) electron beam would be used to project a whole pattern onto the wafer surface. Various experimental arrangements have been described [18–20], either using an emitting surface carrying the design in terms of a patterned photocathode, or taking the pattern from a metal mask on an electron-transparent support. At this point none of these instruments have found application in practical manufacturing. More recently, IBM Deutschland has developed an electron beam proximity printing technique, which uses appropriate electron masks and which promises to

TABLE 8-2 Time Budgets for a 1.0 and 0.5 μm Design Rule Pattern in a Direct Write Facility [15]

	Design Rule	
Operation	1 μm	0.5 μm
	(s)	(s)
Wafer loading	30.000	30.000
Stage motion	69.106	193.045
Alignment	40.350	40.350
Subfield settling	5.077	19.394
Exposure	20.480	85.965
Data transfer	39.913	76.308
Total	184.446	369.654
Exposure time (%)	11	23

combine the precision of the electron beam with a reasonably high throughput (up to 36 wafers of 125 mm diameter per hour) [21]. Figure 8-6 shows a block diagram of the system and part of an electron mask [21].

One factor codetermining exposure time is of course the radiation sensitivity of the electron resist, and instrument designers have sometimes put the onus for improved throughput onto the material scientists. To what extent this may be justified can be judged from the data in Table 8-2, which gives the time budgets of two typical patterning operations [15].

Applications of Electron Beams

The most important application of electron beam lithography is the making of masks, where E-beam is currently the dominant technology. This applies not only to masks for optical lithography, but even more so to x-ray masks and to masks for electron beam and ion beam projection.

In device fabrication, direct-write electron beam lithography is used for the definition of critical features of the smallest size while the rest of the circuit is patterned by optical lithography in a so-called "hybrid" approach [22]. Hybrid lithography is also the technique adopted in the prototype and pilot fabrication of very high speed integrated circuits (VHISICS) where the electron beam plays a crucial role [23].

While these programs are concerned with the large scale manufacture of devices with a 0.5 μm design rule, electron beam lithography is capable of higher resolution and microfabrication in the 10 to 100 nm regime is currently being pursued in a few advanced laboratories [8, 24–27]. In all of these, the exposure tool used is a modified electron microscope. First, this has the advantage of an ultrafine writing spot; the Philips PSEM 500 has a writing spot that can be varied from a nominal 8 to 1000 nm, and the writing spot of the Philips 420 T Scanning Transmission Electron Microscope (STEM) has a smallest spotsize of 2 nm [24, 28].

This kind of resolution opens up the possibility of new devices with completely new architecture, for example, highly connective *neural* networks that simulate the operational structure of the brain [29], or, in another mode, the exploitation of new quantum electronic effects. Electron beams also have important applications in the testing of finished devices and in their customization and repair. Testing of a highly complex IC can no longer be safely accomplished with optical devices. In fact, scanning electron microscopes (SEMs) have become a necessity [30]. By combining an SEM with a computer it is possible to automate IC inspection and detect dimensional anomalies in the workpiece with the help of pattern recognition (template matching) software. This is important for process automation [31].

Electron beam methods are also used in the customization of integrated circuits [32]. Specialized circuits are often not required in very large numbers and would be too expensive if they had to be manufactured from scratch. However, many different specialized circuits can be derived from the same basic circuit pattern by a comparatively small number of new connections. The basic circuits can be manufactured in bulk and later customized to order. It was found that so-called "floating gate nonvolatile switches" can be built into the basic circuits [33] and these later addressed and "set" with a finely controlled electron beam. Electron beam programmable switches make it possible to convert mass produced basic circuits within minutes into custom programmed special integrated circuits. Such application specific integrated circuits (ASICs) are now increasing in commercial importance [34].

A FEW NOTES ON RADIATION CHEMISTRY

Ionization and Its Aftermath

In using an electron beam for the imaging of resists one hopes to exploit the chemical effects of fast electrons. The most important characteristic of fast electrons and of other ionizing radiations is that the energy of the individual quanta is much larger than the energy of the chemical bonds in the material and larger than the ionization potentials of its constituent atoms. Consequently, the primary result of the interaction between the radiation and the material is ionization. A whole sequence of steps follows, leading to a variety of products. The description of these events is the subject of radiation chemistry [35].

Although radiation chemistry and photochemistry are clearly distinct, the boundary between them is rather diffuse. In photochemistry, radiation quanta are absorbed by specific chromophores and promote these into well-defined excited states. As the energy of the quanta increases and starts to exceed some of the bond energies in the substrate, photochemistry leads to fragmentation and occasionally to ionization. In fact, deep-UV photochemistry has many features of radiation chemistry. The distinguishing feature of radiation chemistry is that energy absorption is not any more associated with a particular chromophore, but occurs at random in the material. Radiation chemistry may

TABLE 8-3 Ionization Energies of Some Atoms and Molecules [36]

	E_i (eV)[a]		E_i (eV)[a]
H	13.2	Cl_2	13.2
H_2	15.4	Br_2	12.8
He	24.6	I_2	9.7
He^+	54.4	CH_4	13.1
C	11.27	C_2H_5	11.6
C^+	24.38	C_2H_2	10.5
C^{2+}	47.87	C_3H_6	9.7
N	14.54	$CH_3\cdot$	10.1
N_2	15.6	$C_2H_5\cdot$	8.7
O	13.62	$C_3H_9\cdot$	7.8
O_2	12.2	C_6H_6	9.2

[a] 1 eV = 23.04 kcal/mole.

therefore be defined as being concerned with quantal energies far in excess of the ionization energy of common atoms and molecules (see Table 8-3). The lower energy limit of radiation chemistry is often quoted as 30 eV (700 kcal/mole).

The typical course of a radiation chemical event may be described as follows: The primary step is ionization.

$$M \rightsquigarrow M^+ + e^- \qquad (1)$$

The electron produced in the primary step carries a large amount of kinetic energy that on encounter with neighboring atoms leads to further ionization. This process is repeated over and over again and some of the secondary electrons may recombine with their original partners (geminate recombination) and produce molecules in higher excited states.

$$M^+ e^- \rightarrow M^{**} \qquad (2)$$

These either emit radiation, or fragment into ions or radicals, decay to lower excited states by internal conversion, or finally return to the ground state.

$$M^{**} \rightarrow M + h\nu' \qquad (3)$$

$$M^{**} \rightarrow M^+ + e^- \qquad (4)$$

$$M^{**} \rightarrow R_1^{\cdot} + R_2^{\cdot} \qquad (5)$$

$$M^{**} \rightarrow M^* \qquad (6)$$

Often this sequence leads to final products similar to those that would have been obtained by photochemistry. As the secondary electrons lose energy they become thermalized and may attach themselves either to specific molecules (forming molecular ions), or to the solvent in the form of solvated electrons.

All these events take place in the vicinity of the primary high energy particle and the various product species are formed along its trajectory or in spurs and side branches to it. The distribution of ions and radicals in the system is therefore spatially highly nonuniform. At the electron densities used in electron beam imaging devices, however, the image noise level produced by these nonuniformities remains tolerable.

Radiation Chemical Yield and Dosimetry

Since the energy of the electrons is not absorbed at a specific site or within a particular molecule, the quantum yield of an electron reaction cannot be unambiguously defined. The inherent efficiency of a radiation chemical process is therefore measured by a radiation chemical yield or *G-value*. If a process leads to a certain number of reaction products, the *G*-value of each is defined as the number of moles of that product resulting from 100 eV of energy deposited in the system. In a system of unit volume (1 L) the molarity of the product [P] produced by a dose D is given by

$$[P] = G(P) \times \frac{D}{100} \text{ eV} \tag{7}$$

A determination of the *G*-value requires therefore the measurement of a product quantity and of dose or dose rate.

Dose is often expressed in terms of the unit *rad* (or kilorad, megarad, etc.). One rad is the dose that deposits 100 ergs into 1 g of material. The older unit of *exposure*, the Roentgen, deposits numerically about the same amount of energy into condensed matter, but is by definition a measure of incident, not of absorbed, energy and therefore less useful in radiation chemistry. In the calculation of *G*-values, absorbed dose must be expressed in electron volts. For practical dose calculations see Reference 37.

Since most of the energy deposited in a radiation chemical system is dissipated as heat, the deposited dose can be measured calorimetrically, and that is in fact the method by which the energy content of electron beams is usually calibrated. However, for chemical work in liquids or solids, a chemical dosimeter is used. The *Fricke dosimeter* is now universally accepted. It is based on the conversion of Fe^{2+} ions to Fe^{3+} ions in acid solution. The reaction has a fairly complex mechanism, but it is reproducible and it can be monitored conveniently by the optical density, at 304 nm, of the ferric ions produced.

$$H_2O \rightsquigarrow H\cdot + OH\cdot \tag{8}$$

$$Fe^{2+} + \cdot OH \rightarrow Fe^{3+} + OH^- \tag{8a}$$

$$H\cdot + O_2 \rightarrow HOO\cdot \tag{8b}$$

$$Fe^{2+} + HOO\cdot \rightarrow Fe^{3+} + HOO^- \tag{8c}$$

$$HOO^- + H_2O \rightarrow H_2O_2 + HOO^- \tag{8d}$$

$$2Fe^{2+} + H_2O_2 \rightarrow 2Fe^{3+} + 2OH^- \tag{8e}$$

$$OH^- + H^+ \rightarrow H_2O \tag{8f}$$

The G-value of Fe^{3+} in the process is high, which makes the system a sensitive dosimeter. For the production of trivalent iron by ^{60}Co-gamma radiation or fast electrons in these systems $G(Fe^{3+}) = 15.5$.

In the following we note a few typical processes that illustrate some of the basic principles of organic radiation chemistry [35].

Typical Radiation Chemical Processes

The Radiolysis of Liquid n-Hexane. n-Hexane, C_6H_{14}, is typical of aliphatic hydrocarbons. It contains only C—C and C—H bonds. The fission of C—C bonds leads to hydrocarbon fragments of lower molecular weight, while the fission of C—H bonds leads initially to hexyl radicals that may dimerize, and to hexenes, C_6H_{12}. The list of product yields in Table 8-4 shows that in aliphatic hydrocarbons C—H fission predominates, but that C—C fission products are also formed in appreciable amounts.

The Radiolysis of Benzene. While the G-values of some products in the radiolysis of hexane are quite high, irradiation of benzene is rather ineffective. Electron attachment to benzene is by far the most important process, the radiation yield of benzene anions (solvated electrons) being $G(C_6H_6^-) = 0.8$, while hydrogen is produced only in trace amounts, $G(H_2) = 0.04$. Evidently, benzene is able to dissipate excitation energy without breaking any bonds. This property of aromatics in general is important, for example, in plasma etching, where the presence of aromatic rings is known to confer a degree of radiation protection to the polymer.

TABLE 8-4 Products in the Radiolysis of n-Hexane [38]

Product	G-Value
H_2	5.0
CH_4	0.1
C_2H_4	0.3
C_2H_6	0.3
C_3H_6	0.1
C_3H_8	0.4
C_4H_{10}	0.5
C_5H_{12}	0.3
C_6H_{12}	1.2
C_7	0.2
C_8	0.5
C_9	0.5
C_{10}	0.4
C_{12}	2.0

The Radiolysis of Ethyl Iodide. The radiation chemical decomposition of ethyl iodide illustrates the rule that "weak bonds are broken more frequently than stronger ones." The main products of the reaction are iodine, ethylene, and ethane in almost equal molar quantities.

$$CH_3CH_2-I \rightsquigarrow CH_3-CH_2^{\cdot} + I\cdot \qquad (9)$$

$$2CH_3-CH_2^{\cdot} \rightarrow CH_2=CH_2 + CH_3-CH_3 \qquad (9a)$$
$$(G = 2.1) \qquad (G = 1.9)$$

$$2I\cdot \rightarrow I_2 \qquad (9b)$$
$$(G = 2.1)$$

The Radiolysis of n-Butyl Chloride. This system illustrates the rule that "free radicals assume a configuration of lowest energy more rapidly than saturated molecules." In the radiolysis of n-butyl chloride there is almost only one product, namely, sec-butyl chloride. It must be produced by a chain reaction, since its G-value is 60! The reaction sequence for this process is indicated in Eqs. (10).

$$n\text{-BuCl} \rightsquigarrow n\text{-Bu}\cdot + \text{Cl}\cdot \qquad (10)$$

$$n\text{-Bu}\cdot \rightarrow sec\text{-Bu}\cdot \qquad (10a)$$

$$sec\text{-Bu}\cdot + n\text{-BuCl} \rightarrow n\text{-Bu}\cdot + sec\text{-BuCl} \quad (G = 60) \qquad (10b)$$

The process is driven by the rapid transition of the n-butyl radical to the more stable sec-butyl radical, which is an example of the basic principle just mentioned.

Another important and more general process of the internal rearrangement of a radical is the backbone degradation of polyacrylates in the so-called "Norrish-type I" process. This is a case of backbone disintegration resulting from side-chain scission. It occurs, for example, in the radiolysis (and deep-UV photolysis) of poly(methyl methacrylate) PMMA.

$$\begin{array}{c} \text{CH}_3 \quad\quad \text{CH}_3 \\ | \quad\quad\quad | \\ -\text{CH}_2-\text{C}-\text{CH}_2-\text{C}- \\ | \quad\quad\quad | \\ \text{CO} \quad\quad \text{CO} \\ | \quad\quad\quad | \\ \text{OCH}_3 \quad \text{OCH}_3 \end{array} \rightsquigarrow \begin{array}{c} \text{CH}_3 \quad\quad \text{CH}_3 \\ | \quad\quad\quad | \\ -\text{CH}_2-\text{C}-\text{CH}_2-\text{C}- \\ | \quad\quad\quad | \\ \text{CO} \quad\quad \text{CO} \\ | \quad\quad\quad | \\ \text{OCH}_3 \quad \text{OCH}_3 \end{array} \qquad (11)$$

$$\begin{array}{c} \text{CH}_3 \quad\quad \text{CH}_3 \\ | \quad\quad\quad | \\ -\text{CH}_2-\text{C}=\text{CH}_2 + \cdot\text{C}- \quad + \text{CO}_2 + \cdot\text{CH}_3 \\ \quad\quad\quad\quad\quad\quad | \\ \quad\quad\quad\quad\quad\quad \text{COOCH}_3 \end{array}$$

Once the C—C bond next to the carbonyl has been opened, the resulting unstable radical rearranges to an olefin and a second more stable radical, thereby fragmenting the polymer backbone.

The Radiation Chemistry of Polymers

Backbone Scission and Crosslinking. The radiation chemistry of polymers is mainly concerned with two processes: backbone scission and crosslinking. Whether a polymer will act as a positive or as a negative resist depends on which of the two processes predominates [39].

Backbone scission leads to polymer degradation and to a lowering of the molecular weight of the polymer. It is favored by the inclusion of weak bonds in the backbone (see, e.g., the polyalkene sulfones). In vinyl-type polymers, backbone scission occurs more often if the main chain contains tetra-substituted carbon, that is, carbon atoms that do not carry any hydrogen. Examples are PMMA and poly(tetrafluoroethylene), which both degrade rapidly on irradiation. Vinyl polymers, rubber and polyethylene, where each carbon of the main chain carries at least one hydrogen, tend to crosslink in the solid state.

In view of the importance of scission and crosslinking for the potential use of a polymer in microlithography the radiation yields of scission, G_s and of crosslinking, G_x are important material characteristics.

Determination of the Scission Yield G_s. If crosslinking can be neglected in a polymer, the efficiency of backbone scission may be determined from the change in molecular weight that occurs on irradiation. Before irradiation, the number of molecules of polymer in the sample can be expressed as

$$N^0 = N_A \frac{w}{M_n} \qquad (12)$$

where w is the weight of the sample (in grams), M_n is the number average molecular weight before irradiation, and N_A is Avogadro's number. After exposure to a given dose D, expressed in electron volts per gram, the number of scissions (N_{sc}), by definition of the scission yield G_s, is given by Eq. (13).

$$N_{sc} = \frac{G_s}{100} Dw \qquad (13)$$

Each scission event increases N by one molecular unit so that the molecular weight M_n after exposure to a dose D can be expressed from Eqs. (12) and

TABLE 8-5 Radiation Yields of Scission for a Group of Common Polymers [35]

Polymer	G_s
Poly(α-methylstyrene)	0.3
Poly(isobutene)	4
PMMA	2
Cellulose	11
Poly(α-methylcellulose)	16
Poly(tetrafluoroethylene)	Very high

(13) by an equation of the form

$$\frac{1}{M_n} = \frac{1}{M_n^0} + \frac{G_s}{100 N_A} D \tag{14}$$

Table 8-5 lists values of G_s obtained in this way for a number of common polymers [35].

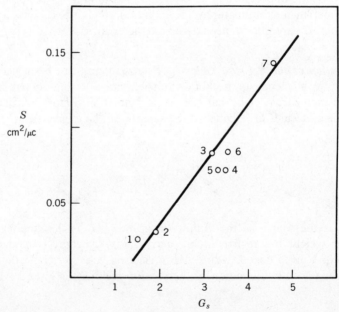

Figure 8-7 The correlation between G_s values and lithographic sensitivity to 15 kV electrons, tested on analogs of PMMA. G-values obtained by exposure to ^{60}Co γ radiation, lithographic sensitivity is in square centimeters per microcoulomb.
(1) PMMA; (2) PMMA-MA Acid; (3) PMMA-MA Anhydride; (4) PMMA-Isobutene; (5) PMMA-co-Cl-Acrylate; (6) PMMA-co-CN-Acrylate; (7) PMMA-MA Acid-MA Anhydride. [Data taken from C. G. Willson, *Introduction to Microlithography*, *ACS Symp. Ser.*, **219**, 126 (1983)].

The determination of scission yield is often accomplished by exposure of the material to γ radiation from a ^{60}Co source. The molecular weight of the polymer is monitored in intervals during the process. These data are of value to the lithographer, because the radiation chemical procedure is simpler than a full lithographic evaluation, and there is a good correlation between G_s and lithographic sensitivity. This correlation is illustrated in Fig. 8-7.

Determination of the Crosslinking Yield G_x. The radiation chemical yield of crosslinking can be determined by monitoring the insolubilization of the polymer. Charlesby and Pinner [40] have shown that the sol fraction, s, as a function of the radiation dose received by the polymer can be described by an equation of the following form.

$$s + s^{1/2} = \frac{G_s}{2G_x} + \frac{9.65 \times 10^5}{M_n^0 G_x} \frac{1}{D} \tag{15}$$

In the experiments, the polymer sample is irradiated and then extracted with a suitable solvent, in which the sol fraction, that is, the fraction of material that is not connected to the gel network, can be determined. A plot of

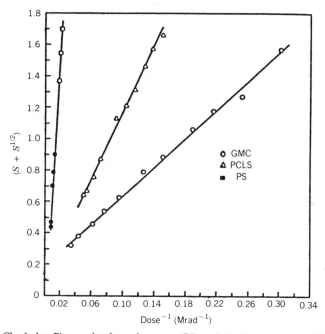

Figure 8-8 Charlesby–Pinner plot for polystyrene (PS), poly(3-chlorostyrene) (PCLS), and the copolymer of 3-chlorostyrene with glycidyl methacrylate (GMC). The G_x values derived from the slope of the plots are 0.03, 0.61, and 1.02, respectively. (Reproduced with permission from A. Novembre and T. N. Bowmer [41].)

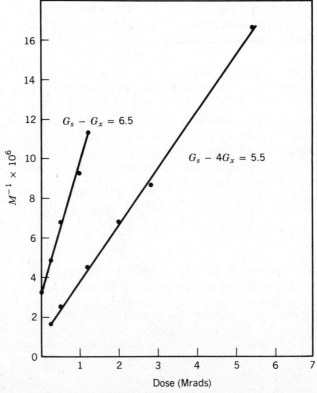

Figure 8-9 Plot of $1/M_n$ and $1/M_w$ versus radiation dose for a poly(butene-1-sulfone). (Reproduced with permission from L. Stillwagon [42].)

$(s + s^{1/2})$ against $1/D$ leads then to the value of G_x. Figure 8-8 shows such a plot for three negative electron resists [41].

If both crosslinking and backbone scission are important in a system, the G-values of both processes can be derived simultaneously by measuring the number average and the weight average molecular weights, M_n and M_w, of the sample under irradiation. The M_n and M_w values can be expressed by two equations of similar form.

$$\frac{1}{M_n} = \frac{1}{M_n^0} + [G_s - G_x]\frac{D}{100 N_A} \tag{16}$$

$$\frac{1}{M_w} = \frac{1}{M_w^0} + [G_s - 4G_x]\frac{D}{100 N_A} \tag{17}$$

Figure 8-9 shows plots of the reciprocal molecular weights of a sample of poly(butene-1-sulfone) as a function of radiation dose [42]. From the slope of the plots, $G_s - G_x$ and $G_s - 4G_x$ are obtained, leading to the values $G_s = 6.83$ and $G_x = 0.33$.

NEGATIVE ELECTRON RESISTS

Resists Containing Epoxy Groups. Most negative working electron resists are based on radiation-induced crosslink formation. Polyethylene, various rubbers, polystyrene, and poly(vinyl chloride) are known to crosslink on irradiation, but from a lithographic point of view their radiation sensitivity is too low. The crosslinking efficiency of these materials can be substantially enhanced by introducing specific radiation sensitive (labile) groups or bonds [43].

Nonogaki and co-workers at Hitachi have found that azido groups and epoxy groups are well suited to this purpose [44]. Epoxy groups are preferred, because they are insensitive to light. An early example of a system with enhanced radiation sensitivity is epoxydized polybutadiene, which crosslinks with a radiation chemical yield of $G_x = 200$. This high yield is the consequence of a chain mechanism in the crosslinking process, initiated by one of the ions or radical ions, $M \cdot ^+$, which are the primary products of irradiation.

$$-CH_2-CH \underset{O}{\overset{}{\diagup}} CH_2 + M^{\cdot+} \longrightarrow -CH_2-\underset{\underset{H}{|}}{\overset{\overset{M}{|}}{\underset{}{\overset{O}{|}}}}C-\overset{\cdot}{C}H_2 \qquad (18)$$

$$\downarrow$$

$$-CH_2-\overset{\overset{OM}{|}}{C}H-CH_2$$
$$|$$
$$O$$
$$|$$
$$-CH_2-CH-\overset{\cdot}{C}H_2$$

The occurrence of a reaction chain is here supported by the fact that in some of these systems polymerization is observed even after irradiation has ceased. Postirradiative polymerization leads to the gradual enlargement of features in the imaged resist and is therefore detrimental to resolution. It can be inhibited

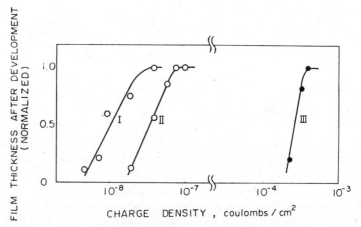

Figure 8-10 Exposure characteristics of two poly(glycidyl methacrylates), I and II, and poly(isobutyl methacrylate), Lucite 45, III. (Reproduced with permission from T. Hirai et al. [44].)

by the addition of radical scavengers such as, for example, 1,1-diphenyl-2-picrylhydrazyl (DPPH) [45].

Successful electron resists were developed from polymers carrying glycidyl groups. The generic structure is poly(glycidyl methacrylate), PGMA. It can be seen in Fig. 8-10 that the presence of the epoxy group makes a large difference in the crosslinking efficiency of the material. Here two poly(glycidyl methacrylate) samples of different molecular weight are compared with their nonepoxydized analog, poly(isobutyl methacrylate) [44].

Taniguchi et al. [46] have investigated the lithographic characteristics of PGMA and report a sensitivity of 0.5 $\mu C/cm^2$ and feature resolution down to 0.5 μm. A copolymer of glycidyl methacrylate with ethylacrylate has similar lithographic characteristics and better film forming properties. It is known under the trade name "COP" [47].

$$\left[\begin{array}{c} CH_3 \\ -CH_2-C- \\ CO \\ | \\ O \\ | \\ CH_2-CH\underset{O}{\underset{\diagdown\diagup}{}}CH_2 \end{array} \right]_m \left[\begin{array}{c} -CH_2-CH- \\ CO \\ | \\ O \\ | \\ C_2H_5 \end{array} \right]_n$$

COP

Copolymers of allyl methacrylate and of propargyl methacrylate with hydroxyethyl methacrylate have been found to function as sensitive electron resists [48, 49].

$$\left[-CH_2-\underset{\underset{\underset{CH_2-CH=CH_2}{|}}{\underset{|}{O}}}{\underset{|}{CO}}- \right]_m \left[-CH_2-\underset{\underset{\underset{CH_2-CH_2OH}{|}}{\underset{|}{O}}}{\underset{|}{CO}}- \right]_n$$

$D(0.5) = 0.064 \, \mu C/cm^2$ for 10 kV electrons $M_w = 48,000$

$$\left[-CH_2-\underset{\underset{\underset{CH_2-C\equiv CH}{|}}{\underset{|}{O}}}{\underset{|}{CO}}- \right]_m \left[-CH_2-\underset{\underset{\underset{CH_2-CH_2OH}{|}}{\underset{|}{O}}}{\underset{|}{CO}}- \right]_n$$

$D(0.5) = 0.42 \, \mu C/cm^2$ for 10 kV electrons $M_w = 26,000$

Resists Based on Polystyrene. While PGMA and its derivatives are highly sensitive and have excellent resolution, they do not withstand plasma treatment and are therefore not suitable for dry etch pattern transfer. This has led to the investigation of the plasma resistant polystyrene and some of its derivatives as electron resists [50, 51].

Polystyrene is known to crosslink slowly under irradiation, $G_x = 0.04$, the principal crosslinking mechanism being centered on the tertiary C atom of the backbone [51a].

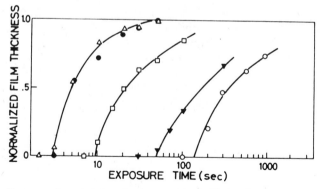

Figure 8-11 Exposure characteristics of polystyrene and of iodated polystyrenes with increasing degree of iodination: ○ 0; ▼ 0.02; □ 0.04; ● 0.28; △ 0.85. (Reproduced with permission from T. Ueno et al. [53].)

314 ELECTRON BEAM LITHOGRAPHY

$$-CH_2-CH(\text{Ph})- \rightsquigarrow -CH_2-\dot{C}(\text{Ph})- + H \quad (19)$$

By substituting some or all of the phenyl rings with halogen, the radiation sensitivity and the crosslinking efficiency can be considerably enhanced. Iodinated polystyrene was introduced by the Hitachi group in 1980 as a negative electron resist [52]. A moderate degree of substitution (some 20%), was sufficient to produce almost the maximum effect (see Fig. 8-11). The key step in the mechanism is the opening of the weaker C—I bond. This produces a phenyl radical that may abstract a hydrogen from the tertiary carbon of the backbone and form a crosslink by radical coupling [53].

$$-CH_2-CH(\text{Ph})-CH_2-CH(\text{Ph-I})- \rightsquigarrow -CH_2-CH(\text{Ph})-CH_2-CH(\text{Ph}\cdot)- + I\cdot$$

$$\searrow -CH_2-CH(\text{Ph-C(Ph)(CH}_2-)(CH_2-))- \quad (20)$$

Iodinated polystyrene has a sensitivity of about 2 $\mu C/cm^2$ and can resolve 1 μm features. The analogous poly(4-chlorostyrene) acts by a similar mechanism and has similar lithographic characteristics [53, 54]. Iodinated polystyrene has been used as a negative resist with the Philips Electron Image Projector [20].

The highest sensitivity in this group of materials is achieved with poly(chloromethylstyrene) for which an exposure requirement of 0.5 $\mu C/cm^2$ is claimed [55, 56].

$$-CH_2-CH(\text{Ph-CH}_2\text{Cl})-$$

Various other resists based on aromatic polymers have been described. For example, Ohnishi introduced poly(vinyl naphthalenes) and poly(chloromethyl-styrene-co-2-vinyl naphthalene) as electron resists with particularly high dry etch resistance [57].

An interesting new development is the introduction of conductive electron resists that eliminate the charging effect that is so troublesome in electron beam lithography. Todokaro et al. [58] used a partially chloromethylated poly(diphenylsiloxane) as the top imaging resist in a bilayer system, and used the ionically conductive ammonium salt of poly(p-styrene sulfonate) as the bottom layer.

$$\left[\begin{array}{c} CH_2-CH- \\ | \\ \bigcirc \\ | \\ SO_3^- \; NH_4^+ \end{array} \right]_n$$

A different approach was taken by Lee and Jopson [59] who found that poly(vinyl pyridine) was fairly insensitive to electrons, but that its sensitivity could be improved dramatically by quaternizing it with methyl iodide (see Fig. 8-12). They assume that on irradiation with electrons crosslinks are generated in two steps: In the first step methyl iodide is split off and is cleaved into radicals by a second electron. The methyl radicals so created produce backbone radicals that eventually lead to crosslinking.

$$-CH_2-CH- \rightsquigarrow -CH_2-CH- \quad (21)$$

$$CH_3I \rightsquigarrow CH_3^{\cdot} + I\cdot \quad (21a)$$

$$-CH_2-\overset{H}{\underset{|}{C}}- \; + \; CH_3^{\cdot} \longrightarrow -CH_2-\overset{\cdot}{C}- \longrightarrow \text{crosslinks} \quad (21b)$$

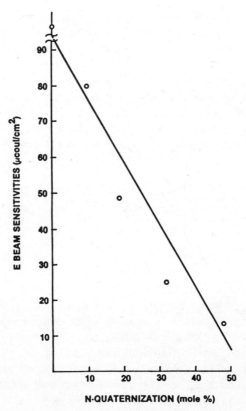

Figure 8-12 Effect of quaternization with methyl iodide on the electron sensitivity of poly(vinyl pyridine). (Reproduced from K. I. Lee and H. Jopson [59].)

Diazoquinone–Novolac Systems as Negative Electron Resists. Conventional positive diazoquinone–novolac photoresists are electron sensitive, but produce negative resist images when exposed to an electron beam. Shaw and Hatzakis [60] obtained quite good results with some commercial resists, the criterion here being image quality rather than photographic sensitivity. Liu et al. [61] have recently described a novolac based negative electron resist that allows a resolution of 0.3 μm at a dose of 4 $\mu C/cm^2$.

Following a suggestion by Pacansky, Oldham and Hieke [62], Mochiji et al. [63], and recently Berker [64] have devised a reversal process leading to good quality negative images. The method is based on the reaction of the ketene intermediate in diazoquinone photolysis (see Chapter 5). In the absence of water, the ketene reacts with the cresol moieties of the novolac to form various esters. This removes the photosensitive component from the system, without producing an alkali soluble acid. In the reversal process the resist is exposed in a vacuum (no water) to an electron pattern, and subsequently flood exposed in

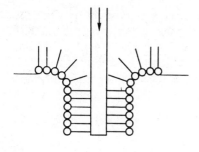

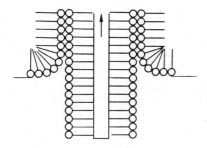

Figure 8-13 Schematic representation of a Langmuir–Blodgett monolayer film being picked up by a glass slide on going down into the water through a monomolecular surface film and deposition of a second layer onto the slide on its way up.

(moist) air to UV radiation that produces photoacid in the nonimage areas. A negative image of the pattern is formed on development in alkali.

Langmuir–Blodgett Films. The use of Langmuir–Blodgett films as resists was explored by Baraud et al. [65, 66] and by Lando [67–69]. Monomolecular films are spread on a water surface by the method pioneered by Blodgett and Langmuir [70]. The films are condensed on the water surface by lateral pressure (using a Langmuir trough) and picked up layer by layer on passing a glass slide or other substrate through the water–monolayer film interface (see Fig. 8-13). Successive monolayers are built up on the glass slide or on a silicone wafer in this way until a film is obtained that is strong enough to serve as resist (400–900 Å) [66].

On irradiation of the closely packed array of olefinic monomers or of diacetylenes, photoinitiated polymerization occurs by a radical chain mechanism. After exposure, the unpolymerized parts of the film are removed with solvent. Baraud et al. used ω-tricosenoic acid as the monomer [65].

$$CH_2=CH+(CH_2)_{20}COOH$$
ω-Tricosenoic acid

With a highly pure preparation (impurities quench the chain reaction), films of high sensitivity (0.5 μC/cm^2) were obtained and pattern with a resolution of 600 Å were written.

Working with acetylene and diacetylene groups containing amphiphiles, Ogawa et al. [71] were able to delineate submicron pattern using a KrF excimer laser scanner and x-rays. Feature sizes of 0.3 to 0.4 μm lines and spaces were reproduced, and the resolution was limited not by the materials but by the optics of the imaging device. With the excimer laser the most successful material was pentacosadiynoic acid,

$$CH_3 \!-\!\!(CH_2)_{\overline{15}} C \equiv C - C \equiv C \!-\!\!(CH_2)_{\overline{8}} COOH$$

with x-rays it was ω-tricocynoic acid.

$$HC \equiv C \!-\!\!(CH_2)_{\overline{20}} COOH$$

While ω-tricosenoic acid functions as a negative resist, Lando and coworkers [67] found that o-octadecyl acrylic acid

$$CH_3 \!-\!\!(CH_2)_{\overline{17}} CH = CH - COOH$$

o-Octadecyl acrylic acid

can be used both as a negative and as a positive resist. When the film is prepolymerized by deep-UV irradiation, exposure to electrons leads to depolymerization and a positive resist image is obtained. Alternatively, a pattern can be written with a high intensity electron beam, which evaporates the material before it can polymerize. The remaining monomer is subsequently polymerized by low intensity electron irradiation. The result of such an experiment is shown in Fig. 8-14.

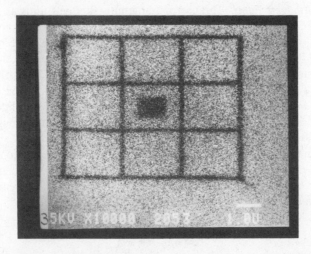

Figure 8-14 Scanning electron micrograph of a grating written into a Langmuir–Blodgett resist film. The film consisted of 10 layers of $CH_3-(CH_2)_{15}-C \equiv C-C \equiv C-(CH_2)_8-COOH$. The lines were written with an intense electron beam, which vaporized the monomer before it could be polymerized. The remaining monomer was subsequently polymerized by a less intense beam. The dark lines are the bare SiO_2 surface.

POSITIVE ELECTRON RESISTS

Positive electron resists rely either on the fragmentation of the polymer chain or on polarity changes to bring about a change in solubility.

Resists Based on Backbone Scission

Poly(methyl methacrylate) and Its Derivatives. Poly(methyl methacrylate), PMMA, which was already described as a deep-UV resist in Chapter 7 is also the classical electron resist. It has excellent lithographic properties: It forms clear, pinhole-free films that adhere well to many substrates, it can resolve a line pattern of submicron size; and it is in general easily and cleanly processed [72, 73]. The material (e.g., Elvacite 2041 or 2010, Du Pont) is coated from solution in ketonic solvents or in cellosolve acetate; after exposure it is developed in 2-ethoxyethanol or in a mixture of 2-ethoxyethanol and ethanol [63, 64].

The mechanism of the radiolytic degradation of PMMA has been extensively investigated [74, 75] (see also Chapter 7). The first step is scission of the C—C bond adjacent to carbonyl, followed eventually by a rearrangement of the primary radical by the Norrish-type I process.

$$\cdots-CH_2-\underset{\underset{OCH_3}{\underset{|}{CO}}}{\overset{\overset{CH_3}{|}}{C}}-CH_2-\underset{\underset{OCH_3}{\underset{|}{CO}}}{\overset{\overset{CH_3}{|}}{C}}-\cdots \rightsquigarrow \cdots-CH_2-\underset{\underset{CH_3O^\cdot}{\underset{|}{\overset{\cdot}{CO}}}}{\overset{\overset{CH_3}{|}}{C}}-CH_2-\underset{\underset{OCH_3}{\underset{|}{CO}}}{\overset{\overset{CH_3}{|}}{C}}$$

$$\cdots-CH_2-\underset{\underset{CH_3OOC\cdot}{+}}{\overset{\overset{CH_3}{|}}{\underset{\cdot}{C}}}-CH_2-\underset{\underset{OCH_3}{\underset{|}{CO}}}{\overset{\overset{CH_3}{|}}{C}}$$

$$\cdots-CH_2-\overset{\overset{CH_3}{|}}{C}=CH_2 + \cdot\underset{\underset{OCH_3}{\underset{|}{CO}}}{\overset{\overset{CH_3}{|}}{C}}-CH_2-\cdots \begin{array}{l}+ CO_2 + CH_3^\cdot \\ + CO + CH_3C\cdot\end{array}$$

(22)

The gaseous components formed in the process play a useful role during development. On escape from the matrix, the gas leaves behind free volume in the exposed areas of the film and in this way selectively speeds dissolution and enhances image discrimination [76, 77].

As an electron resist PMMA has two weak points: It is rather insensitive [its scission yield is $G_s = 1.3$], and it does not stand up to plasma etching. The plasma resistance of PMMA cannot be significantly improved, but efforts to increase its sensitivity were successful. These have centered on the idea of promoting the primary reaction step of side chain scission either by introducing electron-withdrawing substituents at the α carbon of the backbone, by substitution at the ester moiety, or by copolymerization with radiatively more labile components.

Examples of this general idea are copolymers of methyl methacrylate with α-chloroacrylate [75, 78] and with acrylonitrile [79].

$$\left[\begin{array}{c} CH_3 \\ | \\ CH_2-C- \\ | \\ CO \\ | \\ OCH_3 \end{array} \right]_l \left[\begin{array}{c} Cl \\ | \\ CH_2-C- \\ | \\ CO \\ | \\ OCH_3 \end{array} \right]_m \left[\begin{array}{c} CH_3 \\ | \\ CH_2-C- \\ | \\ CO \\ | \\ CN \end{array} \right]_n$$

Both are more sensitive than PMMA alone (see Table 8-6). The electron-withdrawing substituent in the α position of the backbone stabilizes the primary radical, as well as the radical that results from the Norrish rearrangement, and promotes decomposition.

This kind of substitution, however, can have an unwelcome side effect; it may increase the propensity for crosslinking. Figure 8-15 taken from the work of Pittman et al. [80] shows the initial increase in G_s, but later also the steep increase in G_x, which occurs on the gradual substitution of the polymer backbone with fluorine. Willson et al. [81] have attributed this tendency towards crosslinking to the elimination of HF, which leads to unsaturation in the main chain. They proved their point by synthesizing copolymers of methyl methacrylate and methyl-α-trifluoromethyl acrylate and finding an increase in the scission yield from $G_s = 1.3$ to 3.5, and no trace of crosslinking.

Fluorination in the ester moiety has led to useful electron resists, such as poly(perfluorobutyl methacrylate), poly(hexafluoro methacrylate), and poly-(trifluoromethyl-o-chloroacrylate) [82].

$$\left[\begin{array}{c} CH_3 \\ | \\ CH_2-C- \\ | \\ CO \\ | \\ OC_4F_9 \end{array} \right]_n \left[\begin{array}{c} CF_3 \\ | \\ CH_2-C- \\ | \\ CO \\ | \\ OCF_3 \end{array} \right]_n \left[\begin{array}{c} Cl \\ | \\ CH_2-C- \\ | \\ CO \\ | \\ OCF_3 \end{array} \right]_n$$

TABLE 8-6 Lithographic Sensitivity of PMMA and Derivatives

		Sensitivity			
	G_s	E Beam ($\mu C/Cm^2$)	Deep UV (mJ/cm^2)	X-Ray (mJ/cm^2)	Reference
PMMA	1.3	50	3400	4000	1
Poly(fluorobutyl MA)	3	17	480	400	2
PMMA–Methyl-α-chloromethacrylate (38%)	3.5	6			3
PMMA–Acrylonitrile (11%)		4			4
PMMA–Indenone (50%)			40		5
PMMA–3-Oximino-2-butanone (16%)–Methacrylonitrile (15%)			40		6
PMMA–Methacrylic acid (25%)	2	35			7
PMMA–Methacrylic acid–Methacrylic anhydride "Terpolymer"	4.5	8			8
PMMA–Isobutylene (25%)		5			9

1. M. Hatzakis, *J. Electrochem. Soc.*, **116**, 1033 (1969).
2. M. Kakuchi, S. Sugawara, K. Murase, and K. Matsugama, *J. Electrochem. Soc.*, **124**, 1648 (1977).
3. (a) M. Tada, *J. Electrochem. Soc.*, **126**, 1635 (1979). (b) J. N. Helbert, P. J. Caplan, and E. H. Poindexter, *J. Appl. Polym. Sci.*, **21**, 797 (1977).
4. Y. Hatano, H. Morishita, and S. Nonogaki, *Org. Coat. Plast. Chem.*, **35**, 258 (1975).
5. E. Reichmanis, C. W. Wilkins, Jr., and E. A. Chandross, *J. Electrochem. Soc.*, **127**, 2510 (1980).
6. R. L. Hartless and E. A. Chandross, *J. Vac. Sci. Technol.*, **19**, 1333 (1981).
7. M. Hatzakis, *J. Vac. Sci. Technol.*, **16**, 1984 (1979).
8. W. Moreau, D. Merrit, W. Moyer, M. Hatzakis, D. Johnson, and L. Pederson, *J. Vac. Sci. Technol.*, **16**, 1989 (1979).
9. E. Gipstein, W. Moreau, and O. Need, *J. Electrochem. Soc.*, **123**, 1105 (1976).

and in particular poly(2,2,2-trifluoroethyl-α-chloroacrylate [83].

$$\left[-CH_2-\underset{\underset{O-CH_2-CF_3}{|}}{\underset{|}{\overset{\overset{Cl}{|}}{\underset{|}{C}}}}- \right]_n$$

This material is available under the trade name EBR-9. It has outstanding electron sensitivity (0.8 $\mu C/cm^2$ for 20 kV electrons) and more processing latitude than, for example, the electron resist poly(butene sulfone) (PBS) (see p. 324), which has a comparable sensitivity.

The introduction of substituents in the ester part of PMMA has also been successful and so has copolymerization with methacrylic acid [84, 85] with acrylonitrile and with methacrylic anhydride. Moreau et al. haved described a terpolymer of MMA, methacrylic acid and methacrylic anhydride (the opti-

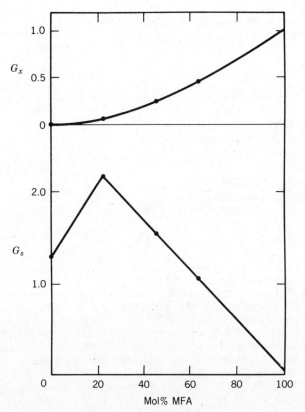

Figure 8-15 G_s and G_x values plotted against mol% of methyl-α-fluoroacrylate in MMA–MFA copolymers. (Reproduced with permission from C. U. Pittman et al. [80].)

mum ratio of the components is 70 : 15 : 15), which is significantly faster than other resists based on PMMA and has other desirable properties [86]. This so-called "Terpolymer" is used in-house by IBM since 1979.

$$\left[\begin{array}{c} CH_3 \\ | \\ CH_2-C \\ | \\ CO \\ | \\ OCH_3 \end{array} \right]_{0.70} \left[\begin{array}{c} CH_3 \\ | \\ CH_2-C \\ | \\ CO \\ | \\ OH \end{array} \right]_{0.15} \left[\begin{array}{c} CH_3 \quad\quad CH_3 \\ | \quad\quad\quad | \\ CH_2-C-CH_2-C \\ | \quad\quad\quad\quad | \\ OC \quad\quad\quad CO \\ \diagdown \;\; \diagup \\ O \end{array} \right]_{0.15}$$

IBM "Terpolymer"

Table 8-6 gives an overview of the radiation sensitivity of PMMA and its derivatives as deep UV, as electron, and as x-ray resist.

Although the T_g of PMMA is quite high (about 105°C), the material tends to soften and sometimes even flow at temperatures that may be required in the processing cycle. Roberts [87] has used thermal crosslinking to increase T_g. For this, small quantities of free methacrylic acid and methacryloyl chloride are introduced into the polymer, and in prebake these produce anhydride crosslinks between the PMMA chains.

$$\begin{array}{c}
\text{CH}_3 \quad\quad \text{CH}_3 \\
| \quad\quad\quad | \\
-\text{CH}_2-\text{C}-\text{CH}_2-\text{C}- \\
| \quad\quad\quad | \\
\text{CO} \quad\quad \text{CO} \\
| \quad\quad\quad | \\
\text{OH} \quad\quad \text{OCH}_3 \\
| \quad\quad\quad \\
\text{Cl} \quad\quad \text{OCH}_3 \quad \xrightarrow{\Delta} \\
| \quad\quad\quad | \\
\text{CO} \quad\quad \text{CO} \\
| \quad\quad\quad | \\
-\text{CH}_2-\text{C}-\text{CH}_2-\text{C}- \\
| \quad\quad\quad | \\
\text{CH}_3 \quad\quad \text{CH}_3
\end{array} \quad
\begin{array}{c}
\text{CH}_3 \quad\quad \text{CH}_3 \\
| \quad\quad\quad | \\
-\text{CH}_2-\text{C}-\text{CH}_2-\text{C}- \\
| \quad\quad\quad | \\
\text{CO} \quad\quad \text{CO} \\
| \quad\quad\quad | \\
\quad\quad\quad\quad \text{OCH}_3 \\
\text{O} \quad\quad\quad\quad\quad\quad + \text{HCl} \\
| \quad\quad\quad \text{OCH}_3 \\
\text{CO} \quad\quad \text{CO} \\
| \quad\quad\quad | \\
-\text{CH}_2-\text{C}-\text{CH}_2-\text{C}- \\
| \quad\quad\quad | \\
\text{CH}_3 \quad\quad \text{CH}_3
\end{array}$$

(23)

On exposure to an electron beam these links as well as the backbone are broken and the polymer becomes soluble in organic solvents. Roberts went one step further; to combine improved flow resistance with improved resistance to plasma etching he designed a resist where methacrylic acid and methacryloyl chloride are copolymerized with styrene. When styrene is the majority component (90% or more of the repeat units) and with a molecular weight of about 20,000, the resulting polymer is a useful positive resist with a good tolerance for plasma treatment [88].

Poly(olefin sulfones). Another way of promoting radiation induced polymer degradation is to introduce a weak link into the backbone. This is the idea behind the use of poly(olefin sulfones) as electron resists. The nature of the poly(olefin sulfones) was recognized in the 1930s by Marvel et al. [89]. The C—S bond has an energy of 62 kcal/mole, compared with the 83 kcal/mole of the average C—C bond. Poly(olefin sulfones) for microlithography were first investigated by Bowden and Thompson at Bell Laboratories [90]. They prepared poly(alkene sulfones) by radical copolymerization of (liquid) SO_2 with a whole range of olefins [91, 92]. Reaction temperatures have to be kept low in synthesis because of the low ceiling temperatures of the poly(alkene sulfones) (see Table 8-7). The resulting copolymers have a regular 1:1 alternating composition.

$$\left[\text{CH}_2-\underset{\underset{\text{R}}{|}}{\text{CH}}-\text{SO}_2 \right]_n$$

TABLE 8-7 Ceiling Temperatures for Polymerization of Olefins with SO_2, [olefin][SO_2] = 27 mole^2/L^2 [95]

Olefin	T_c (°C)
Cyclopentene	102
Butene	64
Hexene	60
cis-2-Butene	46
Cyclohexene	24
2-Methyl-1-pentene	−34

In the exposure phase the primary radiation chemical event was found to be scission of the C—S bond, followed by the spontaneous depolymerization of the alternating polymer (Brown and O'Donnell [93]).

$$\left[CH_2-\underset{R}{CH}-SO_2 \right]_n \rightsquigarrow nCH_2=\underset{R}{CH} + nSO_2 \qquad (24)$$

The first step in the process is the loss of an electron from the molecule [94].

$$R-SO_2-R' \longrightarrow [R-SO_2-R']^+ + e^- \qquad (25)$$

The electron is probably removed from one of the $2p_i$ orbitals of sulfur, but the vacancy there is filled immediately from a higher orbital and appears eventually in a σ(S—C) bonding orbital, thereby transforming C—S into a weak one-electron bond that can be broken by thermal energy or by the excess kinetic energy released in the ionization event.

As a result of the high scission yield the poly(alkene sulfones) are very sensitive resists, both for electrons and for deep-UV radiation. Poly(butene sulfone), which is now the material of choice has an exposure requirement of less than 1 μC/cm^2 at 10 kV. It is commercially available under the trade name PBS and it is widely used in the production of lithographic masks.

After exposure, poly(olefin sulfones) can be developed by two methods: by the usual treatment with solvent [96] or by thermal development. The exposed areas of the resist simply evaporate on heating or in some cases during exposure [95]. This phenomenon has been termed "self-development" and it would seem to be the ultimate in processing convenience. In electron lithography, however, it is a mixed blessing, since the gases released in the process endanger the vacuum of the electron column. Liquid development requires a careful choice of solvent, since the developer is asked to discriminate between two materials, which differ only in molecular weight.

Image Discrimination Based on Molecular Weight Changes. In resists in which the radiation chemical event is backbone scission, image discrimination is

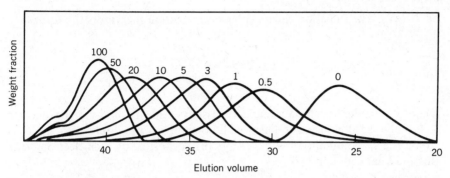

Figure 8-16 Molecular weight distribution as determined by gel permeatian chromatography in samples of poly(butene-1-sulfone) exposed to ^{60}Co radiation. The dose (mrad) is indicated at the peak of each distribution curve. (Reproduced with permission from M. J. Bowden et al. [98].)

based purely on differences of molecular weight. Ueberreiter [97] has shown that the rate of dissolution of a solid polymer does depend on molecular weight and that this dependence can be expressed in the parametric form

$$R = KM^a \quad (26)$$

where K and a are constants. How large a change in molecular weight is required to allow a secure distinction between image and nonimage areas? Figure 8-16 shows the gradual change in the molecular weight distribution (monitored by GPC) that occurs in a sample of PBS on exposure to γ radiation [98].

A reasonable criterion for image discrimination is the condition that the molecular weight distribution of the exposed and unexposed materials do not significantly overlap. In that case, any molecule in the irradiated areas will dissolve faster than any molecule in the nonirradiated areas. This conditions is achieved in the experiment of Fig. 8-15 at a dose of about 3 mrad. At that point the molecular weight ratio of the two samples, M_1 and M_2, is given by

$$\log\left(\frac{M_1}{M_2}\right) = \frac{34}{26} = 1.31$$

$$\frac{M_1}{M_2} = 20 \quad (27)$$

The absorbed dose required to achieve this degree of fission can be derived from a radiation chemical balance. Consider a film of unit area, density d, and thickness r. Its weight is $w = rd$. The number of molecules of monomer units in the sample is $(wN_A)/M_0$ where M_0 is the molecular weight of the repeat unit. If the fraction of bonds to be broken in the sample is p_s, the number of bonds broken is $(wN_A p_s)/M_0$. Let the absorbed dose required to achieve this be D (eV). The radiation chemical balance equation then takes the form

$$D(\text{eV}) = \frac{wN_A 100}{M_0 G_s} p_s \tag{28}$$

where G_s is the radiation chemical scission yield. To convert the dose to Coulombs, as it is usual in electron lithography, we multiply with the electronic charge e and divide by the energy (eV) deposited by each electron. With this the final equation becomes

$$D(\text{coulomb}) = \frac{wN_A 100 \, e}{M_0 G_s E} p_s \tag{29}$$

Charlesby [36] has shown that p_s can be expressed in terms of the molecular weights M_1, M_2.

$$p_s = \left(\frac{M_1}{M_2} - 1\right)\frac{M_0}{M_1} \tag{30}$$

For the experiment shown in Fig. 8-16, M_1 is about 10^6 and M_0 is of the order of 100, the ratio M_1/M_2 having a value of 20. In this system the fraction of bonds to be opened is to a good approximation $p_s = 0.0024$. The dose required to produce this degree of fission in a film of 2 μm thickness using 10 kV electrons (which deposit 10^4 eV) is given by

$$D = \frac{2 \times 10^{-4} \cdot 6 \times 10^{23} \cdot 10^2 \cdot 1.6 \times 10^{-19}}{122 \times 10^1 \times 10 \times 10^3} \times 0.003 = \underline{0.5 \times 10^{-6}} \text{ C/cm}^2$$

and is found to be in good agreement with experiment.

Electron Sensitive Dissolution Inhibitors

A "New Positive Resist". The high radiation sensitivity of poly(butene sulfone), PBS is also the cause of its most serious weakness: it does not resist plasma treatment. This rules out PBS as a practical resist for device production. The difficulty was overcome at Bell Laboratories by an ingenious approach that makes it possible to combine the high radiation sensitivity of poly(olefin sulfones) with the good plasma resistance of novolac. Bowden et al. [99] found that in analogy to the function of diazoquinone in the classical positive AZ resists, poly(olefine sulfones) can act as dissolution inhibitors in novolac resins. Figure 8-17 shows the effect of poly(2-methylpentene-1-sulfone), PMPS,

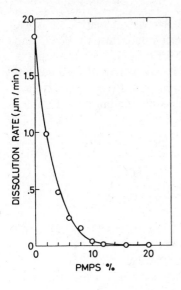

Figure 8-17 Dissolution rate of a novolac resist film as a function of its poly(2-methylpentene-1-sulfone), PMPS, content. (Reproduced with permission from H. Shiraishi et al. [100].)

on the rate of dissolution

$$\left[-CH_2-\underset{\underset{\underset{CH_3}{|}}{\underset{CH_2}{|}}{\overset{CH_3}{\underset{|}{C}}}}-SO_2- \right]_n$$

PMPS

of novolac [100]. On removal of the polysulfone by irradiation and after a short heat treatment, the novolac recovers its original high rate of dissolution in aqueous alkali. For example, with a 10% content of a high molecular weight PMPS [100] in novolac, a resist is obtained with a sensitivity of 3 $\mu C/cm^2$ at 20 kV. This resist has good resolution and the plasma resistance of common positive novolac resists. The system is called "New Photo Resist," NPR. It is commercially available under the trade name REE5000P.

The high sensitivity of NPR is caused by the radiation induced depolymerization (*unzipping*) of the polymer chain of the inhibitor, and for that reason high molecular weight PMPS is being used. In fact, the molecular weight of the inhibitor in NPR is in excess of 500,000. However, poly(olefine sulfones) do not mix well with novolac. In formulating the new resist, workers at Bell Laboratories had to design a proprietary novolac compatible with PMPS, and even then miscibility was restricted to about 10% and the choice of solvent remained critical. The resist is coated from isoamyl acetate, prebaked at 80 to 140°C and developed in 0.4% of tetramethylammonium hydroxide.

It was thought that the discrepancy in molecular weight between novolac and PMPS was in part responsible for the incompatibility problem, and Tarascon et al. [101] have recently prepared PMPS samples of substantially lower molecular weight. As expected, the radiation sensitivity decreased, but pattern quality and resolution (0.25 μm) were improved. Ito et al. [102] have improved compatibility between the base resin and the inhibitor by replacing PMPS with copolymers of the general structure

$$\left[\begin{array}{c}\text{CH}_3\\|\\\text{CH}_2-\text{C}-\text{SO}_2\\|\\\text{CH}_2\text{C}_2\text{H}_5\end{array}\right]_m\left[\begin{array}{c}\text{CH}_3\\|\\\text{CH}_2-\text{C}-\text{SO}_2\\|\\\text{CH}_2-\text{R}\end{array}\right]_n$$

R = $-\text{CH}_2\text{OH}$, $-\text{COOH}$, $\text{CH}_2-\text{CO}-\text{CH}_3$, $-\text{CH}_2-\text{CO}-\text{C}_2\text{H}_5$, and $-\text{O}-\text{C}_2\text{H}_4-\text{O}-\text{C}_2\text{H}_5$

The copolymers are miscible with common cresol novolac resins, the mixtures can be coated from a variety of solvents and they have high glass transition temperatures.

Electron Resists Based on Radiation-Induced Polarity Changes

Most positive electron resists are based on molecular weight changes brought about by backbone scission. More recently, however, materials have been described where image discrimination is based on changes in side chain polarity.

Poly(p-tert-butoxycarbonyl oxystyrene). One of the materials where irradiation causes a change in side chain polarity is the *t*-butoxycarbonyl-protected poly(hydroxystyrene), PBOCST [103].

$$\left[\begin{array}{c}\text{CH}_3\\|\\\text{CH}_2-\text{C}-\\|\\\bigcirc\\|\\\text{O}\\|\\\text{CO}-\text{OBu}\end{array}\right]_n$$

This was described in Chapter 7 as a deep-UV resist. In fact it had been conceived as such, but was later found to respond also to electrons. On heating, or more usefully in the presence of acid, the hydroxyl groups become deprotected, CO_2 and isobutylene escape as gases, and poly(vinyl phenol) is left behind. As a result, the exposed areas of the resist film can be developed either in aqueous alkali, and in that case the system functions as a positive resist, or in an organic solvent that leaves the phenol untouched, but dissolves the unexposed areas, and the system acts then as a negative resist. Poly(*p-tert*-butoxycarbonyl oxystyrene) is a rather sensitive electron resist; exposures of 3 μC/cm^2 are sufficient, resolution is better than 0.5 μm, and the images are fairly resistant to plasma etching.

Poly(ortho-substituted 2-phenylethyl methacrylates). Hatada and co-workers have described polymers of ortho-substituted phenylethyl methacrylates as a new type of electron resist. On exposure to an electron beam these materials split off *o*-methylstyrene, thereby uncovering unprotected free methacrylic acid groups in the side chain.

$$-CH_2-\underset{\underset{\underset{O-CH_2CH_2-C_6H_4-CH_3}{|}}{C=O}}{\overset{CH_3}{\underset{|}{C}}}-$$

$$\rightsquigarrow -CH_2-\underset{\underset{\underset{OH}{|}}{C=O}}{\overset{CH_3}{\underset{|}{C}}}- + CH_2=CH-C_6H_4-CH_3 \qquad (31)$$

The irradiated films are soluble in aqueous alkali and the material can serve as a positive as well as a negative resist, depending on the polarity of the developer. These resists are not particularly sensitive, but they have exceptionally high lithographic contrast [104].

Polystyrene–Tetrathiofulvalene. The ultimate change in polarity occurs in a resist designed by Hofer et al. [105] and based on tetrathiofulvalene (TTF) groups attached in some way to the phenyl groups of polystyrene (PS) [105].

On excitation, the TTF group becomes a strong electron donor and transfers

TABLE 8-8 The Performance of Electron Beam Resists[a]

	Image Type	Trade Name	Manufacturer[b,c]	Sensitivity ($\mu C/cm^2$)	Resolution	Relative Etch Rate[d]	References
Poly(glycidyl methacrylate)	−	OEBR100	TO	0.7	1.5	2.0	4
PGMA-co-poly(ethyl acrylate)	−	COP	MD	0.7	1.5	2.0	5
Poly(chlorostyrene)	−		N	2.0	0.3	1.0	10
Poly(iodostyrene)	−	RE4000N	H	1.8	1.0	1.0	11
Poly(chloromethylstyrene)	−		HP	0.5	1.0	0.8	12
Chloromethylated poly(styrene)	−	CMS	TS	4.0	0.3	0.8	7
Poly(methy methacrylate)	+	PMMA	N	80[e]	0.05	1.9	9
PMMA-co-poly(acrylonitrile)	+	OEBR1030	TO	30[e]	<0.5	1.9	15
PMMA-co-poly(methacrylate anhydride)	+	CP3	TO	15[e]	<0.5	2.0	18
Poly(trifluoromethyl-α-chloroacrylate)	+	EBR9	T	15[e]	<0.5	1.8	19
Poly(fluoroalkyl methacrylate)	+	FBM	DK	?	<0.5	1.4	20
PMMA crosslinked	+	PM	M	40[f]	<0.5	1.9	21
Poly(butene sulfone)	+	PBS	MD	2.0[f]	0.5	∞	24
Novolac–diazoquinone resist	+	AZ1350	AZ	15/50[e]	0.5	1.0	13, 28
Novolac–diazoquinone resist	−			25/60	0.5	1.0	25
Novolac–poly(methylpentene sulfone)	+	RE50000P	H	5.0[f]	<0.5	1.0	26, 27
Poly(styrene)–tetrathiofulvalene	±		I	8.0	<0.5	1.0	32
Inorganic GeSe/AgSe	−		N	500	0.5	0	17
Poly(siloxane)	−		I	1.5	0.5	0	31

[a] Reprinted with permission from Watts [43].
[b] Company codes: TO = Tokyo Okha, MD = Mead, HP = Hewlett-Packard, TS = Toyo Soda, N = Numerous, T = Toray, DK = Daikin Kogyo, M = Microimage, AZ = AZ.
[c] Photoresist Products, I = IBM, H = Hitachi.
[d] Ar sputter etch rate, experimental & calculated, relative to AZ 1350 [18].
[e] HP data.
[f] Interpreted from paper.

an electron to a nearby acceptor, for example, to the halogen of a perhaloalkane.

$$\text{R}\underset{\text{S}}{\overset{\text{S}}{\bigcup}}=\underset{\text{S}}{\overset{\text{S}}{\bigcup}} \xrightarrow[\text{CBr}_4]{h\nu} \left[\text{R}\underset{\text{S}}{\overset{\text{S}}{\bigcup}}=\underset{\text{S}}{\overset{\text{S}}{\bigcup}} \right]^+ \text{Br}^- + \overset{\cdot}{\text{C}}\text{Br}_3 \qquad (32)$$

In the process TTF is transformed into a salt, TTF^+Br^-, which is water soluble. The exposed films can be developed to a positive or negative tone by choosing an appropriate developer. The images do not swell, resolution is better than 0.5 μm, a sensitivity of 8 $\mu C/cm^2$ has been reported for 15 kV electrons and the material has the plasma resistance of polystyrene.

Table 8-8, taken from the work of Watts [43] gives a useful overview of commercially available electron resists.

REFERENCES

1. M. J. Bowden, *J. Electrochem. Soc.*, **128**, 195C (1981).
2. D. Kyser and N. S. Viswanathan, *J. Vac. Sci. Technol.*, **12**, 1305 (1975).
3. D. Kyser and K. Murata, in *Electron and Ion Beams, Science and Technology* (R. Baskish, Ed.), Proc. Ser., p. 205, Electrochem. Soc., Princeton, New Jersey, 1974.
4. L. F. Thompson, *Solid State Technol.*, **17**, 24 (1974).
5. R. K. Watts and J. H. Bruning, *Solid State Technol.*, **24**(5), 99 (1981).
6. M. Nakase and Y. Matsumoto, *SPSE Fall Symp., 18th, Washington, DC*, Prepr. p. 117 (Nov. 1978).
7. D. R. Herriott and G. R. Brewer, in *Electron Beam Technology in Microelectronic Circuit Fabrication*, pp. 141–216, Academic, New York, 1981.
8. P. J. Coane, P. J. Houzego, and T. H. P. Chang, *Solid State Technol.*, **27**(2), 127 (1984).
9. H. C. Pfeiffer, *Solid State Technol.*, **27**(9), 223 (1984).
10. D. R. Herriott, R. J. Collier, D. S. Alles, and J. W. Stafford, *IEEE Trans. Electron Devices*, **ED-22**, 385 (1975).
11. J. L. Mauer, H. C. Pfeiffer, and W. Stickel, *IBM J. Res. Dev.*, **21**, 514 (1977).
12. H. C. Pfeiffer, *J. Vac. Sci. Technol.*, **15**, 887 (1978).
13. N. Kusui, K. Tanaka, Y. Suzuki, and M. Tokita, *Proc. SPIE*, **773**, 111 (1987).
14. T. Matsuzawa, N. Saitou, S. Hosoki, M. Okumura, G. Matsuoka, M. Okayama, and K. Nakamura, *Proc. SPIE*, **773**, 234 (1987).
15. H. C. Pfeiffer, *J. Vac. Sci. Technol.*, **5**, 887 (1978).
16. B. P. Piwczyk and A. E. Williams, *Solid State Technol.*, **26**(9), 145 (1983).
17. R. D. Moore, *Solid State Technol.*, **26**(9), 127 (1983).
18. R. D. Moore, C. A. Caccoma, H. C. Pfeiffer, E. V. Weber, and O. C. Woodward, *Electronics*, **54**(22), 138 (1981).
19. Y. Tarni, *IEEE Trans. Electron Devices*, **ED-27**, 1321 (1980).
20. R. Ward, A. R. Franklin, P. Gould, M. J. Plummer, and I. H. Lewin, *Proc. SPIE*, **393**, 233 (1983).

21. H. Bohlen, U. Behringer, J. Keyser, P. Nemitz, W. Zapka, and W. Kulcke, *Solid State Technol.*, **27**(9), 210 (1984).
22. J. P. Ruskin and J. H. McCoy, *Solid State Technol.*, **24**(8), 68 (1981).
23. W. B. Glendinning, *Solid State Technol.*, **29**(3), 97 (1986).
24. S. Machi and S. P. Beaumont, *Solid State Technol.*, **28**(8), 117 (1985).
25. H. G. Craighead, R. E. Howard, L. E. Jackel, and P. M. Mankiewitch, *Appl. Phys. Lett.*, **42**, 38 (1983).
26. P. M. Mankiewitch, L. E. Jackel, and R. E. Howard, *J. Vac. Sci. Technol.*, **B, 3**, 174 (1985).
27. H. G. Craighead, *J. Electron Microsc. Technol.*, **2**, 147 (1985).
28. C. F. Cook, Jr., T. F. AuCoin, and G. J. Lafrate, *Solid State Technol.*, **28**(10), 125 (1985).
29. E. Kent, *The Brains of Men and Machines*, McGraw-Hill, New York, 1980.
30. M. T. Postek and D. C. Joy, *Solid State Technol.*, **29**(11), 145 (1986).
31. K. Saitch, S. Takeuchi, K. Moriizumi, Y. Watanabe, and T. Kato, *J. Vac. Sci. Technol.*, **B, 4**, 686 (1986).
32. D. C. Shaver, *Solid State Technol.*, **27**(2), 135 (1985).
33. D. C. Shaver, *J. Vac. Sci. Technol.*, **B, 1**, 1084 (1983).
33. Y. Furukawa, Y. Goto, A. Ito, T. Ishizuka, K. Ozaki, and T. Inagaki, *J. Vac. Sci. Technol.*, **B, 3**, 874 (1985).
34. M. Lautner and G. Biegler, *Solid State Technol.*, **29**(8), 111 (1986).
35. J. H. O'Donnell and D. F. Sangster, *Principles of Radiation Chemistry*, American Elsevier, New York, 1970.
36. A. Charlesby, *Atomic Radiations and Polymers*, Pergamon, Oxford, 1960.
37. R. D. Heidenreich, L.F. Thompson, E. D. Feit, and C. M. Meliar–Smith, *J. Appl. Phys.*, **44**, 4039 (1973).
38. E. Menzel and R. Buchanan, *Solid State Technol.*, **28**(12), 63 (1985).
39. M. Dole, *The Radiation Chemistry of Polymers*, Academic, New York, 1972.
40. A. Charlesby and S. H. Pinner, *Proc. R. Soc. London, Ser. A*, **249**, 367 (1959). (b) P. Alexander, R. M. Black, and A. Charlesby, *Proc. R. Soc. London, Ser. A*, **232**, 31 (1955).
41. A. Novembre and T. N. Bowmer, *ACS Polym. Prepr.*, **25**(1), 324 (1984).
42. L. Stillwagon, in *Materials for Electronic Applications* (E. D. Feit and C. Wilkins, Jr., Ed.), *ACS Symp. Ser.*, **184**, 19 (1982).
43. M. P. C. Watts, *Solid State Technol.*, **27**(2), 111 (1984).
44. T. Hirai, Y. Hatano, and S. Nonogaki, *J. Electrochem. Soc.*, **118**, 669 (1971).
45. Y. Ohnishi, M. Itoh, K. Mizuno, H. Gokan, and S. Fujiwara, *J. Vac. Sci. Technol.*, **19**, 1141 (1981).
46. Y. Taniguchi, Y. Hatano, H. Shiriashi, S. Horigone, S. Nonogaki, and K. Naraoka, *Jpn. J. Appl. Phys.*, **18**, 1143 (1979).
47. L. F. Thompson, J. P. Balantyne, and E. D. Feit, *J. Vac. Sci. Technol.*, **12**, 1280 (1975).
48. Z. C. H. Tan, R. C. Daly, and S. S. Georgia, *Proc. SPIE*, **461**, 135 (1984).
49. R. C. Daly, M. J. Hanrahan, and R. W. Blevins, *Proc. SPIE*, **539**, 138 (1985).

50. S. Imamura, T. Tomamura, K. Sugekawa, O. Kogure, and S. Sugawara, *J. Electrochem. Soc.*, **131**, 1122 (1984).
51. E. D. Feit, L. F. Thompson, C. W. Wilkins, Jr., M. E. Wurtz, E. M. Doerries, and L. E. Stillwagon, *J. Vac. Sci. Technol.*, **16**, 1997 (1979).
51a. P. Alexander, R. M. Black, and A. Charlesby, *Proc. R. Soc. London, Ser. A*, **232**, 31 (1955).
52. H. Shiraishi, Y. Taniguchi, S. Horigome, and S. Nonogaki, *Polym. Eng. Sci.*, **20**, 1054 (1980).
53. T. Ueno, H. Shiraishi, and S. Nonogaki, *J. Appl. Polym. Sci.*, **29**, 223 (1984).
54. J. Luitkis, J. Paraszak, J. Shaw, and M. Hatzakis, *SPE Reg. Tech. Conf.*, *Ellenville, NY*, p. 223 (Nov. 1982).
55. E. Feit and L. Stillwagon, *Polym. Eng. Sci.*, **20**, 1058 (1980).
56. H. S. Choong and F. J. Kahn, *J. Vac. Sci. Technol.*, **19**, 1121 (1981).
57. Y. Ohnishi, *J. Vac. Sci. Technol.*, **19**, 1136 (1981).
58. Y. Todokaro, A. Kajiya, and H. Watanabe, *J. Vac. Sci. Technol.*, *B*, **6**, 357 (1988).
59. K. I. Lee and H. Jopson, *Polym. Bull.*, **10**, 39 (1983).
60. J. M. Shaw and M. Hatzakis, *IEEE Trans. Electron Devices*, **ED-25**, 425 (1978).
61. H.-Y. Liu, M. P. de Grandpre, and W. E. Feely, *J. Vac. Sci. Technol.*, *B*, **6**, 379 (1988).
62. W. G. Oldham and E. Hieke, *IEEE Electron Device Lett.*, **EDL-1**, 217 (1980).
63. K. Mochiji et al., *Jpn. J. Appl. Phys.*, **20**, Suppl. *20*-1, p. 63 (1981).
64. T. D. Berker, *Proc. SPIE*, **469**, 151 (1984).
65. A. Baraud, A. Ruaudel-Teixier, and C. Rosilio, *Ann. Chim. (Paris)*, **10**, 195 (1975).
66. A. Baraud, C. Rosilio, and A. Ruaudel-Teixier, *Solid State Technol.*, **22** (6) 120 (1979); *Thin Solid Films*, **68**, 91, 99 (1980).
67. G. Farris, J. B. Lando, and S. R. Rickert, *Thin Solid Films*, **99**, 305 (1983).
68. D. Day and J. B. Lando, *Macromolecules*, **13**, 1478 (1980).
69. V. Enkelmann and J. B. Lando, *J. Polym. Sci., Polym. Chem. Ed.*, **15**, 1843 (1977).
70. K. B. Blodgett and J. Langmuir, *Phys. Rev.*, **51**, 964 (1937).
71. K. Ogawa, H. Tamura, M. Sagaso, and T. Ishihara, *Proc. SPIE*, **771**, 39 (1987).
72. I. Heller, M. Hatzakis, and R. Srinivqasan, *IBM J. Res. Dev.*, **12**, 251 (1968).
73. M. Hatzakis, *J. Electrochem. Soc.*, **116**, 1033 (1968).
74. J. H. Lai and J. N. Helbert, *Macromolecules*, **11**, 617 (1978).
75. J. N. Helbert, P. J. Kaplan, and F. H. Pointdexter, *J. Appl. Polym. Sci.*, **21**, 797 (1977).
76. A. C. Ouano, *Polym. Eng. Sci.*, **18**, 306 (1978).
77. L. E. Stillwagon, *Org. Coat. Plast. Chem.*, **43**, 236 (1980).
78. M. Tada, *J. Electrochem. Soc.*, **126**, 1635 (1979).
79. Y. Hatano, H. Marishita, and S. Nongaki, *Org. Coat. Plast. Chem.*, **35**, 258 (1975).

80. C. U. Pittman, C. Y. Chen, M. Ueda, J. N. Helbert, and J. H. Kwiatkowski, *J. Polym. Sci., Polym. Chem. Ed.*, **18**, 3413 (1980).
81. C. G. Willson, H. Ito, D. C. Miller, and T. G. Teissier, *SPE Reg. Tech. Conf., Ellenville, NY, Tech. Pap.*, p. 207 (Nov. 1982).
82. M. Kaguchi, S. Sugawara, K. Murase, and K. Matsuyama, *J. Electrochem. Soc.*, **124**, 1648 (1977).
83. T. Tada, *J. Electrochem. Soc.*, **126**, 1829 (1979).
84. J. Heller, R. Feder, M. Hatzakis, and E. Spiller, *J. Electrochem. Soc.*, **126**, 154 (1979).
85. M. Hatzakis, *J. Vac. Sci. Technol.*, **16**, 1984 (1979).
86. W. Moreau, D. Merritt, W. Mayer, M. Hatzakis, D. Johnson, and L. Pederson, *J. Vac. Sci. Technol.*, **16**, 1989 (1979).
87. E. D. Roberts, *Org. Coat. Plast. Chem.*, **33**, 359 (1973); **35**, 281 (1975); **37**, 36 (1977).
88. E. D. Roberts, *Polym. Eng. Sci.*, **23**, 968 (1983).
89. D. S. Frederick, H. D. Cogan, and C. S. Marvel, *J. Am. Chem. Soc.*, **56**, 1815 (1934).
90. M. J. Bowden and L. F. Thompson, *J. Electrochem. Soc.*, **120**, 1722 (1973); **121**, 1620 (1974).
91. M. J. Bowden and E. A. Chandross, U.S. Patent 3,884,695 (1975).
92. C. G. Willson, J. M. J. Frechet, and M. J. Farrell, *IBM Res. Rep.*, No. RJ 3259 (1981).
93. J. R. Brown and J. H. O'Donnell, *Macromolecules*, **5** 109 (1971).
94. T. N. Bowmer and J. H. O'Donnell, *Radiat. Phys. Chem.*, **17**, 177 (1981).
95. M. J. Bowden and L. F. Thompson, *Polym. Eng. Sci.*, **17**, 269 (1977).
96. M. J. Bowden and L. F. Thompson, *J. Appl. Polym. Sci.*, **17**, 3211 (1973).
97. K. Ueberreiter, in *Diffusion in Polymers* (J. Crank and G. S. Park, Ed.), p. 218, Academic, New York, 1968.
98. M. J. Bowden, L. F. Thompson, and J. P. Balantyne, *J. Vac. Sci. Technol.*, **12**, 1294 (1975).
99. M. J. Bowden, L. F. Thompson, S. B. Fahrenhotz, and E. M. Doerries, *J. Electrochem. Soc.*, **128**, 1304 (1981).
100. H. Shiraishi, A. Isobe, F. Murai, and S. Nonogaki, in *Polymers in Electronics, ACS Symp. Ser.*, **244**, 167 (1984).
101. R. G. Tarascon, J. Frackoviak, E. Reichmanis, and L. F. Thompson, *Proc. SPIE*, **771**, 54 (1987).
102. H. Ito, L. A. Pedersen, S. A. MacDonald, Y. Y. Cheng, J. R. Lyerla, and C. G. Willson, *Tech. Pap. Reg. Conf. Soc. Plast. Eng., Nevele, NY*, p. 127 (Oct. 1985).
103. J. M. J. Frechet, E. Eichler, H. Ito, and C. G. Willson, *Symp. VLSI Technol., Tokyo, Sept. 1982*.
104. K. Hatada, T. Kitayama, S. Danjo, Y. Tsubokura, H. Yuki, K. Moriwaki, H. Aritome, and S. Namba, *Polym. Bull.*, **10**, 45 (1983).
105. D. C. Hofer, F. B. Kaufman, S. R. Kramer, and A. Aviram, *Appl. Phys. Lett.*, **37**, 314 (1980).

9 X-RAY AND ION BEAM LITHOGRAPHIES

X-RAY LITHOGRAPHY

Optical lithography is limited by diffraction and may not be viable in production below feature sizes of 0.5 μm. At the same time it must be noted that resolution of 0.35 μm is now being achieved with excimer laser projection patterning of some multilevel resists [1]. From the point of view of resolving power electron beam lithography would be the patterning method of choice, but as a scanning procedure it is restricted to low wafer throughput and therefore more suited to mask making than to direct device fabrication. Clearly, there is an area in IC manufacture where neither optical lithography nor electron beam are efficient. X-ray lithography combines in principle the high resolution of electron beam with the large-area exposure mode and good wafer throughput of optical lithography and appears well suited to fill the apparent technology gap.[1]

The original concept of x-ray lithography is summarized in the diagram of Fig. 9-1. A stationary x-ray source is located in a vacuum, some 50 cm vertically above an exposure table, where an x-ray mask and a resist-coated wafer are held in near contact. By opening a shutter, the resist is exposed through the mask to the x-rays. After exposure the wafer is removed and the resist image developed in the usual way [2].

The quality of pattern transfer from mask to wafer is best if mask and wafer are in hard contact. This, however, is not practical because of the

---

[1] Very recently there has been a change in outlook and it is thought that excimer laser lithography might well be able to provide the bulk of the submicron-fabrication production capacity for the next decade.

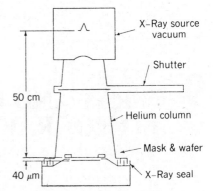

Figure 9-1 Simple schematic of an x-ray exposure tool. (Reproduced with permission from A. Zacharias [2].)

inevitable damage to the expensive mask. A safety gap between mask and substrate is therefore essential, but it introduces a source of image degradation in the form of a "penumbral blurr" [3]. Figure 9-2 shows the exposure geometry in the proximity printing mode of an x-ray exposure tool [4]. It can be seen, that with a finite radiation source of diameter d, at a distance D from the exposure plane, and with a gap s between mask and substrate, the penumbral blurr, δ, is given approximately by the expression

$$\delta = s\left(\frac{d}{D}\right) \qquad (1)$$

With a source distance of 50 cm and a source diameter of 2 mm, together with

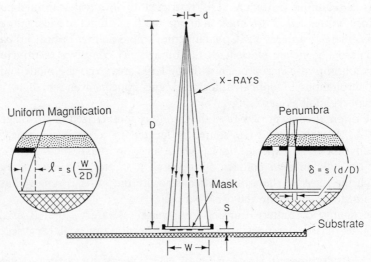

Figure 9-2 Geometry of the proximity printing mode. (Reproduced with permission from H. I. Smith [4].)

a gap $s = 40$ μm (which is standard, although smaller gaps can now be safely maintained) the penumbral blurr is of the order of 0.2 μm. As far as beam geometry is concerned, x-ray lithography is clearly capable of high resolution.

The conceptual simplicity of the arrangement in Fig. 9-1 is to some extent deceptive. All the components of the machine, namely, the x-ray source, the wafer stage, the mask, and the resist are not easily realized and had to undergo considerable evolution before submicron lithography with x-rays could become a pilot-line reality [5].

X-Ray Sources

The earliest sources were static *electron impact x-ray tubes* with Al as the target (wavelength of characteristic radiation 8.3 Å). Bell Laboratories adopted a palladium target as the most useful source [2] (4.4 Å). The palladium radiation can pass almost unattenuated through a fairly thick beryllium window, which makes it possible to keep the source in a high vacuum, while exposure can be made to keep the source in a high vacuum, while exposure can be made at atmospheric pressure under a helium blanket. This simplifies the mechanics of the printing process considerably and makes for easy wafer handling and high throughput.

Even in the best electron impact sources barely 1% of the energy is converted into x-rays [2, 6]. Much better conversion efficiency can be achieved by producing x-rays through the impact of a high energy pulsed laser on a suitable target (*laser plasma sources*); with a neodymium glass laser impinging on a tungsten target conversions of up to 20% have been achieved at a repetition rate of 10 Hz [7]. Furthermore, plasma sources have a relatively small focus, which makes for better imaging. Recently, a plasma x-ray source for lithography has been described, which is based on emission from the collapsing molecular ensemble in a so-called "gas-puff, Z-pinch" plasma [8, 9]. These plasma sources are up to 50 times brighter and better focused than the conventional electron impact x-ray tubes. Figure 9-3 shows the x-ray spectrum of such a plasma source.

The most interesting sources of x-rays for submicron lithography are *storage rings and synchrotrons* [9–11]. While plasma sources are still in the design and development stage, storage rings are by now a mature technology. In the storage ring or the synchrotron, short wave dipole radiation is emitted as the orbiting electrons change their direction in the magnetic field. Under certain conditions, when the electron velocity approaches the speed of light, the radiation pattern changes from a dipole type to a forward lobe and the result is a strong continuous output of an almost parallel (collimated) beam of radiation with a wide range of wavelengths. The storage ring beam is ideally suited for lithography [12, 13]. Being naturally collimated it does not cause a penumbra, but produces deep resist profiles with vertical walls and astonishing aspect ratios (see Fig. 9-8); also, the mask–substrate gap becomes much less critical.

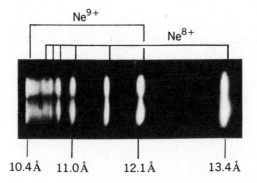

Figure 9-3 X-ray spectrum of a Ne gas plasma. [Reproduced with permission from H. Yoshihara, I. Okada, Y. Saitoh, and S. Itahashi, *Proc. SPIE*, **632**, 133 (1986).]

When the monochromatic characteristic radiation of a static electron impact source is used for lithography, strong Fresnel diffraction patterns appear, which are highly detrimental to image quality. With the polychromatic radiation of the storage ring this effect is largely suppressed (see Fig. 9-4). Furthermore, the high brightness of the storage ring source (it deposits over 100 mW/cm² onto the mask) makes it possible to use less sensitive high resolution resist, such as poly(methyl methacrylate). Okada et al. [14] have recently described the development of a reliable synchrotron radiation lithography beam line.

Electron storage rings are large and costly installations that do not fit into the concept of a commercial manufacturing facility, but the lithographic

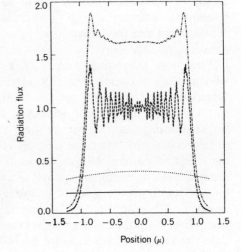

Figure 9-4 Fresnel diffraction from a 2 μm slit at a distance of 50 μm from the resist surface, irradiated with the characteristic $Al_l K_\alpha$ line at 8.3 Å, compared to the image of the same slit obtained with polychromatic synchrotron radiation. (Reproduced with permission from W. D. Grobman [15].)

TABLE 9-1 Cost Comparison of Various Lithography Methods [18] 1986

Lithographic method	($/wafer exposure)
Optical lithography	4
Electron beam	59
X-ray: storage ring	7
compact ring	2
plasma source	25

advantages of the storage ring are so important that very serious consideration is being given to the idea of dedicated storage rings for lithography, and a few firms maintain pilot stations at the national synchrotron facilities in Brookhaven, Stanford [10], and Michigan. Several European companies are now developing a compact storage ring that will occupy a floor space of only 30 m^2 and which will be dedicated specifically to x-ray lithography [16]. The super small storage ring based on superconducting magnets, which is currently being built at the Fraunhofer Institut, Berlin appears to be nearing completion [17].

Wilson has made a careful comparison of the costs of the various lithographic modes [18]. His conclusions are summarized in Table 9-1. The data tell their own story and make the growing interest in synchrotron lithography understandable.

X-Ray Masks

Making an x-ray mask is different from the comparatively simple technology of the chrome-on-glass masks used in optical lithography. The whole range of absorptivities for x-rays between highly absorbing (high atomic number) and poorly absorbing (low atomic number) materials spans only some two orders of magnitude (see Fig. 9-5) [4].

As a result, low atomic number materials that take the place of glass in optical masks have to be used as ultra-thin membranes to be sufficiently transparent to x-rays, while the opaque pattern on the mask must be formed from a comparatively thick layer of a high atomic number material. Gold is now used universally for this purpose. The membrane that carries the pattern can be made from a range of materials; some are organic polymers others are silicon carbide, silicon nitride, boron carbide, and in particular boron nitride. All these materials have excellent tensile strength, which is necessary because the membrane supporting the gold pattern is kept under tension to ensure flatness and reasonable resistance against distortions. Figure 9-6 shows a

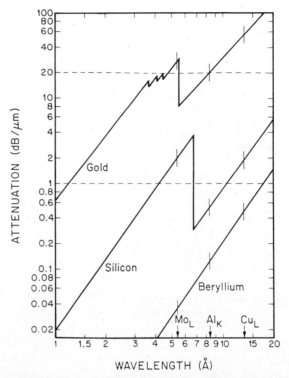

Figure 9-5 The attenuation of x-rays in gold, silicon, and beryllium, three important materials in x-ray lithography. (Reproduced with permission from H. I. Smith [4].)

schematic of a mask structure where the thicknesses of the individual components are indicated [19].

Mask making falls into two operations: fabricating the *mask blank* and applying the gold pattern [20]. Figure 9-7 illustrates a mask blank processing sequence. A sacrificial silicon wafer is coated by chemical vapor deposition, CVD, with hydrogenated boron nitride. This coating is then removed by

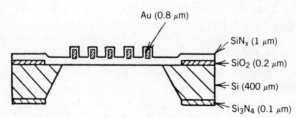

Figure 9-6 Cross sectional view of an x-ray mask. (Reproduced with permission from B. S. Fay and W. T. Novak [19].)

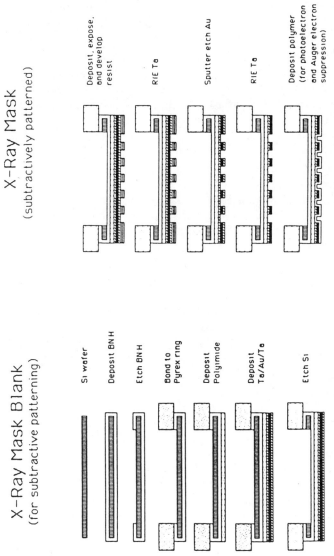

Figure 9-7 Subtractive boron nitride mask blank preparation and patterning sequences. (Reproduced with permission from A. R. Shimkunas et al. [20].)

plasma etching from a large part of one side of the wafer, and the wafer is bonded to a glass or quartz ring. The front side of the wafer is then coated with a thin layer of polyimide and that coating is cured. Three metallic layers are evaporated onto the polyimide: first a thin layer of tantalum (Ta), then a thick (0.8 μm) layer of gold, and finally a second Ta layer. The original silicon wafer is then etched away from the back, leaving the boron nitride as a taut membrane. For an improved boron nitride technology for the making of x-ray masks see Reference 21.

Patterning can be carried out in several ways. The blank just described is patterned by a "subtractive" process. The composite metal layer of the blank is coated with an electron resist and that is patterned in an electron beam exposure tool. The pattern is then transferred by reactive ion etching (RIE) to the Ta layer, and the resulting Ta mask is used to pattern the thick gold deposit by sputter etching. After that, the second Ta layer is patterned by RIE, and finally the gold pattern is encapsulated in polyimide or another protective, and nonabsorbing, polymer. For an alternative method of producing a gold pattern by lithography see Reference 22.

The subtractive method produces a pattern with somewhat sloping sidewalls, which is detrimental to lithographic imaging of very high resolution features. Better resolution is achieved with masks produced by an "additive" patterning process where gold is electroplated onto a suitably prepared blank through the open spaces of a resist stencil. In another method, the resist is patterned with substantial undercut, the patterned surface is metallized, and finally the resist is dissolved from under the metal which is to be removed (lift off). Either of these techniques will produce masks with feature sizes down to 0.2 μm, with distortions of less than 0.1 μm over an area of 20 cm^2 and with about one defect per square centimeter. This defect density seems small enough, but it is still too large for secure production use, since present day devices need between 10 to 15 levels of exposure and the defects of all the exposures accumulate in the device.

An interesting new technique for making very fine line (0.01–0.03 μm) masks is described by Smith and co-workers at MIT. The pattern is defined by the edges of a polyimide relief molding; this is later obliquely shadowed with gold. The fine gold lines that are formed on the side walls of the polyimide mesas are the fine line pattern [23].

This brings us to the question of mask alignment and overlay. The highest quality of pattern definition is not really useful if the masks and the substrate cannot be aligned to even finer tolerances. In most machines, alignment is achieved by optical means, by the matching of interference pattern or Fresnel zone plates. In this respect BN is particularly useful as a blank material because it is transparent to light and allows optical monitoring of the fiducial marks. Some spectacular results obtained by synchrotron lithography with gold on boron nitride masks are shown in Fig. 9-8 [12]. In view of the very high cost of making an x-ray mask, attempts were made to copy the original

Figure 9-8 Poly(methyl methacrylate) structures fabricated using synchrotron radiation. Resist height is 2 μm. (Reproduced with permission from A. Heuberger [12].)

mother-mask into a series of production daughter-masks. This appears now to be feasible [24]; production mask are copied by synchrotron radiation onto single-layer resists.

X-RAY RESISTS

Since the chemical effects of different ionizing radiations are essentially similar, it is expected that electron resists will also be sensitive to x-rays. That is in fact the case; there is good correlation between electron sensitivities and x-ray sensitivities in the group of materials shown in Fig. 9-9. The figure taken from the work of Murase et al. [25] and quoted by Willson [26] refers both to negative and to positive resists. It will be noted from the data in the figure that an electron sensitivity of 1 $\mu C/cm^2$ (which is more than adequate for electron beam lithography) corresponds approximately to an x-ray sensitivity of 100 mJ/cm^2. With the x-ray fluxes available in conventional electron impact sources a resist of this sensitivity requires exposure times in excess of 20 min/wafer, which is not acceptable on a fabrication line. Since the radiation chemical reactivity of a given material is the same in both cases, the sensitivity difference stems from the poor absorption of x-rays in organic materials.

Negative X-Ray Resists

Negative x-ray resists are based on the crosslinking of linear polymers, and as in all crosslinking polymers the gel dose, that is, the radiation dose that brings about incipient gel formation, depends on three material factors: the weight average molecular weight (M_w), the radiation chemical crosslinking yield (G_x), and the mass absorption coefficient (μ_m) of the material for the x-rays to be

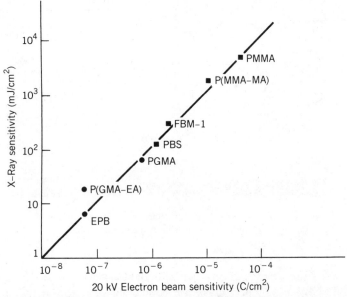

Figure 9-9 Correlation between x-ray sensitivity and electron sensitivity for a group of negative (●) and positive (■) resists. (Reproduced with permission from C. G. Willson [26].)

used in exposure. Charlesby [27] has shown that the general photochemical balance equation for the conditions at the gel point can be adapted in the radiation chemical case as shown in Eq. (2).

$$D_G = D_n^0 = \frac{4.8 \times 10^9}{G_x M_w \mu_m} \qquad (2)$$

The *mass absorption coefficient* of a material is defined by the standard attenuation law.

$$I(x) = I_0 \exp(-\mu_m \rho x) \qquad (3)$$

$I(x)$ is the radiation intensity at depth x in the film, ρ is the density of the material. The absorbing centers, in the case of x-rays, are the individual atoms and the mass absorption coefficient of a material is simply the weighted sum of the coefficients of its constituting elements.

$$\mu_m = \sum \mu_m(i) f_w(i) \qquad (4)$$

Here $f_w(i)$ is the weight fraction of atom i in the system. The mass absorption coefficient of the elements depends very sensitively on the atomic number (Z). This relationship is shown in Fig. 9-10 for the characteristic Al, K_α radiation

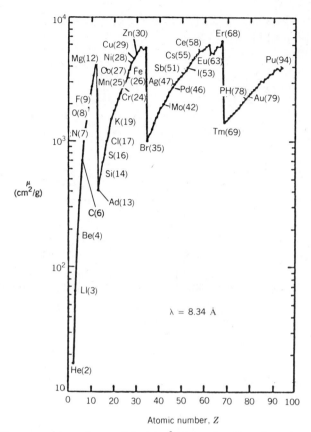

Figure 9-10 X-ray mass absorption coefficient (cm^2/g) as a function of atomic number for the Al, K_α line at 8.34 Å. (Reproduced with permission from A. Eranian et al. [28].)

at 8.34 Å [16]. The curve in Fig. 9-10 has several branches separated by discontinuities, the x-ray absorption edges. Within each branch μ_m is given by the Bragg–Pierce equation.

$$\mu_m = C\rho N_A \lambda^3 \frac{Z^4}{A} \qquad (5)$$

Here C is a constant, ρ is the density of the material, N_A is Avogadro's number, Z the atomic number, A the atomic weight, and λ the wavelength of the radiation.

The obvious way to increasing μ_m is to incorporate into the resist a high proportion of a high atomic number element. Attempts at this have been made as early as 1974 when Thompson et al. [29] prepared poly(vinyl ferrocene) and demonstrated enhanced sensitivity to x-rays. In later developments other metals were incorporated into polymers, but this was found to impair the

TABLE 9-2 X-Ray Mass Absorption Coefficients for the Pd, K_α, Al, K_α and the Cu, L_α Lines [28, 29]

	Pd, K_α 4.37 Å	Al, K_α 8.34 Å	Cu, L_α 13.34 Å
H		1.8	8.9
C	100	718	2714
O	227	1597	5601
F	323	2037	6941
Cl	2013	1023	3596
Br	1500	1021	3101

solubility and processability of the materials. For this reason, efforts to increase x-ray absorption and resist sensitivity to x-rays have centered on the introduction of halogen into radiation sensitive materials. Table 9-2 gives the x-ray mass absorption coefficients of relevant atoms for three important x-ray lines.

It can be seen from these data that fluorine is the element of choice for use with Al, K_α radiation. Chlorine is particularly suited for the Pd, K_α x-ray source because its absorption edge almost coincides with the wavelength of the Pd line (4.37 Å).

Taylor and co-workers [30–33] studied the effect of chlorine on the x-ray sensitivity of a group of acrylates of the type

$$\left[\begin{array}{c} CH_2-CH- \\ | \\ C=O \\ | \\ OR \end{array} \right]_n$$

where R are chlorinated alkyl groups. It was found that not only was the x-ray absorption of the materials much improved by the presence of chlorine, the radiation chemical sensitivity of the materials was also enhanced. This effect depends on the structure of the repeat units of the molecules. Thus, it was found that poly(2,3-dichloro-1-propyl acrylate) is much more sensitive than poly(1,3-dichloro-2-propyl acrylate). The reason for this particular effect was thought to be the presence of secondary C—Cl bonds and vicinal halogen in the more sensitive compound.

Poly(2,3-dichloro-1-propyl) acrylate), DCPA, was considered the most promising of these materials. Its practical lithographic performance (mainly its adhesion) was much enhanced by blending with COP, which is a copolymer of glycidyl methacrylate and ethylacrylate (see Chapter 8).

Chlorinated materials have been further developed as x-ray resists. Choong and Kahn [34] have shown that poly(chloro-methylstyrene), PCMS, which was earlier introduced for electron beam lithography by Imamura [35], functions also as an x-ray resist.

$$\left[\begin{array}{c} -CH_2-CH- \\ \bigcirc \\ CH_2Cl \end{array} \right]_n$$

PCMS

Poly(chloromethylstyrene) has an x-ray sensitivity of 28 mJ/cm^2, it can be exposed in air, and has good dry etch resistance. Another polymer that is a successful electron resist and has since been found useful for x-ray exposures is chlorinated poly(methylstyrene), CPMS. The material that is prepared simply by partial chlorination of poly(methylstyrene) contains the following moieties.

(1) $-CH_2-CH-$ with phenyl-CH$_3$

(2) $-CH-CH-$ with Cl on first CH, phenyl-CH$_3$

(3) $-CH_2-C-$ with Cl on C, phenyl-CH$_3$

(4) $-CH_2-CH-$ with phenyl-CH$_2$Cl

CPMS

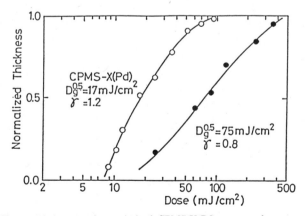

Figure 9-11 Characteristic curve for optimized CPMS-X(Pd) compared to a commercial x-ray resist. (Reproduced with permission from N. Yoshioka et al. [36].)

Yoshioka et al. [36] have recently optimized this material for x-ray exposure and report good sensitivity, $D_n^{0.5} = 17$ mJ/cm^2, and good contrast. Figure 9-11 compares CMPS–XPd with a commercially available material.

Another resist that does not contain halogen or other heavy atoms has recently been reported: copolymers of allyl methacrylate with either 2-hydroxyethyl methacrylate or glycidyl methacrylate [37, 38].

$$\left[\begin{array}{c} \text{CH}_3 \\ | \\ -\text{CH}_2-\text{C}- \\ | \\ \text{CO} \\ | \\ \text{O} \\ | \\ \text{CH}_2 \\ | \\ \text{CH}=\text{CH}_2 \end{array}\right]_l \left[\begin{array}{c} \text{CH}_3 \\ | \\ -\text{CH}_2-\text{C}- \\ | \\ \text{CO} \\ | \\ \text{O} \\ | \\ \text{CH}_2 \\ | \\ \text{CH}_2\text{OH} \end{array}\right]_m \left[\begin{array}{c} \text{CH}_3 \\ | \\ -\text{CH}_2-\text{C}- \\ | \\ \text{CO} \\ | \\ \text{O} \\ | \\ \text{H}_2\text{C} \\ | \\ \text{HC} \diagdown \\ \text{H}_2\text{C}^{\nearrow}\text{O} \end{array}\right]_n$$

These polymers have exceptional electron sensitivity (0.5 μC/cm^2) and reasonable x-ray sensitivity (50 mJ/cm^2), acceptable contrast, and high temperature resistance (Kodak EK 571 and EK 771).

Positive X-Ray Resists

In view of the high resolution patterns that are being defined for x-rays, positive resists with proven stability in the production process are more desirable than negative resists. Most positive electron resists currently available have x-ray speeds too low for practical use; with conventional electron impact x-ray sources they require literally hours of exposure [9]. Interestingly, the most spectacular submicron patterns, such as those shown in Fig. 9-8 have been produced with x-rays and the slowest of all positive resists, PMMA. The resolution, film forming, and processing properties of this material are still unsurpassed after almost three decades of intensive research and development.

Novolac based resists would be highly desirable because of their proven technological stability, but they are in general too slow for routine production if used as x-ray resists. The best results, so far, were obtained with the Olin–Hunt resist HPR-204, which has an x-ray sensitivity of about 500 mJ/cm^2 [39]. Recently Dössel et al. have claimed a novolac resist with chemical amplification, where a dissolution inhibitor is removed catalytically, and which has a sensitivity well below 100 mJ/cm^2 [40].

ION BEAM LITHOGRAPHY

If there is hope that x-ray lithography will become the large scale patterning method of the submicron, or rather subhalf-micron era [18], mask making for feature sizes in the nanometer range may become the realm of ion beam lithography.

Work on the lithographic applications of ion beam has been in progress since the mid 1970s. The potential advantages of ions over electrons are based on the much higher mass of the ions. Because of it, the proximity effects (electrostatic repulsion effects), which degrade the focus of electron beams are virtually absent in ion beam technology [41]. Ions scatter less in the resist material and there is no significant backscatter from the substrate [42, 43]. Also, ions produce secondary electrons of very low energy and therefore much shorter range than the secondary electrons of the electron beam, and finally, because of the much higher energy deposited by ions, resist materials tend to be more sensitive to ion beams than to electron beams [44].

Two types of exposure arrangements have been investigated. Ions have been used as finely focused ion beams (FIB), in analogy to electron beams, or as broad beam sources, illuminating the whole area of a template mask, in analogy to x-rays.

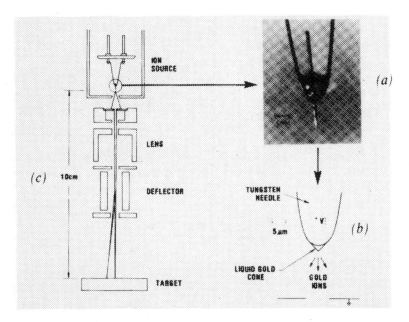

Figure 9-12 Finely focused ion beam system with a liquid metal ion source. (*a*) the heating hairpin and the tungsten needle wet with Au; (*b*) schematic of the tip of the needle with liquid Au pulled into a cone by electric field forces and emitting Au$^+$ ions by field evaporation; (*c*) schematic of the ion-optical column for the formation and deflection of submicron diameter ion beams. (Reproduced with permission from W. L. Brown et al. [43].)

Focused Ion Beams

Figure 9-12 shows a schematic of a simple focused ion beam (FIB) source. It is in essence a fine tungsten needle with a carefully shaped tip that is whetted by a liquid metal [43]. (The two wires, which are connected to the tip, bring in the heating current). The tip is held at some +20 kV against the potential of the base plate of the accelerator. In these conditions the electrostatic field and the surface tension forces in the metal balance and form a fine point, the so-called "Taylor cone" [45]. Figure 9-13 shows the much more developed focusing system of the new Jeol JIBL-200S, which is optimized for Ga$^+$ ions [46].

The ions are focused by electrostatic optics and the beam is scanned over the surface of the wafer. Sources like these have now developed to a stage where they can deliver ion currents of 50 pA [47]. They are being used on an

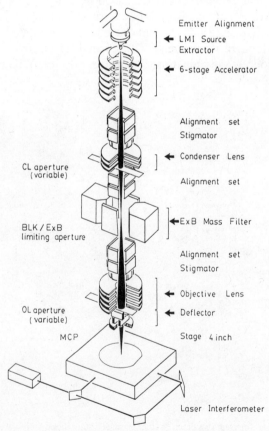

Figure 9-13 Schematic of the ion-optical system of the Jeol JIBL-200S focused ion beam instrument. The ions can be accelerated with potentials from 20 to 200 kV, the beam can focus down to 0.09 µm. (Reproduced with permission from Aihara et al. [46]. © American Institute of Physics, 1988.)

ION BEAM LITHOGRAPHY 351

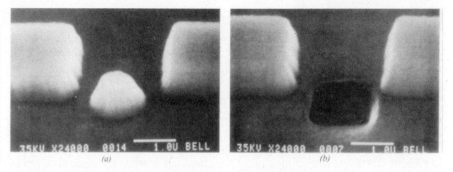

Figure 9-14 Use of focused ion beam milling to remove a submicron isolated defect. (*a*) original pattern, and (*b*) defect removed. (Reproduced with permission from J. Randall et al. [54].)

experimental basis for *maskless ion implantation*, that is, for the direct selective doping of silicon or GaAs wafers [48, 49].

An important application of focused ion beams is the *repair of x-ray masks* and other high resolution patterns [50] where so-called "opaque" defects, for example, improperly deposited metal debris, can be sputtered away by an ion beam. This is illustrated in Fig. 9-14 where a small drop of gold is cleanly removed by focused ion beam milling [51].

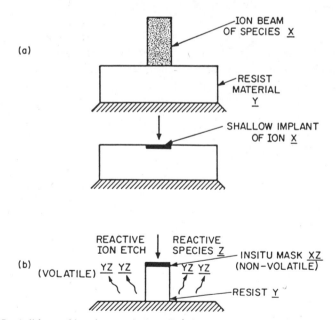

Figure 9-15 A lithographic scheme using an Ga^+ beam to implant a shallow image into the planarizing resist and transferring the pattern into this resist by reactive ion etching. The Ga^+ pattern acts as an *in situ* mask. (Reproduced with permission from T. Venkatesan et al. [54].)

Focused ion beams can be used to *write submicron pattern* into various resists and these can then be developed in a conventional way [52, 53]. An alternative procedure is described by Venkatesan et al. [54] where a high resolution pattern is produced in a thick polymer layer by shallow imagewise implantation of (Ga^+) ions (see Fig. 9-15). Subsequently the implantation pattern is transferred into the thick polymer layer by oxygen reactive ion etching. In the first stages of plasma etching Ga^+ is oxidized to Ga_2O_3, which later acts as an *in situ* plasma resistant etch mask.

In all these procedures scan times are still much too long and exclude the ion beam for direct device making. Also, there is no point in trying to shorten exposure times by using more sensitive resists: Error counts and the continuity of the scan lines are limited by the statistics of the Poisson distribution of the oncoming particle stream, the so-called shot noise limit.

Broad Beam Exposures

Two systems are being developed for the parallel exposure of large areas by ion beam: *Ion Optical Imaging* of Sacher Technik, Vienna, Austria, where the pattern of a mask is projected by ion optics onto the resist substrate [55]. The other is the *Channeled Ion Lithography* developed at the Hughes Research Laboratory [56]. This technique is based on the unexpected observation that correctly oriented single-crystal films of silicon are highly permeable to protons and produce a channeling effect, which makes for an almost perfectly collimated beam of H^+ ions [57] (see the sketch in Fig. 9-16).

Originally, the silicon film was used as a substrate for a gold pattern of submicron resolution, similar to that used in x-ray masks. A significant breakthrough was achieved when gold masks, which suffer from thermal distortion under ion bombardment, were replaced by all silicon masks that are sturdy, thermomechanically homogeneous, and comparatively easy to fabricate [58]. A prototype instrument for masked ion beam lithography was recently built, which incorporates a powerful source of protons, using an accelerating voltage of 300 kV over a distance of 1.6 m, and removing all ions other than H^+ from the beam by a mass separator. The machine is equipped with an optical interference alignment system and with a special high speed–high precision stepper stage. It can provide a throughput of up to 60 four-inch wafers per hour with a limiting resolution of 0.1 μm [56].

Ion Beam Resists

The salient fact about the materials of ion beam lithography is that, in general, polymers are much more sensitive to fast ions than to electrons. For example, Seliger and Fleming [59] reported as early as 1974 that PMMA was about 100 times more sensitive to 60 keV He^+ ions than to electrons, and similar results were obtained by Hall et al. [44a]. Some of their data, relating to the sensitivity of novolac, acting as a negative resist in this case, are shown in Fig. 9-17.

ION BEAM LITHOGRAPHY 353

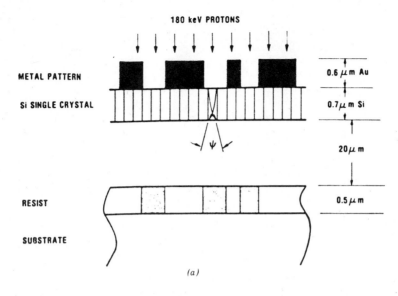

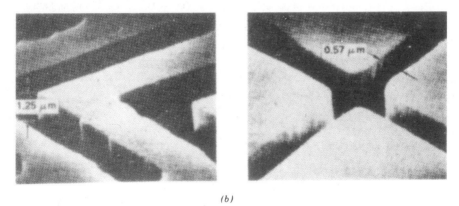

Figure 9-16 (a) Schematic of channeled ion lithography. The 0.7 μm Si single-crystal supports a gold pattern that defines the regions of transmission. Channeling of protons through the crystal reduces multiple scattering and hence the divergence of the emerging beam. (b) Pattern produced in PMMA with 150 keV protons at a dose of 2×10^{13} ions/cm^2. (Reproduced with permission from D. B. Rensch et al. [44b].)

It can be seen that the sensitivity, expressed as a gel dose in terms of the radiation fluency, that is, the number of ions incident per square centimeter, increases with the mass of the ions. The critical dose is evidently a function of the energy deposited in the material, and this energy increases with the ionic mass. However, even in terms of energy deposited per unit volume, oxygen ions are about 10 times more effective than electrons in exposing PMMA. Hall

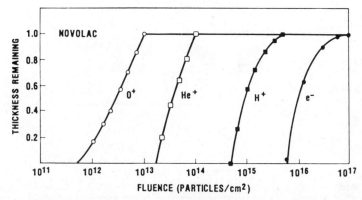

Figure 9-17 Characteristic curves of novolac exposed to electrons, protons, helium ions, and oxygen ions. (Reproduced with permission from T. M. Hall et al. [44a].)

and co-workers [44] interpreted this in terms of the spatial energy spread in the neighborhood of the particle track: Many radiation chemical reactions require the cooperation of two reactive fragments. In the vicinity of an ion track the higher energy density makes it more likely that two neighboring reactive sites are formed.

In crosslinking resists, even under irradiation with ion beams, the general rules of network formation apply. In fact Brault and Miller [58] demonstrated,

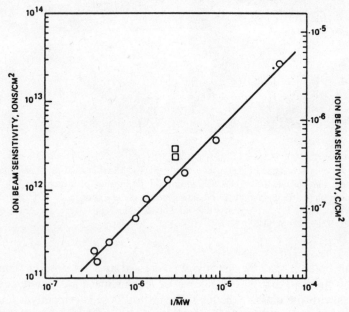

Figure 9-18 The effect of molecular weight on the sensitivity of polystyrene as a proton resist. (Reproduced with permission from R. C. Brault and L. J. Miller [58].)

TABLE 9-3 Proton and Electron Sensitivities of a Group of Resists

Resist[a]	Code	Q_p	Q_e	Q_e/Q_p
Negative Resists				
Poly(vinyl acetate)	PVAc	$9.6\ 10^{-7}$	$4.4\ 10^{-5}$	46
Poly(vinyl cinnamate)	PVCin	$7.5\ 10^{-8}$	$3.0\ 10^{-6}$	40
Poly(methyl siloxane)	PMS	$4.2\ 10^{-6}$	$3.0\ 10^{-4}$	71
Poly(glycidyl MA-co-EA)	COP	$5.4\ 10^{-8}$	$4.2\ 10^{-7}$	7.8
Polystyrene	PS	$4.5\ 10^{-5}$	$4.6\ 10^{-6}$	100
Poly(4-chlorostyrene)	PClS	$2.1\ 10^{-7}$	$4.6\ 10^{-6}$	22
Poly(4-bromostyrene)	PBrS	$1.1\ 10^{-7}$	$1.8\ 10^{-6}$	16
Novolac [44]		$1.4\ 10^{-4}$	$1.3\ 10^{-3}$	9.3
AZ 1350 J [60]		$3.0\ 10^{-6}$	$9.9\ 10^{-6}$	3.3
Positive Resists				
Poly(methyl methacrylate)	PMMA	$4.2\ 10^{-6}$	$1.8\ 10^{-4}$	42
Poly(methylvinyl ketone)	PMVK	$6.4\ 10^{-7}$	$1.7\ 10^{-4}$	270
Poly(*tert*-butyl MA)	PTBMA	$3.2\ 10^{-7}$	$2.3\ 10^{-5}$	72
Poly(methyl-*o*-chloroacrylate)	PMCA	$2.8\ 10^{-7}$	$3.8\ 10^{-5}$	140
Poly(butene sulfone) [44]	PBS	$5.0\ 10^{-7}$	$7.5\ 10^{-7}$	3.4
Poly(trifluoroethyl-*o*-chloroacrylate) [61]		$5.1\ 10^{-7}$	$9.6\ 10^{-7}$	1.9

[a]After Brault and Miller [58] except where indicated otherwise.

in a group of almost monodisperse polystyrenes, the accurate constancy of the product $Q_p \times M_w$, which is demanded by the radiation chemical balance equation (see Fig. 9-18).

Brault and Miller [58] have made a systematic comparison of the response of some common resists and other polymers to the action of 100 keV protons and 20 keV electrons. Some data on ion beam resists [44, 58, 60, 61] are listed in Table 9-3, where it can be seen that there is a definite correlation between proton and electron sensitivities. Proton sensitivities vary a good deal less than electron sensitivities in the group of materials investigated. The logarithm of exposure dose (Q) for electron and proton exposure has been plotted in Fig. 9-19. The linear regression plot of the data has a correlation coefficients of 0.9. The origin of the plot is at $Q_e = 10^{-7}$ and $Q_p = 10^{-8}$, its slope has a value of 1.25. This means that, on the average, the materials are at least 10 times more sensitive to protons than to electrons, and that the ratio Q_p/Q_e increases with the exposure dose.

Geis et al. [62] have used *nitrocellulose as a self-developing ion beam resist*. On exposure to low energy (< 2 keV) Ar^+ ions as well as 193 nm deep UV, nitrocelluose disintegrates and the exposed areas of the films are ablated. At beam currents of 1 mÅ/cm² exposure times of 1 s and less are sufficient to produce high resolution pattern. The material is sufficiently plasma resistant to function as a mask in conventional dry etch processes.

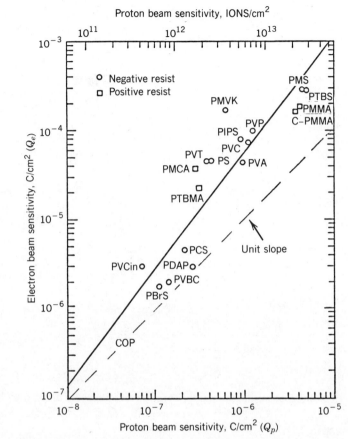

Figure 9-19 Correlation between proton beam (100 keV) and electron beam (20 keV) sensitivities for various resists. The resists are designated by the letter codes in Table 9-3. (Reproduced with permission from R. C. Brault and L. J. Miller [58].)

Ion beam technology is currently under vigorous development in several large laboratories. It is thought that the technology will play a role in microfabrication at the nanometer level, for the realization of quantum effect devices. Present efforts are preliminary approaches to that goal.

REFERENCES

1. M. Rothchild and D. J. Ehrlich, *J. Vac. Sci. Technol.*, B, **6**, 1 (1988).
1a. D. L. Spears and H. I. Smith, *Electron. Lett.*, **8**, 102 (1972).
2. A. Zacharias, *Solid State Technol.*, **24**(8), 57 (1981).
3. L. F. Thompson and M. J. Bowden, in *Introduction to Microlithography* (L. F. Thompson, C. G. Willson, and M. J. Bowden, Eds.), *ACS Symp. Ser.*, **219** 77 (1983).
4. H. I. Smith, *SPIE's Santa Clara Symp. Microlithogr.*, Tutorial T1, Fig. 5 (1987).

5. R. B. McIntosh, G. P. Hughes, J. L. Kreuzer, and G. R. Conti, Jr. (Perkin-Elmer), *Proc. SPIE*, **632**, 156 (1986).
6. A. M. Hawryluk, *Proc. SPIE*, **632**, 256 (1986).
7. J. R. Maldonado, M. E. Paulsen, T. E. Saunders, F. Vratny, and A. Zacharias, *J. Vac. Sci. Technol.*, **16**, 1942 (1979).
8. J. S. Pearlman and J. C. Riordan, *Proc. SPIE*, **537**, 102 (1985).
9. S. Doniach, I. Lindau, W. E. Spicer, and H. Winick, *J. Vac. Sci. Technol.*, **12**, 1123 (1975).
10. P. Pianetta, R. Redaelli, R. Jaeger, and T. W. Barbee, Jr., *Proc. SPIE*, **537**, 69 (1985).
11. I. Okada, S. Saitoh, S. Itabashi, and H. Yoshihara, *J. Vac. Sci. Technol.*, *B*, **4**, 243 (1986).
12. A. Heuberger, *Solid State Technol.*, **29**(2), 93 (1986).
13. W. D. Grobman, in *Handbook on Synchrotron Radiation* (E. E. Koch, D. E. Eastman, and Y. Farge, Eds.), North-Holland, Amsterdam, 1981.
14. K. Okada, K. Fuji, Y. Kawase, and N. Nagano, *J. Vac. Sci. Technol.*, *B*, **6**, 191 (1988).
15. W. D. Grobman, *IEEE Int. Conf. Electron Devices*, p. 415 (Dec. 1980).
16. A. P. Neukermans, *Solid State Technol.*, **27**(9), 185; **27**(11), 213 (1984).
17. A. Heuberger, *J. Vac. Sci. Technol.*, *B*, **6**, 7 (1988).
18. A. D. Wilson, *Proc. SPIE*, **537**, 85 (1985).
19. B. S. Fay and W. T. Novak, *Proc. SPIE*, **632**, 146 (1986).
20. A. R. Shimkunas, J. J. LaBrie, P. E. Mauger, and J. J. Yen, *Proc. SPIE*, **632**, 106 (1986).
21. R. A. Levy, D. J. Resnik, R. C. Freye, A. W. Yanof, G. M. Wells, and F. Cerrina, *J. Vac. Sci. Technol.*, *B*, **6**, 154 (1988).
22. G. E. Georgiou, C. A. Jankoski, and T. A. Palumbo, *Proc. SPIE*, **471**, 96 (1984).
23. S. Y. Chou, H. I. Smith, and D. A. Antoniadis, *J. Vac. Sci. Technol.*, *B*, **3**(6), 1587 (1985).
24. L. Czepregi, H. Seidel, and H. J. Herzol, *J. Electrochem. Soc.*, **131**, 2969 (1984).
25. K. Murase, M. Kakuchi, and S. Sugawara, in *Les Films Sensibles aux Electrons et aux Rayons X, Int. Conf. Microlithogr.*, Paris, June 1977.
26. C. G. Willson, in *Introduction to Microlithography* (L. F. Thompson, C. G. Willson, and M. J. Bowden, Eds.), *ACS Symp. Ser.*, **219**, 138 (1983).
27. A. Charlesby, *Atomic Radiation and Polymers*, Pergamon, Oxford, 1960.
28. A. Eranian, A. Couttet, E. Datamanti, and J. C. Dubois, in *Photodegradation and Stabilization of Coatings* (S. P. Pappas and F. H. Winslow, Eds.), *ACS Symp. Ser.*, **151**, 275 (1981).
29. L. F. Thompson, E. D. Feit, and R. D. Heidenreich, *Polym. Eng. Sci.*, **14**, 529 (1974).
30. G. N. Taylor, G. A. Coquin, and S. Somekh, *Polym. Eng. Sci.*, **17**, 420 (1977).
31. G. N. Taylor and T. M. Wolf, *J. Electrochem. Soc.*, **127**, 2665 (1980).
32. G. N. Taylor, *Solid State Technol.*, **23**(5), 73 (1980).
33. G. M. Moran and G. N. Taylor, *J. Vac. Sci. Technol.*, **16**, 2014 (1980).
34. H. Choong and F. Kahn, *J. Vac. Sci. Technol.*, *B*, **4**, 1066 (1983).

35. S. Imamura, *J. Electrochem. Soc.*, **126**, 1628 (1979).
36. N. Yoshioka, Y. Suzuki, and T. Yamazaki, *Proc. SPIE*, **537**, 51 (1985).
37. Z. C. H. Tan, C. C. Peropoulos, and F. J. Rauner, *J. Vac. Sci. Technol.*, **19**, 1348 (1981).
38. Z. C. H. Tan, R. C. Daly, and S. S. Georgia, *Proc. SPIE*, **469**, 135 (1984).
39. S. Pongratz, H. Betz, and A. Heuberger, *Proc. Kodak Microelectron. Semin.*, San Diego, CA, p. 143 (1983).
40. K.-F. Dössel, H.-L. Huber, and H. Oertel, *Microelectron. Eng. Conf.*, Interlaken, Switzerland Sept. 1985.
41. T. H. P. Chang, *J. Vac. Sci. Technol.*, **12**, 1271 (1975).
42. M. Parikh, *J. Vac. Sci. Technol.*, **15**, 931 (1978).
43. W. L. Brown, T. Venkatesan, and A. Wagner, *Nucl. Instrum. Methods*, **191**, 157 (1981).
44. (a) T. M. Hall, A. Wagner, and L. F. Thompson, *J. Vac. Sci. Technol.*, **16**, 1889 (1979). (b) D. B. Rensch, R. C. Seliger, G. Csonsky, R. D. Olney, and H. L. Stover, *J. Vac. Sci. Technol.*, **16**, 1897 (1979).
45. G. Taylor, *Proc. R. Soc. London, Ser. A*, **280**, 383 (1964).
46. R. Aihara, H. Sawaragi, H. Morimoto, K. Hosono, Y. Sasaki, T. Kato, and M. Hasselshearer, *J. Vac. Sci. Technol.*, B, **6**, 245 (1988).
47. A. Wagner, T. Venkatesan, P. M. Petroff, and D. L. Barr, *J. Vac. Sci. Technol.*, **19**, 1186 (1981).
48. R. L. Kubena, J. Y. Lee, R. A. Jullens, R. G. Brault, P. L. Middleton, and E. H. Stevens, *Proc. IEEE Int. Electron Devices Meet., Washington DC*, p. 566 (Dec. 1983).
49. R. L. Kubena, C. L. Anderson, R. L. Seliger, R. A. Jullens, and E. H. Stevens, *J. Vac. Sci. Technol.*, **19**, 1193 (1981).
50. H. C. Kaufmann, W. B. Thompson, and G. J. Dunn, *Proc. SPIE*, **632**, 60 (1986).
51. J. Randall, D. Ehrlich, and J. Tsao, *Proc. Int. Symp. Electron, Ion Photon Beams*, Tarrytown, NY, *1984*.
52. N. W. Parker, W. P. Robinson, and J. M. Snyder, *Proc. SPIE*, **632**, 76 (1986).
53. R. Aihara, H. Sawaragi, H. Morimoto, and T. Kato, *Proc. SPIE*, **632**, 196 (1986).
54. T. Venkatesan, G. N. Taylor, A. Wagner, B. Wilkens, and D. Barr, *J. Vac. Sci. Technol.*, B, **3**, 1379 (1985).
55. G. Stengl, H. Loeschner, and J. J. Muray, *Solid State Technol.*, **29**(2), 119 (1986); G. Stengel et al., *J. Vac. Sci. Technol.*, **16**, 1883 (1979).
56. J. L. Bartelt, *Solid State Technol.*, **29**(5), 215 (1986).
57. D. B. Rensch, R. C. Seliger, G. Csonsky, R. D. Olney, and H. L. Stover, *J. Vac. Sci. Technol.*, **16**, 1897 (1979).
58. R. C. Brault and L. J. Miller, *Polym. Eng. Sci.*, **20**, 1064 (1980).
59. R. L. Seliger and W. P. Fleming, *J. Appl. Phys.*, **45**, 1416 (1974).
60. Y. Wada, M. Mijitaha, K. Mochiji, and H. Obayashi, *J. Electrochem. Soc.*, **130**, 187 (1983).
61. J. E. Jensen, *Solid State Technol.*, **27**(6), 145 (1984).
62. M. W. Geis, J. N. Randall, R. W. Mountain, J. D. Woodhouse, E. I. Bromley, D. K. Astolfki, and N. P. Economou, *J. Vac. Sci. Technol.*, B, **3**, 343 (1985).

10 MULTILAYER TECHNIQUES AND PLASMA PROCESSING

MULTILAYER TECHNIQUES

Bilayer resists were used in the early 1970s to make fine line metal patterns. For this it was necessary to form strongly undercut resist images, and these were produced by coating the (silicon) substrate first with a fairly thick layer of PMMA and overcoating that with a thin imaging layer of a near-UV sensitive resist [1, 2]. The resist was exposed and developed in the usual way and the pattern then transferred into the underlying polymer by wet etching. With a judicious choice of procedure it was possible to make overcut, vertical, and undercut profiles, as illustrated in Fig. 10-1 [3, 4].

The importance of controlling image profiles in some applications can be easily demonstrated. If an overcut pattern is metallized, the polymer is completely protected against the action of a solvent. With an undercut profile the open spaces of the substrate are again metallized, but a narrow rim of the substrate, which is shielded by the resist overhang, as well as the side walls of the polymer pattern, remain uncovered. In the subsequent development step these uncovered areas allow the solvent to penetrate and lift off the polymer, leaving a clean metal pattern behind (see Fig. 10-2). This so-called *lift-off* technique is routinely used in IC manufacture [6]. The method is remarkably flexible, it has recently been used in conjunction with a Scanning Tunneling Microscope and lines as narrow as 120 nm were produced in a positive resist [7].

As the feature sizes of integrated circuits decreased in the 1970s a whole host of problems started to appear: aspect ratios (i.e., the ratio of height to width) increased, yet the new high-aperture stepper lenses had very shallow

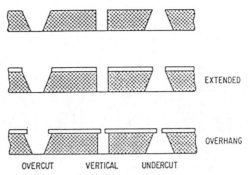

Figure 10-1 Overcut, vertical, and undercut image profiles in a thick resist layer, with and without the top imaging layer retained. (Reproduced with permission from A. W. McCullough [3b].)

depth of focus; linewidth control over nonplanar substrates became more and more difficult and interference patterns formed in resist films coated over highly reflective substrates. Many of these difficulties can be alleviated by the use of multilayer techniques [3, 8].

The component common to all multilayer techniques is a thick *planarizing layer*, which overcoats the wafer and submerges the underlying topography. In early experiments poly(methyl methacrylate) was used as planarizing material. The planarizing layer is overcoated with a thin imaging layer, for example, a novolac–diazoquinone resist. This resist layer is imaged in the usual way, developed, and can function as an *in situ* deep-UV mask (see Fig. 10-3a). The

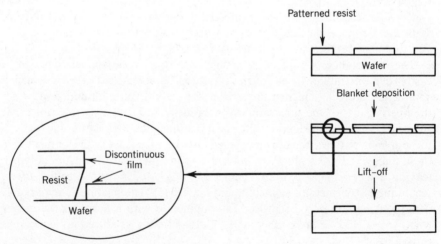

Figure 10-2 The lift-off technique for making a metal pattern. The resist is patterned on the substrate, making sure that it has a substantial undercut. The whole structure is metallized, and finally the resist is lifted off the substrate with solvent. (Reproduced with permission from J. W. Coburn [5].)

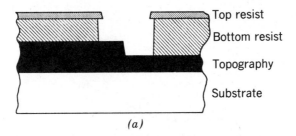

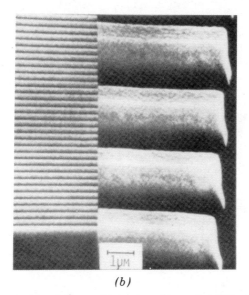

Figure 10-3 (*a*) The principle of the portable conformable mask (PCM). (*b*) Results obtained with a PCM in 1.9 μm PMMA exposed through an image defined in 2000 Å thick AZ 1350 J. (Reproduced with permission from M. Hatzakis [9b].)

pattern is then transferred into the planarizing layer by a deep-UV blanket exposure and subsequent wet development. Since the imaging layer conforms to the planarizing layer and since it is portable with the coated wafer, the system has been called a *portable conformable mask* (PCM) [8, 9]. Figure 10-3*b* shows the result obtained by using a conformable mask in 2 μm PMMA with deep-UV exposure through a mask of 2000 Å of a patterned AZ 1350 J positive resist [9b]. Early on Hatzakis has used a bilayer method for the preparation of metallized patterns [10].

To operate a PCM successfully, the planarizing layer and the imaging layer must not be soluble in each others coating solvent, and indeed in each other. In practice there is often a small interpenetration zone that contains both components. To remove this difficulty and make the interface as sharp as possible, a third "separation" layer is used (Havas, 1976 [11]), which then

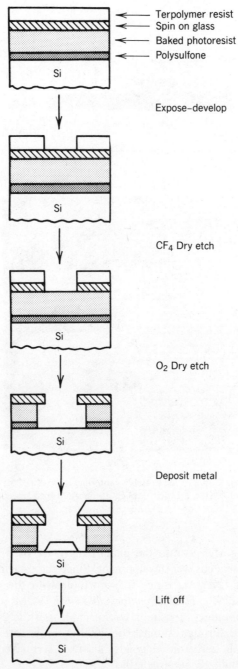

Figure 10-4 The trilevel process for the production of a fine line metal pattern. The image is created by exposure and development in the top imaging resist layer. The pattern is transferred into the "spin-on" glass interlayer by etching with a fluorine containing gas. It is then transferred into the planarizing layer, (baked photoresist) by oxygen plasma etching, which produces an overhang as shown. The whole structure is metallized in a vacuum and finally the resists are removed with solvent. A metal pattern is left on the substrate.

leads to a *trilayer system*. Initially, silicon was used as the interlayer, coated onto PMMA by plasma CVD. More recently various polysilane derivatives have been introduced. These are soluble polymers and can be spun on in the same way as the resist. On exposure to an oxygen plasma they form a thin layer of SiO_2 (glass), hence the term "*spin-on glasses*" [12]. The thickness of the planarization layer is between 1 to 4 µm, the separation layer has a thickness from 0.05 to 0.2 µm, and the imaging layer is usually between 0.3 to 0.5 µm thick. Figure 10-4 shows the flow of operations in the Havas process.

The transfer of the resist pattern into the planarizing layer is achieved either by wet etching, or by plasma etching (dry etching), or by deep-UV exposure and subsequent development. Finally, the pattern formed in the planarizing layer is transferred to the substrate, again by various wet or dry etching methods. In the dry etching transfer to SiO_2 or to silicon, PMMA and similar polymers present a problem because of their poor plasma resistance. In that case, a cap of the original novolac (or inorganic) layer has to be retained on the polymer pattern. Some four different processing strategies can be pursued, as indicated in Fig. 10-5 [3].

The novolac resist cap on the planarizing layer can be retained either by using a solvent for the development of the planarizing layer which does not attack the novolac resist (e.g., chlorobenzene or toluene) or by crosslinking (hardening) the novolac resist. The second solution is much preferred since the crosslinked resist has better deep-UV absorptance and better plasma resistance, and with the hardened resist it is possible to use the fast acting developer methyl isobutyl ketone for the underlying PMMA.

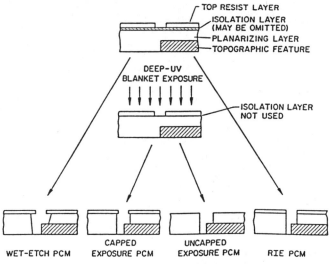

Figure 10-5 Multilayer resist systems of the PCM type. (Reproduced with permission from B. J. Lin [3a].)

Resist hardening is achieved by various radiative exposures (deep UV [13, 14], electrons [15], plasma [16] or photomagnetic curing [17]) in combination with a high temperature bake. The baking operation is invariably accompanied by slight, but increasingly unacceptable changes in image dimensions. To cope with this problem Lin has introduced the idea of *resist hardening in a conformable mold* (RHCM) [18, 19]. The procedure consists of the following steps: The novolac resist image is overcoated with a thick layer of poly(methyl methacrylate) (the mold), and the whole structure is subjected to a high temperature bake (30 min at 200°C). During the thermal process the resist pattern is held by the mold and is crosslinked without changing its geometric dimensions (see Fig. 10-6).

Many polymers have been evaluated as *planarizing materials*, but the original poly(methyl methacrylate), PMMA, is still popular. Poly(methyl isopropenyl ketone), PMIPK,

$$\left[CH_2 - \underset{\underset{CH_3}{\overset{CO}{|}}}{\overset{CH_3}{\underset{|}{C}}} \right]_n$$

is in some respects superior to PMMA; it has most of the desirable physical properties of PMMA and it requires less deep-UV exposure for pattern transfer. More recently an aqueous developable planarization material, poly(dimethyl glutarimide) (PMGI) [4]

$$-CH_2 + \underset{\underset{O=C}{\overset{H_3C}{\overset{|}{C}}}}{\overset{H_3C}{\overset{|}{C}}} \underset{N}{\overset{H_2}{\overset{|}{C}}} \underset{\underset{R}{|}}{\overset{CH_3}{\overset{|}{C}}} +_n$$

PMGI

has been described. Poly(dimethyl glutarimide) is water based and is highly resistant to organic solvents so that no intermixing with conventional solvent coated imaging layers is observed. It has a higher T_g than PMMA or PMIPK and better resistance to plasma etching. It can also be dyed to make it absorb radiation at the exposure wavelength and thus prevent the formation of interference pattern in the imaging layer [20].

The problem of interference pattern has also been dealt with by putting an antireflection (dye) coating, ARC, between the substrate and the planarizing layer. This brings about increased exposure latitude over topography, but it

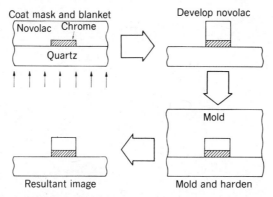

Figure 10-6 The concept of resist hardening in a conformable mold. A self-aligned resist image, obtained by exposing a novolac resist through the quartz support of a chromium mask and developing, is encased in a mold of PMMA and hardened by prolonged baking at 200°C. After removing the mold, the resulting image is found to have retained its nominal dimensions to better than 0.05 μm. (Reproduced with permission from B. J. Lin [18].)

may decrease adhesion and have other minor disadvantages. Much improved images can be obtained by adding a contrast enhancing layer to the system. This is a thin overcoat with a radiation-bleachable dye in a binder. (see Chapter 3). The contrast enhancing layer produces a built in mask on top of the imaging resist and makes for sharper images, although of course, it requires longer exposures [21].

The physics of planarization has been addressed by White [22], by Lavergne and Hofer [23] and more recently by Stillwagon et al. [24]. It was found that the time required for the levelling out of a topographic feature of width, w, depends on the original height, h_0, of the polymer solution film used in coating, on its viscosity, η, and its surface tension, γ, in the following form.

$$t = \frac{w^4 \eta}{h_0^3 \gamma}$$

Planarization for multilevel metallization has been discussed by Saxena and Pramanik [25]. A great number of conventional resists have been evaluated as imaging or top layers in multilayer procedures and a variety of new materials with greater resistance to plasma have recently been described. These will be discussed in a later section.

Multilayer techniques have produced high resolution patterns of astonishing quality, as can be seen by the example shown in Fig. 10-7. It refers to an uncapped resist image obtained with a two layer deep-UV PCM using 2 μm PMMA as the planarizing layer and a 0.2 μm AZ 1350 imaging layer. Note the severe topography of the substrate, which the resist image overcomes without the slightest change in linewidth. Multilayer resists have now become an

Figure 10-7 High resolution pattern obtained with a two layer resist PCM in 2 μm thick PMMA. (Reproduced with permission from J. M. Moran and D. Maydan, *J. Vac. Sci. Technol.*, **16**, 1620 (1979)).

indispensable part of microlithography. Recently, Lee has critically examined the economics of the technique [21].

PLASMA ETCHING

As the feature sizes of integrated circuits decrease, the earlier essentially two-dimensional pattern have given way to three-dimensional structures with high aspect ratio (height to width) and with vertical side walls. Such pattern can be created only by highly directional, anisotropic transfer processes. Wet etching is by its nature isotropic, but some plasma techniques allow directional etching, and that is the principal reason for the importance of plasma processing in microlithography [6, 26–28]. There are other advantages in dry etching as compared to wet etching, namely, greater cleanliness (it is possible to keep particulate contaminations to a lower level), greater ease of production line automation, and lower chemical costs [29, 30].

Nomenclature. The term plasma etching as it is used in the title of this section covers a whole group of processing techniques the most important of which are based on radio frequency (rf) glow discharges. Others are based on large area ion beams. The genealogy and nomenclature of these methods are indicated in the following scheme [31].

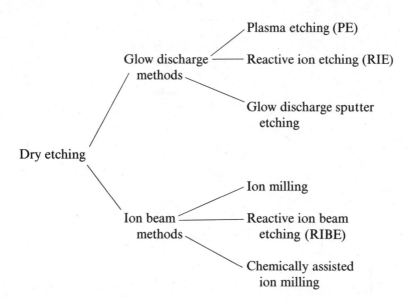

Two of these methods are now in common use: plasma etching in the narrower sense and reactive ion etching, RIE.

The Glow Discharge. A glow discharge occurs in an ionized gas that contains a low concentration of ions (about one in a million of the molecules present are ionized) and free electrons. The electrons can be made to couple into the high frequency electric field of an rf generator where they gain kinetic energy. They pass this on, by collisions, to the neutral molecules, and this collisional process produces ions, radicals, and radical ions. It is the chemical and the physical action of these species that brings about the removal of material called plasma etching.

Early on plasma was used in microelectronics as a cleaning operation (plasma ashing), the removal of traces of organic material from the wafer surface [32]. In plasma ashing the wafers are loaded into a cylindrical vessel, which is connected to an oxygen supply at one end and to a rough vacuum pump at the other. Radio frequency power is applied by an outside coil or by external electrodes. Similar so-called barrel etchers are used to transfer resist pattern to the wafer. The technique is termed plasma etching, and here the etching process is isotropic.

Reactive Ion Etching. Directional, anisotropic, etching is associated with the action of fast moving ions impinging all in one direction on the surface of the material. To have some control over ion acceleration, planar geometry etchers with internal electrodes have been introduced (Fig. 10-8a) [6].

The spatial distribution of potential in a system in which only one of the electrodes is powered by capacitative coupling is shown in Fig. 10-8b. As the discharge is established in the system, the electric field that equalizes the ionic

368 MULTILAYER TECHNIQUES AND PLASMA PROCESSING

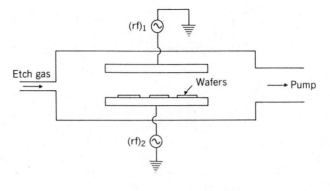

$(rf)_1 \neq 0 \; (rf)_2 = 0$ —— Plasma etching
$(rf)_1 = 0 \; (rf)_2 \neq 0$ —— Reactive ion etching

(a)

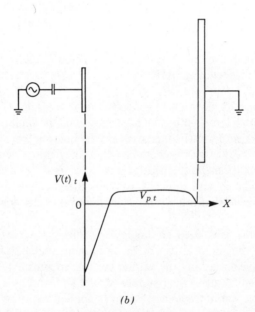

(b)

Figure 10-8 (*a*) Schematic of a planar etcher. (*b*) Distribution of time average potential between the electrodes of a planar etcher. (Reproduced with permission from J. W. Coburn [6].)

fluxes at the electrodes causes a decrease of electron concentration near the powered electrode and forms a so-called plasma sheath. In the plasma sheath that is starved of electrons, the greater part of the voltage drop of the discharge occurs. When the wafers are placed on the grounded electrode they are exposed to neutral radicals and to slow moving ions and etching is isotropic. This case is equivalent to the situation in the barrel etcher. In

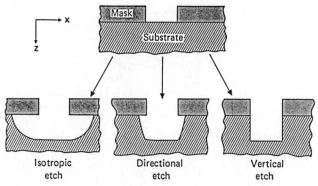

Figure 10-9 Isotropic etching, partly directional and highly directional (anisotropic) etching. (After J. W. Coburn [6].)

reactive ion etching the wafers are placed onto the powered electrode where they are exposed to fast moving ions that have been accelerated in the voltage drop of the plasma sheath.[1] The directional ion bombardment in RIE enhances the etch rate in the vertical, but not in the horizontal direction, and this produces anisotropic etching and leads to the formation of vertical walls in the etch pattern (see Fig. 10-9) [5].

The ratio of ions to neutrals reaching a surface and the relative kinetic energy of the ions determine the directionality of the etching process. Another factor is the overall pressure in the system. The lower the pressure, the fewer are the collisions of the ions on traversing the dark space and the more directional is their action. Below about 0.02 torr pressure, the ions do not suffer any collisions in the accelerating zone. The directionality of the etching process is also affected by the choice of gases that influence the ion–neutrals ratio in the plasma [31]. There are many other conditions that have a bearing on the performance of an etching procedure, indeed the great number of parameters that have to be considered is one of the problems of the method. The reader is referred to the monographs and review articles quoted in the references [6, 26–31, 33].

Paraszak et al. [34] have recently investigated RIE over a wide range of microwave frequencies (from 10 MHz–10 GHz) and found that one of the critical parameters determining the rate of etching and its directionality is the ion energy acquired in the plasma sheath divided by the overall pressure.

Other Techniques. Apart from the two principal plasma etch methods described, a variety of other techniques are being developed. One of these is *glow discharge sputter etching* where the surface of the wafer, placed on the powered electrode, is exposed to nonreactive ions such as Ar^+, which remove the substrate material simply by momentum transfer (sputtering). This method is

---

[1] Magnetic field enhanced RIE has recently received some attention [35].

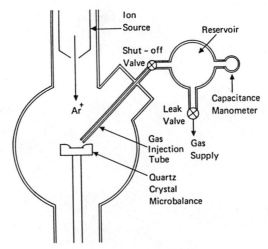

Figure 10-10 Apparatus used in the study of ion-assisted gas surface chemistry. (Reproduced with permission from J. W. Coburn and H. F. Winters [29].)

very general, anything can be sputtered away, but it is also totally nonselective [33].

Several promising techniques are based on wide area ion beams. Ion beam sputtering, also called *ion milling* uses nonreactive ions accelerated through a large dc potential. Ion milling with focused ion beams is used in the repair of masks and in the customization of integrated circuits. In *reactive ion beam etching*, RIBE, a reactive gas, for example, CF_4, is used [36]. In *chemically assisted ion etching*, CAIE, nonreactive ions are used in conjunction with a

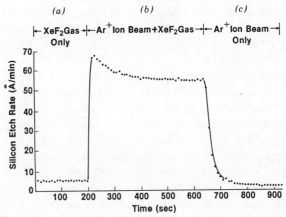

Figure 10-11 The etch rate of silicon (a) in XeF_2 gas; (b) in XeF_2 gas combined with fast Ar^+ ions; (c) under Ar^+ ion bombardment alone. (Reproduced with permission from J. W. Coburn and H. F. Winters [29].)

reactive gas, which is introduced near the wafer surface. Figure 10-10 shows the experimental arrangement that has been used to monitor the etch rate in such a system [37].

The results obtained with an Ar^+ ion beam and with XeF_2 as the reactive gas are illustrated in Fig. 10-11. Both, the ions alone and the XeF_2 gas alone do etch silicon at a measurable rate, but their combination is highly synergistic and the etch rate increases by up to 20 times [37]. The synergistic cooperation of an etchant gas with laser radiation is being explored as a means of laser-assisted dry etching. Some interesting results have been achieved and there is considerable activity in the field [38, 39].

RESISTS FOR USE IN DRY ETCHING

Etch Rate and Resist Composition. To function as an effective mask in plasma etching the developed resist image must withstand the action of the etchant better than the substrate does. Not many conventional resists fulfill this condition; in fact, there are wide variations in the plasma resistance of common polymers and resists. Table 10-1 lists relative rates of removal by an oxygen plasma [32]. It can be seen from the data that the removal rate parallels the radiation chemical scission yield of the materials when exposed to γ radiation (removal rates for 40 polymers are given in Reference 32).

Ohnishi and co-workers at NEC [40] have reported a quantitative correlation between plasma resistance and the elemental composition of polymers containing carbon, hydrogen, and oxygen. In a comprehensive study they found that the rate of RIE, but not that of plasma etching, is determined by the *effective carbon content* of the material. This observation is consistent with the view that RIE is controlled by a sputtering process.

Since carbon has the smallest sputtering yield in these materials, the etching of carbon will be the rate-determining step, and the etch rate will be inversely proportional to the atom fraction of carbon in the repeat unit of the polymer.

TABLE 10-1 Relative Rates of Removal by Oxygen Plasma and G_s Values [32]

Polymer	Etch Rate	G_s
Poly(o-methylstyrene)	1.11	0.3
Poly(phenyl methacrylate)	1.33	
Poly(vinyl methyl ketone)	1.48	
Poly(methyl methacrylate)	2.37	1.2
Poly(methyl methacrylate-co-methacrylonitrile) 94 : 6	2.70	2.0
Poly(isobutylene)	3.56	4.0
Poly(butene-1-sulfone)	7.11	8.0

$$\text{etch rate} \propto \frac{N}{N_C} \tag{1}$$

Here N_C is the number of carbon atoms and N is the overall number of atoms in the repeat unit of the polymer. When Eq. (1) was applied to the experimental data, materials that contained oxygen were always found to have higher etch rates than predicted, which meant that the presence of oxygen in the molecule enhanced the etching process. This idea was incorporated into Eq. (1) by defining an *effective* carbon content ($N_C - N_O$) and replacing Eq. (1) by the following expression.

$$\text{etch rate} \propto \frac{N}{N_C - N_O} \tag{2}$$

Equation (2) correlates rather effectively the etch rates of a variety of polymers exposed to an Ar^+ ion beam (300 eV) or to an oxygen ion beam, as can be seen in Fig. 10-12.

Watanabe and Ohnishi [41] have extended their steady state approach to the oxygen reactive ion etching of silicon containing polymers. Assuming that the steady state etch rate of these polymers is controlled by the sputtering rate of the SiO_2 that is formed in the oxygen plasma, they have expressed the etch rate in the form

$$R = \frac{\rho(\text{Si in oxide})}{\rho(\text{Si in polymer})} \times \text{sputtering rate of } SiO_2 \tag{3}$$

where $\rho(\text{Si in oxide})$ and $\rho(\text{Si in polymer})$ are the mass densities of silicon in the oxide layer and in the bulk of the polymer, respectively. Figure 10-13 taken from the work of Jurgensen et al. [42] shows a plot of the steady state etch rate as a function of the prevailing sputtering rate of SiO_2 for a range of conditions of self-bias, oxygen pressure, and power density. The straight line correlation confirms the steady state sputtering hypothesis of Watanabe and Ohnishi [41].

In general, aromatic components increase the plasma resistance of a material, and for that reason novolac resists and materials derived from polystyrene have been widely used as imaging top layers in bilayer resist applications. However, the etch ratio (selectivity) between, say, polystyrene and the planarizing PMMA is not large (1 : 2.5) and the pattern transfer step is often a critical operation. There is a need for greater selectivity, and one way of achieving it is by using a trilayer technique where the pattern is transferred from the imaging top layer into a thin separation layer of silicon or SiO_2, which then forms the plasma resistant template for pattern transfer into the planarization layer.

Trilayer lithography is costly because it adds several processing steps to device fabrication. There is therefore great interest in devising resists that produce highly plasma resistant images and are able to combine the role of the radiation sensitive imaging layer and that of the plasma resistant interlayer in

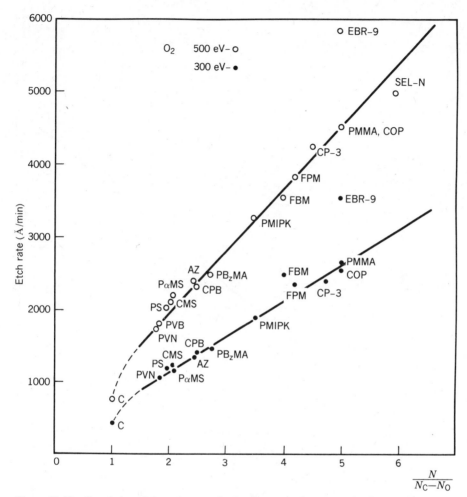

Figure 10-12 Correlation of the etch rates of polymers exposed to oxygen ion beams of 300 and 500 eV with their elemental composition. N_C is the number of C atoms, N_O the number of O atoms, and N the overall number of atoms in the repeat unit. Key to figure: PMMA Poly(methyl methacrylate); COP poly(glycidyl methacrylate-co-ethyl ascrylate); CP-3 poly(methyl methacrylate-co-t-butyl methacrylate); EBR-9 poly(o-chloro-trifluoroethyl acrylate); PBZMA poly(benzyl methacrylate); FBM poly(hexafluorobutyl methacrylate); FPM poly(fluoropropyl methacrylate); PS polystyrene; CMS chloromethylated polystyrene; PoMS poly(o-methylstyrene); PVN poly(vinyl naphthalene); AZ 1350 J is a common diazoquinone–novolac resist; CPB is cyclized poly(butadiene). (Reproduced from H. Gokan et al. [40], with permission from the Electrochemical Society, Inc.)

a single coating. Intensive efforts have been made in all the leading laboratories to design resist materials with increased plasma resistance that would be compatible with the current processing and pattern transfer routines.

One way of increasing plasma resistance is to treat the developed resist images with metal containing solutions. Ito et al. [15a] could demonstrate improved resistance after treatment with hexamethyldisilazenes, Stillwagon

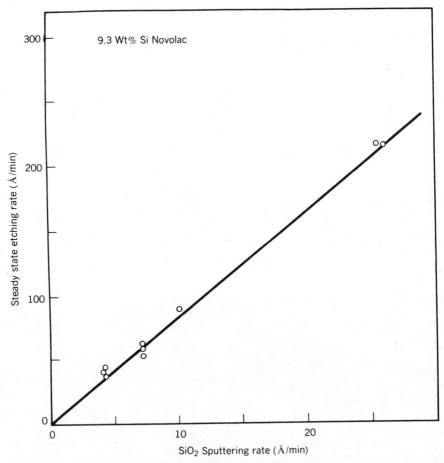

Figure 10-13 Correlation of the steady state etch rate of a 9.3% Si containing novolac in a variety of etching conditions (oxygen pressure, electrode bias, frequency) with the sputtering rate of SiO_2 in these conditions. (Reproduced from Jurgensen et al. [42].)

and co-workers [15b] have successfully used $TiCl_4$; more recently considerable improvement was observed after treatment with dibutyl magnesium dissolved in hexane [15c]. McColgin et al. [43] have described a process, (SABRE) where the developed resist image is silylated in a vacuum oven at 90°C at 100 torr for some 15 min, with a bifunctional, crosslinking silane, such as, for example, dichloromethylsilane

$$\begin{array}{c} H \\ | \\ Cl-Si-Cl \\ | \\ CH_3 \end{array}$$

TABLE 10-2 Effect of Iodination on the Etch Resistance of Polystyrene [44]

% Iodination	Relative Etch Rate
0	1.0
2	0.89
4	0.86
13	0.72
27	0.51
61	0.10
85	0.04

In the subsequent reactive etching step the pattern is etched with an etching selectivity of 15 : 1.

Another method used to achieve higher plasma resistance is the inclusion of heavy atoms directly into the material. Ueno et al. [44] have shown, for example, that incorporation of iodine in polystyrene, by iodination in the para position of the phenyl ring, spectacularly increases the resistance of the polymer towards oxygen plasma (see Table 10-2).

Taylor et al. [45] demonstrated the utility of including elements that form refractory oxides. Thus organic silicon derivatives rapidly produce SiO_2 in an oxygen plasma, and the thin glass layer so formed is an ideal etching barrier with practically infinite etch resistance. Other organometallics (e.g., organotin derivatives) have been explored [46].

Silicon Containing Negative Resists. By far the most successful solution to the plasma resistance problem is the inclusion of silicon in the resist material. In part this is due to the fact that the organic chemistry of silicon is well established and that there is a wealth of synthetic experience in the field.

MacDonald et al. [47] developed a plasma resistant material with useful sensitivity and with good physical properties by copolymerizing *trimethylsilylmethyl styrene* with chlorostyrene.

Suzuki et al. [48] increased the sensitivity of a similar system by copolymerization of silylated polystyrene with chloromethylstyrene. A sensitive material, which is glassy at room temperature, was prepared by Morita and others [49] by the *chloromethylation of poly(diphenyl siloxane)*.

$$\left[\begin{array}{c} \text{Ph} \\ | \\ -\text{Si}-\text{O}- \\ | \\ \text{Ph} \end{array}\right]_m \left[\begin{array}{c} \text{Ph} \\ | \\ -\text{Si}-\text{O}- \\ | \\ \text{C}_6\text{H}_4 \\ | \\ \text{CH}_2\text{Cl} \end{array}\right]_n$$

Hartney and Novembre [50] chlorinated block copolymers of methylstyrene and dimethylsiloxane. With a silicone content of 15.5% by weight and with a chlorine content of 4.9% these resists could be patterned by exposure doses of less than 1 $\mu C/cm^2$.

Brominated poly(1-trimethylsilyl propyne) was recently described as a sensitive deep-UV resist [51]. The polymer was synthesized by $NbCl_5$ catalyzed polymerization from trimethylsilylpropene and subsequently brominated. With about 10 mol% bromination the sensitivity of the polymer to 254 nm radiation is 25 mJ/cm^2.

Ohnishi et al. [52] have taken a two component approach to negative plasma resistant materials. They prepared the unsaturated base polymer poly(triallyl phenylsilane) by radical chain polymerization from triallyl phenylsilane.

$$\text{PhSi}(\text{CH}_2\text{CH}=\text{CH}_2)_3 \xrightarrow[\text{PhH}]{\text{BPO}}$$

$$\left[\begin{array}{c}\text{Si} \\ \text{Ph}\end{array}\right]_{0.47} \left[\begin{array}{c}\text{Si} \\ \text{Ph}\end{array}\right]_{0.44} \left[\begin{array}{c}\text{Si} \\ \text{Ph}\end{array}\right]_{0.04} \quad (4)$$

This was used together with bis-azides, for example, 2,6-bis(4′-azidobenzal)-methylcyclohexanone as a negative resist. The sensitivity of the system in the near UV is about 3 mJ/cm^2. A similar idea was realized by MacDonald et al. [48] when they used poly(trimethylsilylmethyl styrene) as the base resin in combination with radical generating additives such as 1,2,4-trichlorobenzene and 3,3′-diazidodiphenyl sulfone.

Silicon Containing Positive Resists. In fabricating pattern with feature sizes of the order of 1 µm or less, negative resists are often less useful because of the danger of swelling in the development step. For this reason plasma resistant positive resists are of interest.

Hatzakis et al. [53] have described electron resists based on *poly(dimethylsiloxane)* and *poly(phenylmethylsiloxane)*

$$\left[\begin{array}{c} CH_3 \\ | \\ -Si-O- \\ | \\ CH_3 \end{array}\right]_n \qquad \left[\begin{array}{c} CH_3 \\ | \\ -Si-O- \\ | \\ Ph \end{array}\right]_n$$

Poly(dimethylsiloxane)   Poly(phenylmethylsiloxane)

These materials produce a highly etch resistant pattern but they have low glass transition temperatures and poor radiation sensitivity.

Reichmanis et al. [54] have investigated the possibility of making the classical PMMA resist more etch resistant by the introduction of silicon. They polymerized the following *siloxane substituted propyl methacrylates*.

$$\left[\begin{array}{c} CH_3 \\ | \\ -CH_2-C- \\ | \\ C=O \\ | \\ O \\ | \\ (CH_2)_3 \\ | \\ CH_3-Si-CH_3 \\ | \\ O \\ | \\ CH_3-Si-CH_3 \\ | \\ CH_3 \end{array}\right] \left[\begin{array}{c} CH_3 \\ | \\ -CH_2-C- \\ | \\ C=O \\ | \\ OH \end{array}\right]_n$$

and obtained materials of reasonable etch resistance, but low sensitivity. The sensitivity could be increased by introducing a third component into the copolymers, namely, 3-oximino-2-butanone methacrylate. This work demonstrates in some detail the effect of silicon incorporation on etch rate. In agreement with the results of others, Reichmanis and Smolinsky found that a silicon content of approximately 10% by weight is sufficient to produce a useful oxygen–plasma etch barrier.

A more successful and more versatile two component system was synthesized from the usual cresols and *p*-trimethylsilylmethylphenol [55].

The copolymers obtained were used as base resins in positive resists by adding diazoquinones as dissolution inhibitors for UV resists and polysulfones for electron resists. Sensitivities were similar to those of the nonsilylated resins, namely, 120 mJ/cm^2 for the photoresist and 8 μC/cm^2 for the electron resist version. A similar system, where trimethylsilyl groups were introduced into novolac was described by Ohnishi and co-workers [56]. In yet another modification of the same idea poly(vinyl phenol), molecular weight 300 to 7000, was derivatized at the phenolic oxygen with trimethylsilyl groups [57].

Organotin Polymers. Organotin polymers were synthesized by Labadie and co-workers and were found to lead to electron resists of low sensitivity. With a tin content of over 10% by weight the polymers are resistant not only to oxygen reactive ion etching, but also to attack by fluorine containing gases, which is of interest for some applications [58, 59].

TABLE 10-3 Organosilicon Resists for Use in Bilayer Systems

Structure	Type	Company	Reference
$\left[\begin{array}{c} Me \\ | \\ Si-O \\ | \\ CH=CH_2 \end{array}\right]\left[\begin{array}{c} Me \\ | \\ Si-O \\ | \\ Me \end{array}\right]$	Negative EB	IBM	53
$\left[\begin{array}{c} Me \\ | \\ Si-O \\ | \\ CH=CH_2 \end{array}\right]\left[\begin{array}{c} Ph \\ | \\ Si-O \\ | \\ Ph \end{array}\right]$	Negative deep UV	IBM	60
$\left[\begin{array}{c} CH_2-CH \\ | \\ C_6H_4-SiMe_3 \end{array}\right]\left[\begin{array}{c} CH_2-CH \\ | \\ C_6H_4-CH_2Cl \end{array}\right]$	Negative EB, deep UV	NEC	48
$\left[\begin{array}{c} CH_2-CH-S \\ | \quad \| \\ Si \quad O \\ R_3 \end{array}\right]$	Positive EB	RCA	61
$\left[\begin{array}{c} CH_2-CH \\ | \\ C_6H_4-SiMe_3 \end{array}\right]\left[\begin{array}{c} CH_2-CH \\ | \\ C_6H_4-Cl \end{array}\right]$	Negative EB, deep UV	IBM	62
$\left[\begin{array}{c} Ph \\ | \\ Si-O \\ | \\ C_6H_4-CH_2Cl \end{array}\right]\left[\begin{array}{c} Ph \\ | \\ Si-O \\ | \\ Ph \end{array}\right]$	Negative EB, XR, deep UV	NTT	49
$\left[\begin{array}{c} R \\ | \\ Si \\ | \\ R' \end{array}\right]\left[\begin{array}{c} R \\ | \\ Si \\ | \\ R' \end{array}\right]$	Positive deep UV, mid UV	IBM	63
$\left[\begin{array}{c} Et \quad Ph \\ | \quad | \\ Si-Si-C_6H_4 \\ | \quad | \\ Ph \quad Et \end{array}\right]$	Positive deep UV	Hitachi	64

TABLE 10-3 Organosilicon Resists for Use in Bilayer Systems

Structure	Type	Company	Reference
(structure 1)	Negative near UV (+bisazide)	NEC	65
(structure 2)	Negative EB, deep UV	NTT	66
(structure 3)	Negative near UV (+bisazide)	NTT	66
(structure 4)	Positive deep UV	AT & T, Bell	67
(structure 5)	Positive near UV (+quinonediazide)	AT & T, Bell	68
(structure 6)	Positive near UV (+quinonediazide)	NEC	56

Table 10-3, taken from the work of Ohnishi [52] gives an overview of recently reported silicon containing resists.

Polysilanes. An interesting newer class of plasma resistant materials are the polysilanes. While the earlier aromatic polysilanes discovered by Kipping [69] and the permethylpolysilanes prepared by Burkhard [70] were crystalline and insoluble, Wesson and Williams [71] discovered soluble polysilanes that were later developed in particular by West and co-workers [72]. These polysilanes have mixed and preferably bulky aliphatic substituents, and they can be prepared by the condensation of substituted methylsilyl dichlorides with a dispersion of sodium in toluene or other common solvents [73–75].

$$\text{R}-\text{CH}_2-\underset{\underset{\text{R}}{|}}{\overset{}{\text{Si}}}-\text{Cl}_2 \xrightarrow[\text{toluene}]{\text{Na (5\%)}} \left[\underset{\underset{\text{R}}{|}}{\overset{\overset{\text{CH}_3}{|}}{\text{Si}}}\right]_n + 2\text{NaCl} \qquad (5)$$

The alkyl-substituted silanes absorb radiation at around 310 nm. When a phenyl ring is attached directly to the Si backbone the absorption maximum is red shifted to about 340 nm (see Table 10-4).

Furthermore, the polysilanes have the remarkable property that for short chain lengths (degrees of polymerization of less than 20), the 310 nm absorption peak depends on molecular weight and with decreasing chain length shifts to shorter wavelength [74]. Figure 10-14 shows a plot of λ_{max} as a function of chain length for two representative polysilanes. Qualitatively this phenomenon is understood to be the result of increased electron delocalization in the σ–d molecular orbitals, which contribute to Si—Si bonding in the backbone as well as to the molecular orbitals of the excited state [76].

The polysilanes are radiation sensitive. They undergo backbone scission and become soluble by virtue of a molecular weight change [63]. The mecha-

TABLE 10-4 Absorption Characteristics of High Molecular Weight Polysilanes [75]

Polymer	M_w	λ_{max} (nm)
$(\text{Me}_2-\text{Si})_n$		304
$(\text{PhMe}-\text{Si})_n$	190,000	341
(p-Tolyl Me—Si)$_n$	75,000	337
(3-Phentyl Me—Si)$_n$	286,000	303
(n-Propyl Me—Si)$_n$	644,000	306
(n-Butyl Me—Si)$_n$	110,000	304
(n-Dodecyl Me—Si)$_n$	483,000	309
(Cyclohexyl Me—Si)$_n$	804,000	326

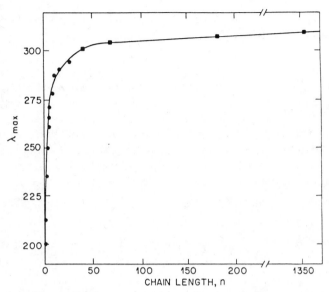

Figure 10-14 Ultraviolet absorption maxima of poly(alkyl silanes) plotted as a function of chain length n. (●) represents $Me(Me_2Si)_n$, (■) represents n-dodecyl$(Me_2Si)_n$. (Reproduced with permission from D. C. Hofer et al. [63].)

nism of the degradation process is unusual and it is thought that a silylene fragment is extruded from within the chain [77]. This either results in chain contraction or it leaves two macro radicals behind. These are stabilized by hydrogen abstraction and continue being fragmented by photoextrusion [72].

$$\begin{array}{c}A\\|\\-Si-\\|\\B\end{array}\begin{array}{c}A\\|\\Si-\\|\\B\end{array}\begin{array}{c}A\\|\\Si-\\|\\B\end{array}\xrightarrow{h\nu}\begin{array}{c}A\\|\\Si:\\|\\B\end{array}+2\cdot\begin{array}{c}A\\|\\Si-----\\|\\B\end{array}\begin{array}{c}A\\|\\SiH\\|\\B\end{array}$$

$$CH_3-\begin{array}{c}A\\|\\Si-H\\|\\B\end{array}\qquad H\begin{array}{c}A\\|\\Si-----\\|\\B\end{array}\begin{array}{c}A\\|\\SiH\\|\\B\end{array}$$

In solution, the quantum yield of chain scission in polysilanes is high ($\phi = 0.97$ for poly(methyl phenylsilane). It is much lower (0.017) in solid

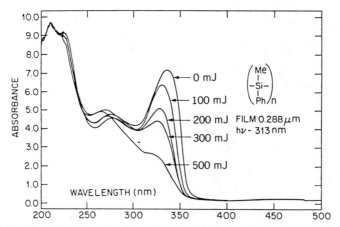

Figure 10-15 Photobleaching of a film of poly(methyl phenyl silane) exposed to 313 nm radiation. (Reproduced with permission from D. C. Hofer et al. [63].)

films, presumably because here chain repair within the reaction cage is highly favored [75].

Since the absorption maximum shifts to shorter wavelength as the length of the polymer chain decreases, the spectrum of the polymer film changes in the last stages of irradiation (see Fig. 10-15). This means that the absorption of the resist film will gradually bleach on exposure and that light will be able to penetrate right down to the substrate. As a result the resist images develop cleanly and with vertical walls. The polysilanes are well suited as imaging resists in bilayer processes. They have reasonable sensitivity (50–100 mJ/cm^2) and have been used to define features of 0.8 μm size.

The polysilanes can also be operated in a self-developing mode [78]. When irradiated at short wavelength (e.g., at 245 nm, with a KrF excimer laser) where the absorption does not bleach and where therefore every Si—Si bond is eventually broken, the material is directly volatilized (photoablated) on exposure.

PLASMA DEVELOPMENT

One of the fundamental problems of negative polymeric resists is the swelling of the resist image that occurs in the developing solvent. An interesting strategy by which this problem can be avoided is plasma development, a procedure whereby the unexposed, or the exposed, areas of the resist are removed not by solvent, but by treatment with plasma or RIE.

In order for plasma development to be possible, a way has to be found to make the exposed areas of the resist either more plasma resistant, or less

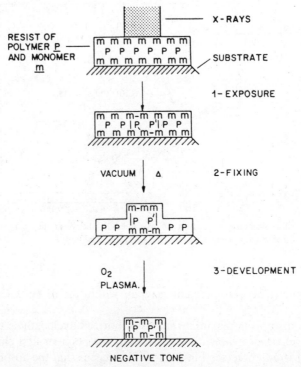

Figure 10-16 Schematic of the procedure for dry development by photochemical grafting. (Reproduced with permission from G. N. Taylor et al. [80].)

resistant than the unexposed material. Two different routes to this end have been explored. In the first, the system is so formulated that a monomer, which confers increased plasma resistance, is photochemically grafted onto a base resin. In the second method, a plasma differentiating agent is brought into the resist film from the gas phase after image exposure, or even during exposure.

Photochemical Grafting of Etch Resistant Monomers. Taylor and Wolf [79] have given an excellent description of the development of the first methodology, originally used to provide a "dry" developable x-ray resist. The general idea is outlined in the following scheme (Fig. 10-16).

The resist is formulated from a polymer P that has a high plasma etch rate, and a moderately volatile monomer M that is considerably more etch resistant. On exposure, the monomer is locked into the polymer either by grafting or by network formation. After exposure, the unreacted monomer in the unexposed as well as the exposed areas is removed in a vacuum and the *fixed* image is developed by oxygen plasma or RIE. This produces a negative resist pattern. Positive pattern can be obtained in a similar way by using other plasmas.

For x-ray exposures Taylor et al. used poly(2,3-dichloro-1-propyl acrylate), DCPA, which etches rapidly in an oxygen plasma, just as PMMA does.

$$\left[\begin{array}{c} \text{CH}_2-\text{CH}- \\ | \\ \text{C}=\text{O} \\ | \\ \text{O} \\ | \\ \text{CH}_2 \\ | \\ \text{CHCl} \\ | \\ \text{CH}_2\text{Cl} \end{array} \right]_n$$

DCPA

The photoreactive monomer additive was N-vinylcarbazole (NVC), which as an aromatic compound is more etch resistant.

N-Vinylcarbazole

The resist consisted of 80% by weight of DCPA and 20% NVC. It was exposed to Pd(L_α) x-rays with a dose of 12 mJ/cm^2, and fixed by evacuation for 1 h at 90°C and 0.1 torr. Lines and spaces of 1 μm could be produced in a 0.65 μm thick resist coating.

To improve etch selectivity Taylor and co-workers experimented with silicon containing monomers and settled finally for bis(acryloxybutyltetramethyldisiloxane), BABTDS [81].

$$\text{CH}_2=\text{CH}-\overset{\text{O}}{\overset{\|}{\text{C}}}-\text{O}(\text{CH}_2)_4\overset{\text{CH}_3}{\underset{\text{CH}_3}{\overset{|}{\underset{|}{\text{Si}}}}}-\text{O}-\overset{\text{CH}_3}{\underset{\text{CH}_3}{\overset{|}{\underset{|}{\text{Si}}}}}(\text{CH}_2)_4\text{O}-\overset{\text{O}}{\overset{\|}{\text{C}}}-\text{CH}=\text{CH}_2$$

BABTDS

With a combination of DCPA and BABTDS a sensitive resist was obtained that could be developed by oxygen RIE to which small concentrations of fluorinated gases had been added (CF$_4$ or CF$_3$Cl) [80].

Electron beam exposable and plasma developable systems were described by Yoneda et al. [82]. They used poly(isopropyl methacrylate) and poly(butyl methacrylate) as base resins and triphenylsilane, triphenylsilanol, triphenylsilazide, triphenylvinyl silane, and others as additives. On exposure to electrons the silicon containing additives are locked into the base resin and after removal of the surplus by a vacuum, the resist pattern is developed by oxygen RIE in a negative mode. The processing sequence is shown in Fig. 10-17a.

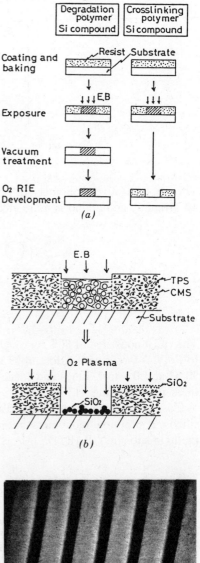

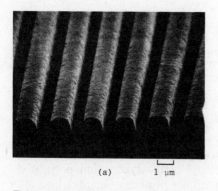

Figure 10-17 (*a*) Processing steps in the dry development technique after Yoneda. (*b*) Detail of positive tone development. (Reproduced with permission from Y. Yoneda et al. [82].)

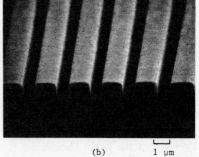

Figure 10-18 Resist pattern made by electron exposure and dry development in positive tone and subsequently treated (*a*) with gaseous $CF_4 + O_2$; (*b*) with a 10% aqueous solution of HF and NH_4F 1 : 9. (Reproduced with permission from Y. Yoneda et al. [82].)

If a crosslinking electron resist, such as chloromethylated polystyrene, CMS, is used as base resin with the same additives, the system works in a positive mode. As the base resin is crosslinked by the exposure, the additives become insoluble in it and precipitate out as a separate phase. In that form they do not confer plasma resistance to the resin; the system is not evacuated in this case and on subsequent plasma etching the exposed (crosslinked) areas of the pattern are etched away more rapidly. A particulate and easily removable SiO_2 residue is left on the substrate (Fig. 10-17b). A pattern of good image quality and quite high aspect ratio can be produced. Figure 10-18 shows a positive resist pattern obtained in this manner, and where the SiO_2 debris was removed either by a short treatment with $CF_4 + O_2$ plasma or by treatment with aqueous HF + NH_4F.

The same principle can be applied to UV sensitive photoresists. Taylor et al. have sensitized the DCPA + N-vinylcarbazole system by adding various aromatic quinones [83]. Optimum sensitivity was achieved with about 3.5% by weight of phenanthrenequinone. Tsuda et al. [84] have designed a plasma developable photoresist based on locking aromatic moieties via radiation sensitive azide functionalities into a base resin. These resists are sensitive to UV, to electrons, and to x-rays.

Gas Phase Functionalization. The general concept of gas phase functionalization as a preliminary to plasma development is indicated schematically in Fig. 10-19 [85].

The resist film contains a reactant (A), which on irradiation is transformed into the product P. After irradiation the film is exposed to a metal-containing gas (MR). If the gas reacts preferentially with A, the unexposed areas of the film will be metallized and, provided the metal will react with the plasma to form a protective refractory oxide, the exposed and hence not metallized areas will etch faster and the system will function as a positive resist. If the gas reacts preferentially with P, the exposed areas will be protected and the system will work with negative tone.

Examples of photofunctionalization were described by Yamada et al. [86] who selectively grafted dichloromethylvinyl silane onto electron beam exposed PMMA. In the subsequent oxygen RIE treatment the graft was converted into SiO_2. A similar procedure was adopted by Wolf et al. [87] who used dimethyldichlorosilane as the grafting agent, hydrolyzed he graft with water, and then developed the exposure pattern by oxygen RIE.

Taylor and co-workers [85] used diborane (in N_2) as the metal-containing gas and polyisoprene and later, and more successfully, poly(chloroacrylate) and its copolymers with allylacrylate as base resins. Better results were obtained with azide containing photoresists such as, for example, Selectilux N-60. On exposure of these (normally negative working) resists to an optical pattern and subsequent treatment with $SiCl_4$ or $SnCl_4$ positive images were obtained on plasma development.

388 MULTILAYER TECHNIQUES AND PLASMA PROCESSING

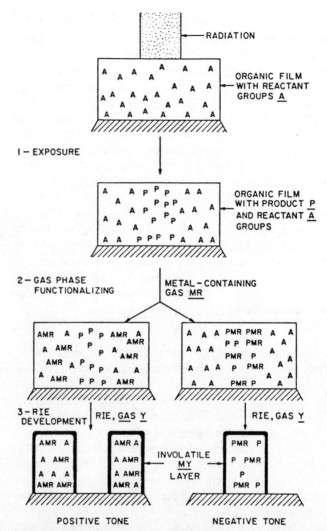

Figure 10-19 Schematic of gas phase functionalization procedure. (1) Exposure of material with reactive groups A; (2) conversion of A to product P; (3) selective functionalization of either A or P with a metal containing reagent MR, where M is the metal and R is a reactive or inert group; (4) reactive ion etch development using a gas, Y, which forms a nonvolatile compound MY, and volatile compounds AY, PY, and so on. (Reproduced from G. N. Taylor et al. [85], with permission from the Electrochemical Society, Inc.)

An interesting system using gas phase silylation has been described by Coopmans and co-workers [88, 89]. It is *the DESIRE process*. This process is based on the observation that the incorporation of diazoquinone into novolac inhibits the penetration of silylating agents into the film, and that this inhibition effect stops on exposure of the films to near-UV radiation. If a novolac–diazoquinone film is exposed to a radiation pattern in the usual way, and subsequently exposed in a vacuum oven at up to 160°C to a gaseous silylating agent such as, for example hexamethyldisilazene,

$$(CH_3)_3-Si-NH-Si-(CH_3)_3$$
Hexamethyldisilazene

silylation occurs only in the radiation-exposed areas. Although the differentiation between image and nonimage areas is based on the rate of permeation of the silylating agent into the film and not on selective functionalization, image discrimination is very sharp indeed. The penetration of the agent is limited to the top 100 to 250 nm of the resist layer. There, silicon is chemically incorporated into the layer by reaction with the hydroxyl groups of the novolac.

Figure 10-20 The 0.6 μm lines and spaces produced with the DESIRE process. The lines cross 1 μm high aluminum steps without any distortion. [Reproduced with permission from B. Roland, J. Vandendriesche, R. Lambaerts, B. Denturck, and C. Jakus, *Proc. SPIE*, **920** (1988).]

The pattern is transferred into the bulk of the resist film by oxygen reactive ion etching as indicated. In the early stages of the plasma treatment silicon is transformed into SiO_2, which subsequently protects the image areas from further erosion. Figure 10-20 shows the result of applying the DESIRE process to a single layer (2.3 μm thick) novolac resist where 0.5 μm lines with an aspect ratio of 4.6 : 1 are seen to cross faultlessly over a 1 μm oxide step.

Other systems based on gas phase functionalization have been described in the literature [90, 91] and analytical and mechanistic studies have been reported [92, 93].

REFERENCES

1. J. R. Franco, J. R. Havas, and H. A. Levine, U.S. Patent 3,873,361 (1973).
2. K. Grebe, I. Ames, and A. Ginzberg, *J. Vac. Sci. Technol.*, **11**, 458 (1974).
3. B. J. Lin, in *Introduction to Microlithography* (L. F. Thompson, C. G. Willson, and M. J. Bowden, Eds.), *ACS Symp. Ser.*, **219**, 287 (1983).
4. M. P. de Grandpre, D. A. Vidusek, and M. W. Legenza, *Proc. SPIE*, **539**, 103 (1985); A. W. McCullough, D. A. Vidusek, M. W. Legenza, M. de Grandpre, and I. Imhof, *Proc. SPIE*, **651**, 316 (1986).
5. J. W. Coburn, *Solid State Technol.*, **29**(4), 117 (1986).
6. J. W. Coburn, *Plasma Etching and Reactive Ion Etching*, American Vacuum Society Monograph, American Institute of Physics, New York, 1982.
7. M. A. McCord and R. F. W. Pease, *J. Vac. Sci. Technol.*, B, **6**, 293 (1988).
8. B. J. Lin, *Proc. SPIE*, **174**, 114 (1979); *J. Electrochem. Soc.*, **127**, 202 (1980).
9. (a) K. L. Tai, W. R. Sinclair, R. G. Vadimsky, J. M. Moran, and M. J. Rand, *J. Vac. Sci. Technol.*, **16**, 1977 (1979). (b) M. Hatzakis, *Solid State Technol.*, **24**(8), 74 (1981).
10. M. Hatzakis, U.S. Patent 4,024,293 (1977); *J. Vac. Sci. Technol.*, **26**, 2987 (1979).
11. J. R. Havas, *Electrochem. Soc. Ext. Abstr.*, **76**(2) (1976).
12. A. D. Butherus, T. W. Hou, C. J. Mogab, and H. Schonhorn, *J. Vac. Sci. Technol.*, B, **3**, 1352 (1985).
13. H. Hiraoka and J. Pacansky, *J. Vac. Sci. Technol.*, **19**, 1132 (1981).
14. (a) R. Allen, M. Forster, and Yung-Tsai Yen, *J. Electrochem. Soc.*, **129**, 1379 (1982). (b) K. J. Orvek and M. L. Dennis, *Proc. SPIE*, **771**, 281 (1987).
15. (a) H. Ito, S. A. MacDonald, R. D. Miller, and C. G. Willson, U.S. Patent 4,552,833 (1985). (b) L. E. Stillwagon, P. J. Silverman, and G. N. Taylor, *Tech. Pap. Reg. Tech. Conf. SPE, Ellenville, NY*, p. 87 (Nov. 1985). (c) H. Hiraoka, *Proc. SPIE*, **771**, 174 (1987).
16. J. M. Moran and G. N. Taylor, *J. Vac. Sci. Technol.*, **19**, 1127 (1981).
17. P. A. Ruggiero, *Solid State Technol.*, **27**(9), 165 (1984).
18. B. J. Lin, *Proc. SPIE*, **771**, 180 (1987).
19. F. S. Lai, B. J. Lin, and Y. Vladimirsky, *J. Vac. Sci. Technol.*, B, **4**, 426 (1987).
20. M. P. C. Watts, *Proc. SPIE*, **468**, 2 (1984).

21. J. Lee, *Solid State Technol.*, **29**(6), 143 (1986).
22. K. L. White, *J. Electrochem. Soc.*, **132**, 1543 (1983).
23. D. B. Lavergne and D. C. Hofer, *Proc. SPIE*, **539**, 115 (1985).
24. L. E. Stillwagon, R. G. Larson, and G. N. Taylor, *Proc. SPIE*, **771**, 186 (1987); L. E. Stillwagon and R. G. Larson, *Proc. SPIE*, **920**(17), (1988).
25. A. N. Saxena and D. Pramanik, *Solid State Technol.*, **29**(10), 95 (1986).
26. J. A. Mucha and W. D. Hess, in *Introduction to Microlithography* (L. F. Thompson, C. G. Willson, and M. J. Bowden, Eds.), *ACS Symp. Ser.*, **219**, 215 (1982).
27. D. L. Flamm, V. M. Donelly, and D. E. Ibbotson, in *VLSI Electronics Microscience* (N. G. Einspruch, Ed.), Vol. 8, p. 190, Academic, New York, 1984.
28. A. T. Bell, in *Techniques and Applications of Plasma Chemistry* (J. R. Hollahan and A. T. Bell, Eds.), p. 1, Wiley-Interscience, New York, 1974.
29. J. W. Coburn and H. F. Winters, *Annu. Rev. Mater. Sci.*, **13**, 91 (1983).
30. C. J. Mogab, in *VLSI Technology* (S. M. Sze, Ed.), McGraw-Hill, New York, 1983.
31. H. H. Sawin, *Solid State Technol.*, **28**(4), 211 (1985).
32. G. N. Taylor and T. M. Wolf, *Polym. Eng. Sci.*, **20**, 1087 (1980).
33. S. J. Fonash, *Solid State Technol.*, **28**(1), 150 (1985).
34. J. R. Paraszak, E. D. Babich, J. E. Heidenreich, R. P. McGouney, L. M. Ferreiro, and N. J. Chou, *Proc. SPIE*, **920**(33) (1988).
35. E. R. Lory, *Solid State Technol.*, **27**(11), 117 (1984).
36. J. J. Hannon and J. M. Cook, *J. Electrochem. Soc.*, **131**, 1164 (1984).
37. J. W. Coburn and H. F. Winters, *J. Appl. Phys.*, **50**, 3198 (1979).
38. P. D. Brewer, J. M. Recksten, and R. M. Osgood, Jr., *Solid State Technol.*, **29**(4), 273 (1985).
39. W. M. Holber and R. M. Osgood, Jr., *Solid State Technol.*, **30**(4), 139 (1987).
40. H. Gokan, S. Esho, and Y. Ohnishi, *J. Electrochem. Soc.*, **130**, 143 (1983).
41. F. Watanabe and Y. Ohnishi, *J. Vac. Sci. Technol.*, B, **4**, 422 (1986).
42. C. W. Jurgensen, A. Shugard, N. Dudash, E. Reichmanis, and M. J. Vasile, *Proc. SPIE*, **920**(34) (1988), and *J. Vac. Sci. Technol.*, **A6**, 2938 (1988).
43. W. C. McColgin, R. C. Daly, J. Jech, and T. B. Brust, *Proc. SPIE*, **922**(35) (1988).
44. T. Ueno, H. Shiraishi, T. Iwayanagi, and S. Nonogaki, *J. Electrochem. Soc.*, **132**, 1168 (1985).
45. G. N. Taylor, T. M. Wolf, and J. M. Moran, *J. Vac. Sci. Technol.*, **19**, 872 (1981).
46. S. A. MacDonald, J. W. Labadie, and C. G. Willson, *ACS Polym. Rep.*, **26**(2), 343 (1985).
47. S. A. MacDonald, H. Ito, and C. G. Willson, in *Microelectronic Engineering*, Vol. 1, p. 269, 1983.
48. M. Suzuki, K. Saigo, H. Gokan, and Y. Ohnishi, *J. Electrochem. Soc.*, **30**, 1962 (1983).
49. M. Morita, A. Tanaka, S. Imamura, T. Tamamura, and O. Kogure, *Jpn. J. Appl. Phys.*, **1-659**, 22 (1983).
50. M. A. Hartney and A. E. Novembre, *Proc. SPIE*, **539**, 90 (1985).

51. A. S. Godz, G. L. Baker, C. Klausner, and M. J. Bowden, *Proc. SPIE*, **771**, 18 (1987).
52. Y. Ohnishi, M. Suzuki, K. Saigo, Y. Saotome, and H. Gokan, *Proc. SPIE*, **539**, 62 (1985).
53. M. Hatzakis, J. Paraszczak, and J. Shaw, *Proc. Microcircuit Eng. '81, Lausanne, Switz., Sept. 1981*.
54. E. Reichmanis, G. Smolinsky, and C. W. Wilkins, Jr., *Polym. Prepr.*, **26**(1), 341 (1985).
55. R. G. Tarascon, A. Shugard, and E. Reichmanis, *Proc. SPIE*, **631**, 40 (1986).
56. Y. Saotome, H. Gokan, K. Saigo, M. Suzuki, and Y. Ohnishi, *J. Electrochem. Soc.*, **33**, 563 (1986).
57. W. C. Cunningham and Chan-Eon Park, *Proc. SPIE*, **771**, 32 (1987).
58. J. W. Labadie, S. A. MacDonald, and C. G. Willson, *J. Imaging Sci.*, **30**, 169 (1986).
59. J. W. Labadie, S. A. MacDonald, and C. G. Willson, *Macromolecules*, **20**, 10 (1987).
60. J. M. Shaw, M. Hatzakis, J. Liutkus, and E. Babich, *Tech. Pap. Reg. Tech. Conf. SPE, Ellenville, NY*, p. 285 (1982).
61. K. B. Killchowski and R. R. Pampalone, U.S. Patent 4,357,369 (1982).
62. S. A. MacDonald, F. Steinmann, H. Ito, M. Hatzakis, W. Lee, H. Hiraoka, and C. G. Willson, *Int. Symp. Electron, Ion Photon Beams, Los Angeles, CA, 1983*.
63. D. C. Hofer, R. D. Miller, and C. G. Willson, *Proc. SPIE*, **469**, 16 (1984).
64. K. Nate, H. Sugiyama, and T. Inoue, *Electrochem. Soc. Fall Meet., New Orleans, LA, Ext. Abstr.*, Vol. 84-2, No. 530 (1984).
65. K. Saigo, Y. Ohnishi, M. Suzuki, and H. Gokan, *J. Vac. Sci. Technol.*, **B3**(1), 331 (1985).
66. A. Tanaka, M. Morita, and O. Kogure, *Proc. SPSJ Int. Polym. Conf.*, *1st*, p. 63 (1984).
67. E. Reichmanis and G. Smolinsky, *Proc. SPIE*, **469**, 38 (1984).
68. C. W. Wilkins, Jr., E. Reichmanis, T. M. Wolf, and B. C. Smith, *Int. Symp. Electron, Ion Photon Beams, Tarrytown, NY, 1984*.
69. F. S. Kipping, *J. Chem. Soc.*, **125**, 229 (1924).
70. C. A. Burkhard, *J. Am. Chem. Soc.*, **71**, 963 (1949).
71. J. P. Wesson and T. C. Williams, *J. Polym. Sci., Polym. Chem. Ed.*, **18**, 959 (1980).
72. R. West, L. D. David, P. I. Durovich, K. L. Stearly, K. S. Srinivasan, and H. Yu, *J. Am. Chem. Soc.*, **103**, 7352 (1981).
73. P. Trefonas, III, P. I. Durovich, H.-Y. Zhang, R. West, R. D. Miller, and D. C. Hofer, *J. Polym. Sci., Polym. Lett. Ed.*, **21**, 819 (1983).
74. P. Trefonas, III, R. West, R. D. Miller, and D. C. Hofer, *J. Polym. Sci. Polym. Lett. Ed.*, **21**, 823 (1983).
75. R. D. Miller, D. C. Hofer, D. R. McKean, C. G. Willson, R. West, and P. T. Trefonas, III, in *Materials for Microlithography* (L. F. Thompson, C. G. Willson, and J. M. J. Frechet, Eds.), *ACS Symp. Ser.*, **266**, 93 (1984).

76. A. R. Neureuther, P. K. Jain, and W. G. Oldham, *Proc. SPIE*, **275**, 110 (1981).
77. M. Ishikawa and M. Kumada, *Adv. Organomet. Chem.*, **19**, 51 (1981).
78. J. M. Zeigler, L. A. Harrah, and A. W. Johnson, *Proc. SPIE*, **539**, 166 (1985).
79. G. M. Taylor and T. M. Wolf, *J. Electrochem. Soc.*, **127**, 2665 (1980).
80. G. N. Taylor, M. Y. Hellman, M. D. Feather, and W. E. Willenbrock, *Polym. Eng. Sci.*, **23**, 1029 (1983).
81. G. N. Taylor, T. M. Wolf, and J. M. Moran, *J. Vac. Sci. Technol.*, **19**, 872 (1981).
82. Y. Yoneda, M. Miyagawa, S. Fukuyama, T. Narusawa, and H. Okuyama, *Proc. SPIE*, **539**, 145 (1985).
83. G. N. Taylor, T. M. Wolf, and M. R. Goldrick, *J. Electrochem. Soc.*, **128**, 361 (1981).
84. M. Tsuda, S. Oikawa, W. Kanai, K. Hashimoto, A. Yokata, K. Nuino, I. Hijikata, A. Uehara, and H. Nakane, *J. Vac. Sci. Technol.*, **19**, 1351 (1981).
85. G. N. Taylor, L. E. Stillwagon, and T. Venkatesan, *J. Electrochem. Soc.*, **131**, 1658 (1984).
86. M. Yamada, J. Tamano, K. Yaneda, S. Morita, and S. Hattori, in *Proc. Dry Processing Symp., Tokyo*, p. 90 (Nov. 1982).
87. T. M. Wolf, G. N. Taylor, T. Venkatesan, and R. T. Kraetsch, *J. Electrochem. Soc.*, **131**, 1664 (1984).
88. F. Coopmans and B. Roland, *Proc. SPIE*, **631**, 34 (1986).
89. B. Roland and A. Vranken, Eur. Patent Appl. 0184567 (1986).
90. H. Ito, S. A. MacDonald, R. D. Miller, and C. G. Willson, U.S. Patent 4,552,833 (1985).
91. K. N. Chiong, B. L. Yang, and J. Yang, U.S. Patent 4,613,398 (1986).
92. T. B. Brust and S. R. Turner, *Proc. SPIE*, **771**, 102 (1987).
93. R.-J. Visser, J. P. W. Schellekens, M. E. Reuhman-Huisken, and L. J. van Ijzendoorn, *Proc. SPIE*, **771**, 111 (1987).

INDEX

ABC Crosslinker, 36
Absorption spectra of diazoquinones, 188
Acetone pinacol, 40
Acetophenone derivatives, as photoinitiators, 108, 111
Acid, as catalyst, 278
Acid catalyzed depolymerization of phthalaldehyde, 290
Acid catalyzed thermolytic depolymerization, 283
Acid hardening resins, 184
Acid sensitive carbonates, 280
Acridine derivatives, 150
Acridinium dyes, 120
Acriflavin, 120
Acrylated cellulose acetate, 160
Acrylates, tetrafunctional, 157
Activation energy, of polymerization, 136
Acyloximino groups, 270
Acylsilanes, 39
Additives, 154
Adhesion, 18
  modulation of, 15, 169
Adhesion promotors, 154, 155
Aerial image, 18, 231, 232, 245
AIBN, 2,2'-azobis(butyronitrile), as photoinitiator, 107
Airy disc, 234
Alcohol soluble nylons, 157
Alcohol soluble polyamides, 160
Alkali soluble cellulose derivatives, 157

Alkoxybenzyl radicals, 108
Alkyl ethers of benzoin, as photoinitiators, 108
Amines, as antioxydants, 154
4-Aminomethyl benzoate, as coinitiator, 116
Ammonia vapor, 184
Anisotropic pattern transfer, 20
Anodized alumina, 162
Anthracene, 283
  energy migration, 76
Anthraquinone, as photoinitiator, 115, 160
Anthraquinone sulfonate, as photoinitiator, 115
Antioxydants, 127, 154
Application specific integrated circuits (ASICs), 302
Aquation, of phenolate, 215
Araldites, 149
Argon ion laser, 123
Aromatic ketones, n^*-transitions, 88
Aryldiazonium salts, 148, 149
Aryl nitrones, 248
Arylsulfinates, 122
Ascorbic acid, 119
Autoacceleration effect, 131, 132, 133
Automatic IC inspection, 302
AZ 1300, 187
AZ 1300 J, 190
AZ 1300 series of resists, 205
AZ 1350 J, 189, 192
AZ 2400, 189, 190

395

AZ 4000, 187
AZ 24110, 262
Azide photolysis, spectral sensitization of, 88
Azide resists, 18
Azides, 32, 38
    spectral sensitization, 90
4-Azidochalcone, 39
Azido group, 33, 311
2,2'-Azobis(butyronitrile), AIBN, as photoinitiator, 107
Azo compounds, as photoinitiators, 107
Azo dye, 181, 182
Azoplate resist AZ 1350 J, 192
AZ resists, 18

Backbone degradation, of polyacrylates, 306
    by C—C bond scission, 269
Backbone fragmentation, 265
Backbone scission, 269, 307, 310
    resists based on, 319
Backscatter of electrons, 294
Bar pattern, 236, 237
Beam geometry, 337
Beam shapes, 297
Benzene, the radiolysis of, 305
Benzene-chromium-tricarbonyl, as photoinitiator, 126
Benzil ketals, as photoinitiators, 110
Benzoin, ketoxime esters of, as photoinitiators, 112
Benzoin derivatives, as photoinitiators, 108, 160
Benzophenone, 40, 42, 70, 147
    H-abstraction by, 113, 115
    as photoinitiator, 115, 117
1-Benzoylcyclohexanol, 111
Benzoyl peroxide, as photoinitiator, 107
Benzoyl radicals, 107, 108
Benzyl radical, 108
Benzyltrimethylstannane, 122
BF_3, as Lewis acid, 147
Bifunctional epoxides, 156
Biimidazole(s), 170
    as photoinitiators, 113, 123
    as sensitizer, 124
Binders, 133, 154, 155
    oligomeric, 154
    polymeric, 154
    reactive, 156
Bisazide-phenolic resin, 259
Bisazide resists, 258
Bisazide-rubber systems, 195
Bisazides, 18

2,6-Bis(4-azidobenzal)-4-methylcyclohexanone, ABC, 36, 38
4,4'-Bis(dimethylamino) benzophenone, Michler's ketone, 117
Bis(ketocoumarins), as photoinitiators, 119
Bisphenol A, 29
Bitumen of Judea, 1, 13, 22
Bleachable dye, 248
Block copolymers, diacetylene containing, 145
Boron carbide, 339
Boron nitride, 339
Boron trifluoride, 147, 149
Bragg relation, 199
Broad beam exposures, 352
    sources, 349
Bronsted acids, 148, 150
Butting errors, 297
n-Butylmercaptan, 146

Cage recombination, 105
Carbenes, 32, 39, 182, 183
Carbocations, 147, 148, 150
Carbon tissue process, 8, 9
Caruthers' equation, 145
Cascading, principle, 240
    the modulation transfer function, 242
Case II mass transfer, 211, 212
Cation size effect, 215, 216, 218
Ceiling temperature, 285, 286, 324
Cellulose acetate, 160
Cellulose derivatives, alkali soluble, 157
Chain length, kinetic, 130
Chalcones, 28
Changes in mechanical wet strength, 159
Changes of index refraction, 159
Changes of permeability, 159
Changes of solubility, 159
Changes of tackiness, 159
Channeled ion lithography, 352
Characteristic curves, 47, 202
    of bisazide-phenolic resists, 259
    of commercial resists, 202, 204
    of novolac-diazoquinone resists, 214
    of resists, 215
Characteristic ratio, 130, 133
    values of, 131
Charlesby–Pinner plot, 309
Chemical dosimeter, 304
Chemical image, 231
Chemical milling, 16
Chemical vapor deposition, 258
Chlorinated poly(methylstyrene), CPMS, 347

Chlorobenzene, 66
1-Chloronaphthalene, 66, 67
2-Chlorothioxanthone, as photoinitiator, 118
Cholic acid, 265
Cinnamic acid, 11
Cinnamoyl group, triplet energy of, 86
Circuit boards, 16, 164
Cis-trans isomerization:
  of azobenzenes, 94
  of spirobenzopyrans, 94
  of stilbenes, 94
Coating aids, 154
Coating of resist, 56
  on spinner, 57
Coherence properties, 256
Coherent radiation, 169
Coinitiators, 103, 116
Color coupler, 43
Color developer, 43, 44
Color proofing, 158
  materials, 168
  off-press, 168
Color reproduction, 7, 168
Color separation, 7
  negatives, 168
Compact storage ring, 339
Composite systems, 231
Computer chips, 16
Computer generated pattern, 163
Computer simulation of polymerization, 135, 137
Contact frame, 16, 17
Contact gauge, 204
Contact printing, 18, 242
  mode, 257
Contrast, 227, 228
  of resist, 52
  photographic, 49, 51
Contrast enhancement, 46, 246, 249
  by computer simulation, 250
  layer (CEL), 248
  by oxygen inhibition, 247
  technique of Griffin and West, 248
Coordination center, 24
Coordination numbers, in amorphous systems, 82
COP, 312, 346
COP Copolymers, 157
Copolymers of alkyl methacrylate, 348
Copolymers of allyl methacrylate, 312
Copolymers of methyl methacrylate and methacrylic acid, 265
Copolymers of MMA and indenone, 274

Copolymers of styrene and 1-vinyl-naphtalene, 78
Correct working exposure, 229
Corrugated surfaces, printing on, 161
Cresoles, isomeric composition of, 192
Critical base concentration, 212
  degree of deprotonation, 218
  dimension of mask, 229
  dimensions, 203, 204
  exposure, 204
  MTF of the resist, CMTF, 246
Cr_2O_7, photoreduction of, 22
Crolux, engineering film, 169
Cronavure, gravure film, 15
Crosslink density, ρ, 49, 51
Crosslink formation, quantum yield of, 49
Crosslinkers, 133, 155
Crosslinking, 22, 24, 307, 310
  agent, 43
  mechanism of, 39
  physical chemistry of, 46
  polymerization, 135, 156
  quantum yield of, 34, 55, 56, 140, 144
  by radicals, 40
  resists, development of, 57
  yield G_x, determination of, 309
Crosslinks, 11, 182
  formation of, 42
Crystal reactivity, 143
Crystalline acrylates, polymerization of, 141
Cumylphenol, sulfonyl ester of, 187
Customization of integrated circuits, 302
Cyanine-type dyes, 123
2-Cyanoisopropyl radicals, 108
Cyclization, 37
Cyclized poly(cis-isoprene), 38, 258
Cycloaddition, 26, 27
Cyclobutane ring formation, singlet and triplet path, 86
Cyclohexene oxide, polymerization, 151
Cyclohexylene, as soft segment, 144

Deep UV, 252
  photochemistry, 302
  resists based on PMMA, 277
Deep-UV lithography, 20
  physics of, 252
Defect densities, 18
Degreasing of substrate, 56
Delayed fluorescence, 78
Depolymerizable polycarbonates, 282
Depolymerization, spontaneous, 286, 287, 324

Deprotonation of phenol, 212, 215
Depth of focus, 250
Desirability curves, 210
Detector response, 226
Determination of G-value, 304
Determining the MTF of an imaging material, 241
Deuterium lamp, 253
Developer, 22
 Shipley 2401, 192
Development, 57
 puddle-, 57
 spray-, 57
Development curves, 192, 198 199, 200
Development end-point detector, 199, 200
Diacetylene containing block copolymers, 142, 145
Diacetylene-bis(p-toluenesulfonate), 143
Diacetylene crystals, 141, 143
Diacetylenes, 317
 crosslinking mechanism, 143
 polymerization of, 141
Diacrylates, phenylene-, 28
Dialkyl-4-hydroxylphenyl sulfonium salts, 152
Dialkylphenacylsulfonium salts, 151
Diallylamine, 146
Diaryliodonium salts, 150, 283
1,2-Diazepins, 45
4,4'-Diazidobenzalacetone, 36
4,4'-Diazidobenzophenone, 36
4,4'-Diazidodiphenyl sulfide, 258
3,3'-Diazidodiphenyl sulfone, 259
4,4'-Diazidostilbene, 36
4-Diazidostilbene-poly(vinyl pyrrolidone), 246
2-Diazodimedones, 264
Diazo papers, 13
Diazo-Meldrum's acid, 262
5-Diazo-Meldrum's acid, 263
Diazonaphthoquinone(s), 178, 181, 189
 decomposition of, 191
 substituted in the 5-position, 185
 substituted in the 4-position, 185
Diazonium salts, 181
 extended spectral range, 149
 photolysis of, 46
Diazopiperidine dione, 263
Diazopyrazolidine dione, 263
Diazoquinone, 13, 180, 181, 183
 photodecomposition of, 44
 resists, 262
 spectra, 188, 190
Diazoquinone-novolac resist, 18, 179, 326

Diazoresin, 45
Diazotetramic acid, 263
Diazotype copying system, 181
Dibutylphthalate, as plasticizer, 139
Dichromated colloids, 3, 4, 181
 gelatin, 4
 poly(vinyl alcohol), 4
 poly(vinyl butyral), 4
 poly(vinyl pyrrolidone), 4
 starch, 3
Dichromated gelatin, 13, 22
Dichromate-poly(vinyl alcohol) system, 248
Dielectric interlayers, 158
Dienes, 145
2,2'-Diethoxyacetophenone, DEAP, as photoinitiator, 111
Differential scanning photocalorimeter, 133
Diffraction efficiency, of holographic material, 170, 172
Diffusion transfer, 234
Dill equation, 206
Dill equation parameters A,B,C, 207
Dimedone, 122
2,2'-Dimethoxy-2-phenylacetophenone, DMPA, 110
4-(Dimethylamino) benzaldehyde, as coinitiator, 116
4-(Dimethylamino) methylbenzoate, as coinitiator, 116
4-(Dimethylamino)phenyldiazonium chloride, 46
4-Dimethylaminostyrene, 41
Dimethylaniline and biphenyl, exciplex fluorescence, 71
Dimethylmaleimide, 30
Dioxycyclopentadiene, 149
Diphenylbenzofuran, 129
Diphenylcyclopropene, 30
Diphenyliodonium hexafluoroarsenate, 282
1,1'-Diphenyl-2-picrylhydrazyl, DPPH, 312
Diphenylsulfide, as initiator, 104
Dipole resonance transfer, 71, 72
Direct-write electron beam lithography, 301
Dispersity of molecular weight, 53, 55
Dissolution curves, 180
Dissolution inhibitors, 46, 179
 of novolac, 280
Dissolution of novolac resist, 327
Dissolution promoter, 180
Dissolution rate, 179, 193, 214, 215
 function of inhibitor concentration, 209
 monitor, 199
 parameters E_1, E_2, E_3, 208
Distribution, lognormal, 51

Distribution of site reactivities, 92, 93
  and free volume, 94
  in PCP, 93
  in poly(vinyl cinnamate), 93
  in PPDA, 93
Dithiols, 145
Ditoluyliodonium hexafluoroarsenate, 150, 152
Domains in photopolymerization, 137
Dose unit, rad, 304
Dosimetry, 304
Drying, of resist, 205
Dry film solder mask, 166
Dry resist, 162, 163
Dry resist film, 163
Dry resists, 15
Dycryl plate, 15
Dyes, 154
  photoreducible, 120
Dye sensitization, with photochemical pumping, 124
Dye sensitized biimidazoles, 170
Dye sensitized photopolymerization, 122

EBES machine, 298
EBR-9, 321
Edge acuity, 230
Edge profile of a negative resist image, 233
Effect of inhibitor structure, 220
Efficiency of initiation, 139
Electron beam, transmission mask, 300
  variable shaped, 299
Electron beam lithography, 20, 294
  tools, 298
Electron beam mask writing system, 298
  proximity printing technique, 300
Electron beam resists, performance of, 330
Electron donors, 121
Electron exposure tool, 295
Electron gun, 295
Electronic pattern generation, 294
Electronics, 158
Electron impact x-ray tubes, 337
Electron microscope, 295
Electron optics, 295, 297
Electron projection, 300
Electron resists, 315, 320
  negative working, 311
Electron scattering, 294, 295
Electron sensitive dissolution inhibitors, 326
Electron sensitivities, 343
Electron transfer, in spectral sensitization, 88
Ellipsometry, 200
Elvacite 2041, 2010, 319

Emission of excimer, 68
Encapsulants, 158, 167
Ene-compounds, 146
Energy migration, 32, 75, 76
  antenna effect, 79
  in anthracene crystal, 76
  in copolymers, 78
  down-chain, 78
  excitons, 78
  in N-isopropyl carbazole, 78
  jump numbers, 82
  in photoreactive polymers, 80
  in polymers, 78
  in poly(N-vinylcarbazole), 79
  in PPDA, 79
  range of, 81
Energy transfer, 65, 70, 73
  critical distance, 72
  by electron exchange, 73
  Perrin formula, 75
  quenching sphere, 75, 76
  rate of, 72
  thermodynamics of, 75
Epoxides, bifunctional, 156
Epoxy derivatives, 147
Epoxydized polybutadiene, 311
Epoxy groups, 145, 311
Epoxy monomer, 150
Epoxy resins, 147
Esters of poly(vinyl benzoic acid), 280
Etchants, 15
Etching, by reactive ions, 20
Etching of printed circuits, 163
Ethylacrylate, 346
2-Ethylanthraquinone, as photoinitiator, 115
Ethylcinnamate glasses, 97
Ethyl cinnamates, photoproducts of, 98
Ethylene oxide, 147
Ethyl iodide, the radiolysis of, 306
Excimer, 67
Excimer binding energy, 68
Excimer fluorescence:
  of perylene, 68
  of pyrene, 68
Excimer laser projection printer, 257
Excimer lasers, 254, 255, 257
Excimer laser scanner, 318
Excimers, emission of, 68
  as exciton traps, 79
  Hirayama's rule, 79
Exciplex, 41, 70, 115
Exciplex fluorescence, 71
Excitons, trapping of, 32
  in energy migration, 78

Exposure, 205
Exposure dose, 47, 57
Exposure latitude, 204, 230
Exposure unit, Roentgen, 304

Far UV, 252
Feature sizes in printed circuits, 166
Ferrocenelike iron–arene complexes, 152
Ferrocenes, as photoinitiators, 127
Ferrocinium ion, 127
Fickian behaviour in diffusion, 211, 212
Fillers, 155
Finely focused ion beams, 349
Flare curve, 231
Flat bed presses, 7
Flexographic plates, 15, 147, 159, 160, 161
Flexography, 158
Floating gate nonvolatile switches, 302
Flory function, 52, 55
Fluorescein, 119, 120
Fluorescence
  delayed, 78
  of p-phenylenediacrylic acid, 80
Fluorophosphate glasses, 257
Focused ion beams, 350, 352
Fountain solution, in lithography, 6, 7
Fourier transform, 239
Fourier transform infrared spectrometry, FTIR, 132
Free acrylic acid, as adhesion promoter, 155
Free energy of charge transfer, 89, 284
Free volume distribution, 94
Free volume effect, 180
Frequency response, 237
Fresnel diffraction, of x-rays, 338
Fricke dosimeter, 304
Functionality changes, 178
Functionalized oligomers, 156
Furans, 149

Ga^+ ions, implantation, 352
"Gas-puff, Z-pinch" plasma, 337
Gelatin, 4
  dichromated, 13, 22
Gel curve, 48
  initial slope of, 49, 51, 52
Gel dose, 48, 58
Gel fraction, 51
Gel layer, 211
Gel point, 47, 48
Gel point exposure, 48, 228
Generic photoinitiators, 109
Glass transition, 138, 168
Glass transition temperature 140, 158, 197

Glycidyl methacrylate, 346, 348
  copolymer with ethylacrylate, 312
  copolymers, 270
Gold masks, 352
Gold pattern, 340
Grafting reaction, 35
Graphic arts, 16, 158
Graphic arts transfer, 169
Gravure cells, 161
Gravure films 15, 167
Gravure plate, 9
  variable area-variable depth, 161
Gravure printing, 8, 10, 16, 158
Gravure printing plates, 167
Gravure, variable depth-constant area, 167
Gravure screen, 161
G_s values and lithographic sensitivity, 308

H-abstraction, by benzophenone, 113
Halftone, 8
  dots, 7
  reproduction, 7
Harmonic analysis, 236
Heat of polymerization, 133, 134, 136
  cyclic ethers, 146
Heat resistant materials, 29
Heliography, 3
Heliogravure, 3
Helium-neon laser, 123
n-Heptene, 146
Heteroexcimers, 70
Hexamethylene diisocyanate, as soft segment, 144
1,6-Hexanediol diacrylate, 135
n-Hexyloxydiazonium hexafluorophosphate, 279
Hirayama's rule, 79
Hologram(s), 16, 169
  recording, 171
  with diacetylenes, 143
Holographic material(s), 169, 170
  diffraction efficiency, 172
Holography, 125, 158
Hughes Research, 352
Hunt HPR 204, 187
Hunt resist WX-159, 187
Hunt WX-118, 187
Hybrid lithography, 301
Hybrid photopolymer system, 42
Hybrid silver halide–photopolymer system, 162
Hydrogen abstraction, 103, 113
Hydrogen donors, 103, 115, 121
  as coinitiators, 114

tertiary amines, 115, 116, 121
Hydrophobic components, 218
Hydroquinone, 119
2-Hydroxyethyl methacrylate, 348

I-line at 365 nm, 252
IBM EL system, 298
IBM exposure tool EL-3, 297
Illuminance, 226
Illuminance distribution, 232
Image, aerial, 18
Image amplification, 102
Image degradation, 234, 294, 336
Image discrimination based on molecular weight changes, 324
Image resolution, 242
Image reversal, 39, 182, 184
Imaging chain, 231
Imidazole 113, 184
N-Iminopyridinium ylides, 45
Incident charge density, 228
Incident energy density, 228
Indene, 182, 183
Indene carboxylic acid, 13, 178, 180, 183, 189
Indene carboxylic acid, decarboxylation, 182
Indene dimer, 182, 183
Index of refraction, 158
 changes, 159
 modulation, 169
Induction period, in photopolymerization, 127, 128
Infrared analysis, 132
Inhibition of dissolution, 178, 180, 218
Inhibitor functionality, 222, 223
Initiation, 128
Initiation of polymerization, 103
Initiation step, 127
Initiation:
 dye sensitized, 118
 efficiency, 105, 139
 by onium salts, 147
 quantum yields, 105
 rate, 130
 temperature dependence, 106
 viscosity dependence, 106
Initiators, 155
Initiator with surfactant properties, 110
Initiator systems, practical, 115
Ink acceptance, 162
Intaglio plate, 3
Intaglio printing, 16
Integrated circuit, 16, 19
Integrated optics, 16

Integrated semiconductor devices, 18
Interference, 234
Interference pattern, 169
Internal conversion, 66
Intersystem crossing, 66
Iodated polystyrenes, 313, 314
Iodonium salts, 149, 150
Ion beam lithographies, 20, 349
Ion beam resists, 352
 data on, 355
Ion implantation, 20, 197, 205, 351
Ionization, 302
Ionization energies of atoms and molecules, 303
Ionization potentials of atoms, 302
Ion optical imaging, 352
Irgacure 184, 111
Iron-arene complexes, 152, 155
Isopropanol, 40
 as hydrogen donor, 113
Isobenzofurane, 122
Isocyanates, multifunctional, 157
Isomer effect of cresoles, 219
Isomeric composition, 193
Isomeric novolac resins, 219
2-Isopropylthioxanthone, as photoinitiator, 118

Ketene, 182, 183, 216
Ketene scavengers, 182
3-Ketocoumarins:
 as photoinitiators, 118
 as sensitizers, 88
Ketones:
 $\pi\pi^*$ triplets of, 114
 n-π^* triplets of, 114
 quantum yield of scission, 275
Kinetic chain length, 130, 133, 139, 145, 146
Knife edge, 235
Kodak Micropositive Resist 809, 187
Kodak Micro Positive Resist 820, 205
Kodak resists, 50
Kodak Thin Film Resist (KTFR), 18, 38

Langmuir-Blodgett films, 144, 317
Laser(s), 123
 argon ion, 123
 helium-neon, 123
Laser direct imaging, 164
Laser driven pattern generator, 258
Laser-exposable lithographic plates, 162
Laser exposure, 20
Laser interferogram, 211
Laser interferometers, 199

Laser interferometry, 197, 198, 200
Laser lines, 255
Laser plasma sources, 337
Laser plates, 158
Laser-sensitive photoresist, 164
Laser scanner, 123, 162, 164
Latent image, 231, 278
Lenses, replication of, 16
Letterpress, 158
Letterpress plates, 159, 160, 162
Letterpress printing, 16
Lewis acids, 148, 150, 184
Lifetime, of triplet state, 88
Lift-off procedures 182, 184
Ligand exchange reaction, 126, 153
  radical formation by, 125
Ligands, in coordinative bonding, 23
Light scattering, 234
Line spectra (Shpolskij method), 96
Line spread function, 234, 235, 239, 294
Link configuration, 195
Liquid development, of poly(butene sulfone), 324
Liquid n-hexane, the radiolysis of, 305
Listing of excimer lasers, 256
Lithium acrylate, 170
Lithographic fountain solution, 6
Lithographic offset plates, 162
Lithographic plate, 140, 159, 162
  laser-exposable, 162
Lithographic sensitivity, 321
Lithographic stone, 6
Lithography, 158
  deep UV, 20
  electron beam, 20
  halftone, 8
  invention of, 5
  ion beam, 20
  offset printing, 7
  x-ray, 20
Lithography methods, cost comparison of, 339
Litho-offset, 16
Litho-offset press, 8
Living polymers, 148
Lloyd Jones diagram, 232
Lognormal distribution, 51
Low pressure mercury lamp, 254

Maleic anhydride, 29
Mask, optical, 18
Mask aligner, 18
Mask blank, 340
Masked ion beam lithography, 352

Mask making, 301
Mask making devices, 298
Mask repair, 20
Mass absorption coefficient, 344
Mechanism of dissolution, 215
Melamine, 184
Meldrum's diazo, 263
Membrane model of Arcus, 216, 217
Membranes, ultra-thin, 339
Mercaptoacetic acid, 145
Mercury G-line at 437nm, 252
Methacrylic acid, 323
Methacrylic chloride, copolymerized with styrene, 323
N-Methyldiethanolamine, as coinitiator, 116
N,N'-Methylenebisacrylamide, 170
Methylene blue, 120, 122, 170
Methylene linkages, position, 192
4-Methylmercapto-α, α-dimethyl-morpholino acetophenone, 111
Methylmethacrylate, 131
Methyl methacrylate:
  copolymers with acrylonitrile, 320
  copolymers with o-chloroacrylate, 320
Methyl vinyl ketone–methyl methacrylate copolymer, 268, 272
Michler's ketone, 4,4'-bis(dimethylamino) benzophenone 117
  as coinitiator, 117
Michler's ketone as photoinitiator, 26, 117
"Micraligner", Perkin-Elmer, 18
Microcomponents, 16
Microfabrication, 16, 20
Microgel particles, formation, 136
Microlithography, optical, 18
Microposit 111, 187
Microposit 1300, 187
Microposit 1400, 187
Mid UV, 252
Migration of energy, 75, 76
Minimum line width, 242
Models of novolac dissolution, 215
Modulation, 237
  of adhesion, 15, 169
  of index of refraction, 169
  of permeability, 15, 167
  of refractive index, 15
  of tackiness, 15, 168
Modulation transfer (MT) factor, 238
Modulation transfer function, MTF, 239, 295
Molar heat of reaction, 132
Molecular oxygen, triplet ground state of, 106

Molecular weight:
　effect on dissolution rate, 193
　weight average, 50
Molecular weight distribution, 53
　dispersity of 53, 55, 192, 193
Monazoline, 184
MRS-1, 259, 260
MTF of an electron beam instrument, 296
　of imaging systems, 236, 241
　of printer lens, 244
　of Zeiss optics, 240
Multifunctional isocyanates, 157
Multilevel resists, 20
Multiplicity of excited state, effect of, 106

Nanometric electron lithographic system, 298
Naphtalene-2-carboxylic acid-tert-butyl ester, 280
Near UV, 252
Negative color proofing system, 170
Negative electron resists, 311, 316
Negative resist, 13, 228
Negative X-ray resists, 343
Network, three-dimensional, 47
Neural networks, 302
"New Positive Resist", NPR, 326
Newspaper production, 161
Nitrene(s), 32, 33, 260
　reaction with oxygen, 39
o-Nitrobenzyl chemistry, resists based on, 264
　reaction, 287
　substituted polyether, 265
Normalized film thickness, 47, 227
Norrish type I process, 269, 306
Norrish type II process, 111, 275
Novolac, 13, 178, 191, 192, 265
　as ion resist, 352
　dissolution, 211, 212
　methylene links, 191
　resist with chemical amplification, 348
　resists, 195
Nucleophiles, 148
Numerical aperture, 243
Nylon derivatives, 160
Nylons, alcohol soluble, 157
Nyloprint, 15, 102, 160

o-Octadecylacrylic acid, 318
Off-press color proofing, 168
Off-press proofing materials, 168
Offset, litho-, 16
Offset printing, 7

OFPR, Tokyo Oka, 187
Olefinic monomers, 317
Oligomeric binders, 154
Olin-Hunt resist HPR-204, 348
Onium salts, 276
　initiation by, 147
　as radical polymerization initiators, 284
Optical density, 226
　image transfer in projection printer, 244
　microlithography, 18
　parameters of resist, 207
　printer, 295
Organometallics, 122
Overlay, 297
Oxidation potential of sensitizer, 284
3-Oximino-2-butanone mathacrylate, OMMA, 271
Oxiranes, 147
Oxycylohexene, 149
Oxygen inhibition, 162
　of bisazide reaction, 246
Oxygen scavengers, 162
Ozalid process, 181

Panchromatic material, 122
Paracresol-based novolacs, 217, 218
o-Particle barriers, 158
Pattern generated by computer, 163
Pattern generation, 299
Patterning, additive, subtractive, 342
Pattern transfer:
　anisotropic, 20
　by wet etching, 205
PBS, Poly(butene sulfone), 324
PCP, distribution of site reactivities, 93
Peak rate of polymerization, 136
"Peel apart" systems, 169
Pentacosadiynoic acid, 318
Pentaerythritol triacrylate, 170
n-Pentylmercaptan, 146
Penumbral blurr, in x-ray lithography, 336
Percolation model, of polymerization, 135
Perkin-Elmer, Micralign projection printer, 253
Perkin-Elmer "Micraligner", 18
Permanent protective encapsulant, 167
Permeability changes, 159
　modulation of, 15, 167
Peroxides, as photoinitiators, 107
Peroxy radicals, 128
Perrin quenching sphere, 76
Perylene, 68, 150
Phase changes in photopolymerization, 136, 138

Phenol ester, 183
Phenolic base resin, 178, 191
Phenolic coupler, 181
2-Phenylacetophenone:
  as photoinitiator, 108
  derivatives of, 109
Phenylene diacrylates, 28
p-Phenylene diacrylic acid, 28, 80
Phenylhydrazine, 119
1-Phenyl-1,2-propanedione-2-O-benzoyloxime, PPO, 112
Phosphines, 147
Photoactive component (PAC), 185, 188
Photocalorimetric curve, 134
Photocalorimetry, 132
Photochemical balance equation, 49
Photochemistry, 20
Photochromic polymer, kinetics of inversion, 95
Photochromic system, 278
Photocrosslinking, 26
Photocycloaddition, 12, 26, 94
Photodimerization, 11
Photofabrication, 16, 157
Photofragmentation, 103, 107
Photo-Friess rearrangement, 268
Photogeneration:
  of acids, from onium salts, 278
  of a catalyst, 278
  of radicals, 103
Photographic contrast, 49, 51, 227, 265
Photographic process, 231
Photographic sensitivity, 48
Photography, 6, 122
Photogravure, 4
Photoinitiated catonic polymerization, 147
  condensation polymerization, 144
  radical polymerization, 14, 15, 103
Photoinitiation, of radical polymerization, 103
Photoinitiator:
  anthraquinone, 115
  anthraquinone sulfate, 115
  benzophenone, 115, 117
  benzoyl peroxide, 107
  2-chlorothioxanthone, 118
  2,2'-diethoxyacetophenone, (DEAP), 111
  2,2'-dimethoxy-2-phenylacetophenone (DMPA), 110
  2-ethylanthraquinone, 115
  2-isopropylthioxanthone, 118
  2-tert-butylanthraquinone, 115
Photoinitiators, 103, 125
  acetophenones, 108
  alkyl ethers of benzoin, 108
  2,2'-azobis(butyronitrile), AIBN, 107
  azocompounds, 107
  benzil ketals, 110
  benzoin derivatives, 108
  biimidazoles, 113
  derivatives of acetophenone, 111
  generic, 109
  3-ketocoumarins, 118
  ketoxime esters of benzoin, 112
  peroxides, 107
  2-phenylacetophenone, 108
  thioxanthones, 118
  triazines, 112
Photolithography, halftone reproduction by, 7
  invention of, 7
Photomagnetic curing, 195, 197
Photomechanical reproduction, 4
Photopolymer, hybrid system, 42
Photopolymerization, 102, 159
  in crystalline phase, 140
  of 1,6-hexanediol diacrylate, 135
  induction period, 128
  in solid state, 140
Photopolymerization systems, components, 155
Photoproduct concentration, 221
Photoreactive polymers, energy migration in, 80
Photoreduction of Cr(VI), 22
Photoresist, laser sensitive, 164
Photosensitization of onium salts, 283
Photostationary state, 130
Phthalaldehyde, copolymerized with o-nitrobenzaldehyde, 287
Planar technology, 16, 18
Planography, 16
Plasma, resistance to, 195
Plasma ashing, 57
Plasma etching, 205
Plasma resistance, of PMMA, 320
Plasma resistance of polystyrene, 313
Plastic flow, 197
Plasticizer, 125
Plate, intaglio, 3, 4
Plates, presensitized, 12
Plating-up, of pattern, 163
PMMA and derivates, 321
  as ion resist, 352
PMPS, poly(2-methylpenten-1-sulfone), 327
Point spread function, 234
Polarity changes, 24
  photoinduced, 44

INDEX 405

Poly(p-acetooxystyrene), 268
Poly(acrylamide), 38
Poly(alkene sulfones), 323
Poly(N-alkyl-o-nitramides), 267
Polyamide plate, 160
Polyamides, 160
Polybutdiene, as soft segment, 144
Poly(butene sulfone), 276, 321, 324
Polycarbonate, 282
Poly(chloromethylstyrene), PCMS, 314, 346, 347
Poly(chloromethylstyrene-co-2-vinyl-naphthalene), 315
Poly(4-chlorostyrene), 314
Poly(cis-isoprene), 58
  cyclized, 36
Polychrome resist, 192
Polydiacetylenes, 143
Poly(1,3-dichloro-2-propyl acrylate), 346
Poly(2,3-dichloro-1-propyl acrylate), 346
Poly(diphenylsiloxane), partially chloromethylated, 315
Polyenes, 145
Polyesters, unsaturated, 156
Poly(fluorobutyl methacrylate), 270
Poly(p-formyloxystyrene), 268
Polyfunctional inhibitors, 220
Polyfunctional reactive prepolymers, 157
Poly(glycidyl methacrylate), PGMA, 270, 312
Poly(glycidyl methacrylates), exposure characteristics, 312
Polyhalogen compounds, as coinitiators, 125, 127
Poly(hexafluoro methacrylate), 320
Polyhydroxybenzophenone, 187
Polyimide, photoreactive, 41, 42
Polyimides, 29, 276
Polymer degradation, 307
Polymer gravure plate, 161
Polymer plate, 13
Polymeric binders, 154
Polymeric gravure plate, electron microgram of, 161
Polymeric ketones, 272
Polymerization inhibitors, 129
Polymerization, activation energy of, 136
  addition-, 102
  in amorphous solids, 137
  cationic, 102, 147
  condensation-, 102
  crosslinking, 156
  of crystalline diacetylenes, 142
  of epoxides, 152
  heat of, 134, 136
  peak rate of, 136
  photoinitiated, 14, 15
  postirradiative, 151, 311
  radical-, 103
  of tetrahydrofurane, 151
  topotactic, 143
Poly(methacrylic acid), 266
Poly(methyl methacrylate) (PMMA), 269, 319, 348
Poly(methyl methacrylate-co-indenone), 272
Poly(methylisopropenyl ketone), PMIPK, 274
Poly(2-methylpentene-1-sulfone), PMPS, 326, 327
Polynuclear aromatic hydrocarbons, 150, 151
Poly(olefin sulfones), 324
  as dissolution inhibitors, 326
  as electron resists, 323, 324
Poly(ortho-substituted 2-phenylethyl methacrylates), 329
Poly(perfluorobutyl methacrylate), 320
Poly(phthalaldehyde), 286, 288
  and diphenyliodonium salt, 290
  as dissolution inhibitor, 290
  as self-developing resist, 288
Poly(phthalaldehyde) films, 289
Poly(phthalaldehyde-co-2-nitrobenzaldehyde), IR spectrum, 289, 290
Polysilanes, 249
Polystyrene, 313
Polystyrene-Tetrathiofulvalene, 329
Polysulfones, 286
  ceiling temperature, 286
Poly(p-tert-butoxycarbonyl oxystyrene) PBOCST, 267, 279, 328
Polythiols, 145
Poly(2,2,2-trifluoroethyl-α-chloroacrylate), 321
Polyurethanes, 157
Poly(vinyl alcohol), 4, 23, 40
Poly(vinyl butyral), 4, 23
Poly(vinyl cinnamate), 12, 18, 24, 25
  distribution of site reactivities, 93
  effect of temperature, 87
  quantum yield of cycloaddition, 86
  quantum yield function, 92
  reactivity distribution, 93
  spectral sensitization of, 82
  triplet sensitization, 84
Poly(vinyl cinnamylidene acetate), 27
Poly(vinyl ferrocene), 345
Poly(vinyl naphthalenes), 315
Poly(vinyl phenol), 38, 259, 268, 277

Poly(vinyl pyridine), 40, 315
  electron sensitivity, 316
Poly(vinyl pyrrolidone), 4, 23, 38, 170, 249
Poly(N-vinylcarbazole), energy migration in, 79
Position of methylene link, effect of, 194, 219
Positive color proofing system, 169
Positive photoresist, 13, 178, 179, 256, 282
Positive polymer plate, 13
Positive x-ray resists, 348
Postexposure bake, 57, 196, 205
Postexposure thermal treatment, 197
Postirradiative polymerization, 151, 311
PPDA, distribution of site reactivities, 93
  energy migration in, 79
  site reactivity distribution in, 95
Practical dose calculations, 304
Practical electron exposure tool, 298
Preexposure bake, 57, 196, 197, 204, 205
Prepolymers, polyfunctional, 157
Presensitized plates, 12
Press, litho-offset, 8
Primary cage effect, 105
Primary radicals, generation of, 128
Printed circuit boards, 149, 158, 162, 163
Printed circuits, 10, 16, 17
Printing:
  contact, 18
  gravure, 16
  intaglio, 16
  letterpress, 16
  plates, 157
  process, 337
  relief, 16
  silk screen, 10
Printing on corrugated surfaces, 161
Printing plate:
  Dycryl, 15
  flexographic, 15
  gravure, 9
  Nyloprint, 15
  positive, 181
Processing of resists, 56, 197
Profiles of resist images, 233
Projection alignment, 18
Projection printing, 243
Projection systems, 257
PROLITH program, 208
Propagation:
  activation energy of, 133
  rate of, 130
Propagation of chain reaction, 103

Propargyl methacrylate, copolymers with hydroxyl MA, 312
Properties of alkali ions, 217
PROSIM program, 208
Protein substrates, 23
Protonic acids, 184
Protonic Bronsted acids, 147
Proton sensitivities, 355
Proximity printing, 242, 243, 257, 336
Puddle development, 57
Pyrene, 68
1-Pyrenebutyric acid methylester, 81
Pyridinium salts, 29

Quantum electronic effect, 302
Quantum yield function, poly(vinyl cinnamate), 92
Quantum yield:
  of crosslinking, 55, 139, 140, 144
  of initiation, 105
  of radical formation, 107
  of scission, 275
  of triplet formation, 88
Quartz crystal microbalance, 200
Quenching sphere, in energy transfer, 75

Radiation chemical yield, G-value, 304
Radiation chemistry, 20, 269, 302
  low energy limit of, 303
Radiation dose, 228
-Radiation from ^{60}Co source, 309
Radiation induced depolymerization, 285, 327
Radiation sources, 252
Radiation yield of crosslinking, G_x, 307
Radiation yield of scission, G_s, 307, 308
Radical(s):
  crosslinking by, 40
  mobility of, 36
  photogeneration of, 103
Radical formation:
  by ligand exchange, 125
  quantum yield of, 107, 139
Radical occlusion, 139
Radical scavengers, 154
Radiolysis of n-hexane, 305
Radiolysis of poly(methyl methacrylate), PMMA, 306
Radiolytic degradation of PMMA, 319
Raleigh criterion, 243, 244, 250
Rare gas/monohalide systems, 254
Raster scan mode, 298
Rate of dissolution, 178

RD 2000N, 259
Reactant sites, 96
  site geometry, 96
  site selection spectroscopy, 96
Reactive binders, 135, 156
Reactive ion etching, 20
Reactive prepolymers, 135
Reactivity distribution, poly(vinyl cinnamate), 93
Reactivity histograms, 93
Reciprocity failure, 258
Recording of x-rays, with diacetylenes, 143
Red azodye, 183
Reduction potential of onium salt, 284
Refractive index, modulation of, 15
Relief printing, 14, 16, 159
Repair of x-ray masks, 351
Resist coating, 56
Resist:
  development of, 57
  photographic speed, 48
  postbake of, 57
  prebake of, 57
  term introduced, 3
Resist contrast, 52, 245
Resist dissolution, "Stone wall" model, 193
Resist processing, 56
Resist profile, 183, 210
Resists:
  AZ, 18
  azide, 18
  diazoquinone-novolac, 18
  dry, 15
  Kodak, 50
  multilevel, 20
  thick film, 50
  water processable, 29
Resists based on polystyrene, 313
Resist swelling, 234
Resist systems with chemical amplification, 278
Resoles, 192
Resolution limits, 245, 252
Resolving power, 240
Response surface, 209
Reversal process, 316
Reversible swelling, 59
Riboflavin, 120
Riston, 163
Riston LUV, 164
Riston LV, 164

Rose Bengal, 119, 120, 129
Rotary whirler, 56
Rubber, 18

Sacher Technik, Vienna, 352
Salt, TTF^+Br^-, 331
SAMPLE program, 208
Scanning electron microscopes, 302
Scattering of secondary electrons, 234
Schiemann reaction, 149
Scission yield G_s, determination of, 307
Screen printing, 10, 16
Secondary cage effect, 105
Secondary electrons, 294
Secondary structure model, 217, 218
Secondary structures of novolacs, 220
Self development, of poly(butene sulfone), 324
Semiquinone radicals, 119
Sensitivity ratings, 227
Sensitization, 272
  of photopolymerization, 144
  spectral, 26
Sensitizer, UV absorbing, 124
Sensitizers, ketokoumarins, 88
Sensitometric curve, 48, 226, 228
  of negative material, 232
  of a negative photoresist, 228
  of a photographic negative, 227
  of photographic paper, 232
  of typical positive resist, 230
Sensitometry of imaging materials, 226
Serigraphy, 10, 16
Shadow printing, 242
Shallow imagewise implantation, 352
Shapes, three-dimensional, 158
Shpolskij spectra, 96
Signal amplification, by a chain reaction, 143
Silicon, 18
Silicon carbide, 339
Silicon dioxide, 18
Silicon masks, 352
Silicon nitride, 339
Siloxycarbenes, 39
Silver halide photography, 43, 88, 227, 278
Silylated poly(phthalaldehyde), 290
Singlet oxygen, 107, 129
Singlet-triplet splitting, 88
Site geometry, 97
Site reactivities:
  distribution of, 92
  distribution in PPDA, 95
Site selection spectroscopy, 96, 97

Solder mask, 157, 158, 165
Solder wave technology, 165, 167
Solid state photopolymerization, 140
Solubility, modulation of, 159
Solubility changes, 159
Solubility parameters, 59
  tabulation, 60
Solvated electrons, 303
Solvent extraction, 132
Sources of electrons, 297
Spectral mismatch, 82, 83
Spectral sensitization, 26, 84
  of azide photolysis, 88
  of azides, 90
  by electron transfer, 88
  of poly(vinyl cinnamate), 82
  triplet energy transfer, 83, 84
  by triplet energy transfer, 85
  in silver halide photography, 88
Speed, 227
Spin coating, 57, 205
Spin speed, and film thickness, 203
Spontaneous depolymerization, 324
Spray development, 57
Square wave, 237
Stabilizers, 127, 155
Standing wave effects, 197
Starch, 3, 23
$n\pi^*$ States of ketones, 114
Step function, 245
Stone, lithographic, 6
Storage rings, 337
Stripping of resist, 57
Styrene, 29
Styrene-maleic anhydride copolymers, 157
Styrylpyridine, 29
Styrylpyridinium, 30
Styrylquinolinium, 30
Substrate, degreasing of, 56
Substrate preparation, 56
Sulfonium salts, 149
Swelling:
  in developer, 58
  reversible, 59
Synchrotons, 337

Tackiness, modulation of, 15, 159, 168
"Taylor cone" in focused ion beam sources, 350
Termination of radical chain, 103
Termination, rate of, 130
Terpolymer, containing methacrylonitrile, 271, 273
"Terpolymer" of IBM, 322

p-Tertbutylanthraquinone, as photoinitiator, 115
p-Tertbutylbenzoic acid, 272
Tertiary alcohols, as coinitiators, 115
Tertiary amines, 70, 162
  as antioxydants, 129
  as coinitiators, 115
  as hydrogen donors, 116, 121
Tetracene, 97
Tetrafunctional acrylates, 157
Tetrahydrofuran, as coinitiator, 115
  polymerization, 154
  polymerization of, 151
Tetrahydrofuryl radical, 115
Thermalized electrons, 303
Thermogravimetric analysis, 132, 279, 288
Thiazines, 120
Thick film resists, 50
Thin film devices, 16
Thiol-Ene Systems, 144, 147
Thiols, 144
Thionine, 120
Thioxanthone, 32, 81, 147
Thioxanthones, as initiators, 118
Three-dimensional shapes, 158, 185, 186
p-Toluenesulfonic acid, 267
p-Toluenesulfonic acid esters of diacetylenes, 143
Tonable adhesive layer, 168
Tone (negative or positive), 229
Toner, colored, 125
Topotactic polymerization, 143
Transfer constants, 128
Transition metal carbonyls, photoreactivity of, 126
  as photoinitiators, 125, 140
Trapping of exciton, 32
Triacrylate monomer, 123
Trialylsulfonium salts, 150
Triazines, as photoinitiators, 112
Trichlomethyl radical, 126, 127
ω-Tricosenoic acid, 317, 318
Triethanolamine, 184
Triethylamine, as coinitiator, 116
Triphenylsulfonium ion in polymer backbone, 276
Triplet energy transfer, spectral sensitization, 83, 84, 85
$n\pi^*$-Triplets of ketones, 114
Triplet sensitization, poly(vinyl cinnamate), 84
Triplet sensitizers, table of, 88
Triplet state:
  and cage effect, 106

lifetime, 88
  quantum yield of formation, 88
  singlet-triplet splitting, 88
Tris(trichloromethyl-s-triazine) as initiator, 123
Throughput of items, 230

Unsaturated polyesters, 156
Urethanes, 143
  functionalized, 157
UV absorption spectrum, 142
UV curing, 16, 147, 151
UV sensitizer, 124

Variable area-variable depth gravure plates, 161
Variable shaped electron beam, 299, 300
Vector scan, 298
Very high speed integrated circuits (VHISICS), 252, 301
Videodisc copying, 158
Videodiscs, copying of, 16
Vinylbenzophenone, 41
Viscous media, photopolymerization in, 131
Visible light laser-sensitive resist, 123

Wafer, 18
Wafer stepper, 19
Water processable resists, 29
Weller's equation, 89
Wet strength, changes in, 159
Wetting agents, 154
Wire boards, 16, 123, 157, 158, 162, 165
Wolff rearrangement, 182
Wooden plates, 160
Woodward–Hoffman rule, 26
Work point, 48, 204, 229
Working exposure, 229

Xanthenes, 120
Xerographic method, 162
X-ray exposure tool, 336
X-ray lithography, 20, 335, 337
X-ray masks, 339
  repair of, 351
X-ray mass absorption coefficient, 346
X-ray resists, 343
X-rays, 318
X-ray sensitivities, 343
X-ray sources, 335, 337